Parasitism & Ecosystems

Parasitism and Ecosystems

EDITED BY

Frédéric Thomas
François Renaud
National Centre for Scientific Research, France

Jean-François Guégan
Institut de Recherches pour le Développement, France

OXFORD
UNIVERSITY PRESS

OXFORD

UNIVERSITY PRESS

Great Clarendon Street, Oxford OX2 6DP

Oxford University Press is a department of the University of Oxford.
It furthers the University's objective of excellence in research, scholarship,
and education by publishing worldwide in

Oxford New York

Auckland Cape Town Dar es Salaam Hong Kong Karachi
Kuala Lumpur Madrid Melbourne Mexico City Nairobi
New Delhi Shanghai Taipei Toronto

With offices in

Argentina Austria Brazil Chile Czech Republic France Greece
Guatemala Hungary Italy Japan Poland Portugal Singapore
South Korea Switzerland Thailand Turkey Ukraine Vietnam

Oxford is a registered trade mark of Oxford University Press
in the UK and in certain other countries

Published in the United States
by Oxford University Press Inc., New York

British Library Cataloguing in Publication Data

Data available

Library of Congress Cataloging in Publication Data

Data available

Typeset by Newgen Imaging Systems (P) Ltd., Chennai, India
Printed in Great Britain
on acid-free paper by
Antony Rowe, Chippenham

ISBN 0 19 852986 4 (Hbk)
ISBN 0 19 852987 2 (Pbk)

10 9 8 7 6 5 4 3 2

In the memory of Louis Thaler

Professor of Palaeontology at the University of Montpellier (France), Louis Thaler was convinced that a multidisciplinary approach to the study of evolution would benefit all biological sciences. To this aim, he incessantly encouraged collaborations between disciplines, and promoted the need to combine available techniques and methodologies. His communicative enthusiasm, his sustained guidance, his grasp of new ideas fashioned the present research landscape in Montpellier. His involvement as President of scientific committees in many French research institutions (University of Montpellier, CNRS, INRA, IRD, CIRAD, BRG, . . .), inflected the thinking of the whole French research community. He was once the mentor of the co-editors of this book; no doubt, its topic would have enthralled him.

Nicole Pasteur
Directeur de l'Institut des Sciences de l'Evolution,
founded by Louis by Thaler in 1980

Contents

Conclusion—Parasites, communities, and ecosystems: conclusions and perspectives 171
Gary G. Mittelbach

References 177

Index 217

Contributors

Jean-Baptiste André Laboratoire Génome, Populations, Interactions, Adaptation, UMR 5171, USTL—CC. 105, Bât. 24, Place Eugène Bataillon, 34095 Montpellier Cedex 5, France.

Michael B. Bonsall Department of Biological Sciences, Imperial College London, Silwood Park Campus, Ascot Berks, SL5 7PY, UK.

Thierry Boulinier Laboratoire d'Ecologie, CNRS—UMR 7625, Université Pierre & Marie Curie, 7 Quai St Bernard, F-75005 Paris, France.

Sam P. Brown Department of Zoology, University of Cambridge, Downing Street, Cambridge CB2 3EJ.

Andy P. Dobson Department of Ecology and Evolutionary Biology, Eno Hall, Princeton University, Princeton, NJ 08544-1003.

Jean-Baptiste Ferdy Laboratoire Génome, Populations, Interactions, Adaptation, UMR 5171, USTL—CC 105, Bât. 24, Place Eugène Bataillon, 34095 Montpellier Cedex 5, France.

Bernard Godelle Laboratoire Génome, Populations, Interactions, Adaptation, UMR 5171, USTL—CC 105, Bât. 24, Place Eugène Bataillon, 34095 Montpellier Cedex 5, France.

J.-F. Guégan GEMI, UMR IRD-CNRS 2724, Centre IRD de Montpellier, 911 Avenue Agropolis BP 64501, 34394 Montpellier Cedex 5, France.

Alexander D. Hernandez Department of Ecology, Evolution and Natural Resources, Rutgers University, Cook College, New Brunswick, NJ 08901.

Robert Holt Department of Zoology, 223 Bartram Hall, P.O. Box 118525, University of Florida, Gainesville, Florida 32611-8525.

Peter Hudson Center for Infectious Disease Dynamics, Biology Department, Penn State University, University Park, PA 16802.

Armand M. Kuris Department of Ecology, Evolution and Marine Biology, Marine Science Institute, University of California, Santa Barbara California.

Kevin D. Lafferty USGS Western Ecological Research Center. Marine Science Institute, University of California, Santa Barbara California, USA.

Michel Loreau Laboratoire d'Ecologie, UMR 7625, Ecole Normale Supérieure, 46 rue d'Ulm, F–75230 Paris Cedex 5, France.

Thierry De Meeüs Génétique et Evolution de Maladies Infectieuses GEMI/UMR CNRS-IRD 2724, Equipe: "Evolution des Systèmes Symbiotiques", IRD, 911 Avenue Agropolis, B.P. 5045, 34032 Montpellier Cedex 1, France.

Gary G. Mittelbach W. K. Kellogg Biological Station and Department of Zoology Michigan State University, Hickory Corners, MI.

Anders Pape Møller Laboratoire de Parasitologie Evolutive, CNRS UMR 7103, Université Pierre et Marie Curie, Bât. A, 7ème étage, 7 quai St. Bernard, Case 237, F-75252 Paris Cedex 5, France.

S. Morand CBGP, UMR INRA-IRD-CIRAD-Agro.M., Campus International de Baillarguet,

CS-30016, 34988 Montferrier sur
Lez cédex, France.

R. Poulin   Department of Zoology, University of
Otago, P.O. Box 56, Dunedin, New Zealand.

Andrew F. Read   Institute of Evolution,
Immunology and Infection Research, University of
Edinburgh, EH 93 JT Edinburgh, Scotland, UK.

François Renaud   Génétique et Evolution de
Maladies Infectieuses GEMI/UMR CNRS-IRD
2724, Equipe: "Evolution des Systèmes
Symbiotiques", IRD, 911 Avenue Agropolis,
B.P. 5045, 34032 Montpellier Cedex 1, France.

Jacques Roy   Centre d'Ecologie Fonctionnelle
et Evolutive, UMR 5175, CNRS
F-34293 Montpellier Cedex 5, France.

Michael V.K. Sukhdeo   Department of Ecology,
Evolution and Natural Resources, Rutgers
University, Cook College, New Brunswick,
NJ 08901.

Frédéric Thomas   Génétique et Evolution de
Maladies Infectieuses GEMI/UMR CNRS-IRD
2724, Equipe: "Evolution des Systèmes
Symbiotiques", IRD, 911 Avenue Agropolis,
B.P. 5045 34032 Montpellier Cedex 1, France.

David Tilman   Department of Ecology, Evolution
and Behavior, University of Minnesota,
St. Paul, MN 55108.

Richard C. Tinsley   School of Biological
Sciences, University of Bristol,
Bristol BS8 1UG, UK.

Parasites, diversity, and the ecosystem

Peter Hudson

The dualities of parasitism

Dualism is a dominant theory of life that considers reality to be a balance between two independent and fundamental principles: good and evil, mind and matter, nature and nurture. In the same manner we see the thread of dualism run through the ecology of parasitism: they can generate diversity but cause extinction, they may castrate a host but increase its growth rate, and they can stimulate an immune response but at the same time encourage a secondary chronic infection. Parasites inhabit individual hosts that are distributed as discrete patches, much like a metapopulation but these hosts are also nested within a spatially structured metapopulation and these within a meta-community of competent hosts. They often divert the host's resources to themselves and away from other consumers and so change energy flow patterns, the use of critical resources, and so influence ecosystem functioning. The majority of living organisms are parasitic and their role as specialist consumers and their influence on biodiversity may well make them important players in many ecosystems.

Given the rather special and probably pivotal role parasites may play in many ecosystems, it is somewhat surprising that few workers have considered the role of parasitism at the ecosystem level. Probably the central question is to ask, how do parasites influence ecosystem functioning? Or more specifically, what are the consequences of parasite removal for the community and energy

Center for Infectious Disease Dynamics, Biology Department, Penn State University, University Park, PA 16802.

flow in the ecosystem? What is the biomass of parasites within the ecosystem and how does this compare with other natural enemies? How do the parasites influence the flow of specific chemicals and minerals through the system? How do parasites influence biodiversity? And how does biodiversity influence parasitism? Questions that we are only just starting to get vague answers to but nevertheless the questions that are the underlying driving force behind the production of this book. There is a common assumption that parasite biomass is negligible (e.g. Polis 1999) but is this assumption really correct? To illustrate the sheer significance of parasites in an ecosystem let me tell you about a comment made by my friend and colleague, Armand Kuris. He once asked me what I thought was the biomass of parasites on the Carpinteria salt marsh (about 70 ha) where he and Kevin Lafferty have studied the trematodes of the gastropod and bird fauna for many years. I had not a clue, looked deep into my glass of wine and fumbled with kilograms. 'Our provisional estimate is in the order of several elephants (if they weigh 3 tons maybe as many as 7–10) with a reproductive rate equivalent of several babies per year (maybe as high 1–2 baby elephants per day) for the 200 warmest days of the year' he replied. Astonishing, absolutely amazing and as a card carrying parasitologist, I was embarrassed by my lack of comprehension. Just imagine a small herd of elephants on a wetland in Southern California, they would be considered a dominant feature and if they were just consuming vegetation they would have a fantastic impact on the environment, especially in that small area of habitat. But these parasites are living off snails and birds and with their high reproductive rate they must be having a huge impact on the growth rate of their hosts, influencing the flux of energy to other trophic levels and shaping community structure by reducing competitive abilities of their hosts and vulnerability to predation. Of course, this is only part of the potential impact of the parasites on the ecosystem since Kuris and Lafferty have not estimated the biomass of the plant pathogens, many of the parasites in the crustacean or in the tertiary consumers.

Asking good questions and making estimates of parasite biomass can help us to get the role of parasites in perspective, but the answers to the questions are far from easy and we should appreciate the amount of hard work that has gone behind the studies of parasites in the Carpenteria salt marshes of Californian. To make some of the questions answerable, we may need to restructure them into a form that can be answered, perhaps by starting at the level of the individual and then using this foundation of understanding to explore issues at population, community, and ecosystem level. Hence we may ask, if parasites have an impact on the individual host and what are the emergent properties we may observe at the population and community level? How does the parasite interact with other natural enemies and then what are the consequences of these interactions to ecosystem functioning? This approach is based on undertaking insightful experiments at the lower level, monitoring changes in the intensity of parasites and age related effects and then integrating our understanding through models and identifying the patterns and emergent features we would predict to observe at the higher levels. Another approach is to examine an ecosystem that is subject to an epidemic. For example, what happened to the marine ecosystem of the North sea when Phocine distemper virus reduced the population of seals? Did fish survival change? Did seabirds compensate with improved breeding production and survival or were such events so transitory as to have little influence? Again modelling and understanding can provide insights. Another dominant approach for examining ecosystems effects is to apply the comparative method to identify patterns and

then dissect the data to propose the putative mechanisms. The chapters in this book use all of theses techniques and together provide an integrated and clear examination of parasites at the ecosystem level.

This short chapter serves as an introduction. I shall try to lay the scene for the role of parasites in ecosystems and in doing this I have to admit I face the tensions of my own inner duality. On the one hand I find the task daunting, our knowledge vague, and the scale of the issue massive, and I am aware that focusing on the parasite component of an ecosystem may inadvertently trivialize other critical components. On the other hand, the task is exciting and a challenge that we should rise to: pathogens and parasites have not been included in the theories of trophic structure and are frequently ignored from ecosystem ecology (Polis and Strong 1996; Polis 1999), we have a growing understanding of the role of parasites, some excellent modelling approaches and the time is ripe for a book like this. So to set the scene for the book I will look at a specialist parasite and a generalist parasite in two contrasting food webs to examine how they influence community structure and the ecosystem. I shall pick some fundamental questions that examine the role of parasites in ecosystems and illustrate those with a few examples, selecting some examples that are not used by others in the book, and of course, shamelessly referring to my own work.

Specialist and generalist parasites in the ecosystem

While the study of parasites was once the sole domain of the specialist parasitologist, often focused on the difficult and challenging task of working out the life cycle and taxonomic position of parasites, it is now apparent that this fundamental biological knowledge has allowed parasitology to come of age so that a wide range of scientists and disciplines are now addressing parasitological questions. The issues scale from the molecular to the ecosystem and to my mind, the challenge for the future is to ensure that the discipline becomes integrated vertically so that an understanding of the processes of infection and persistence at the molecular level can be incorporated in making predictions about the temporal and spatial spread of diseases and in identifying how parasitism influences ecosystem functioning. For example, parasites often weaken their hosts, making them morbid, and thus susceptible to predation (Hudson *et al.* 1992a; Packer *et al.* 2003). How does this influence the way energy flows through the food web? the consequences on the demographics of the host, the predator, the parasite, and the competitors in the system?

Understanding the importance of species and groups within an ecosystem is one of the central challenges to ecology and so if we are to investigate the role of parasites, the design of the question is important. Probably one of the most important questions would be to ask: if we were to remove the parasite from the system what would be the consequences? You may predict that host population growth rates would rise and would then lead to an increase in the growth rate of the other consumers of that host but then maybe they would suppress our host through the classic process of the paradox of enrichment. Alternatively, you may predict that the parasites keep the prey unhealthy allowing the predators to catch and obtain a feed so the removal of parasites makes the prey healthy and the predators starve so the host population would rise to be regulated by some other factor such as food availability. On the other hand, some other parasite may invade the niche and perhaps one that used the predator as an obligate host and then changed the dynamics of the whole community. So this apparently simple question is not trivial but we can quite quickly see there are a suite of mutually exclusive hypotheses to test in the wild. I suspect the true answer to the role of the parasite depends on how the parasite–host relationship is embedded in the food web and more specifically whether the parasite is a generalist, shared between species, or a specialist. In a simple food web when we have a single host infection that shapes the population dynamics of the host then this can be dominating and have far-reaching repercussions to the whole community. I will try to illustrate using studies on red grouse and their caecal nematode, *Trichostrongylus tenuis*. If grouse lived in a more complex ecosystem

then the effects of this specialist parasite may well be buffered by the other interactions. Much to my dismay, grouse do not live in the Serengeti, one of the more fascinating ecosystems in the world and one with a relatively complex food web. I know of no detailed studies on a specialist parasite in this system but there have been studies on a number of interesting generalist parasites. Rinderpest infected many of the ungulate species and appeared to have led to a dramatic and far-reaching change in the ecosystem, so we shall examine this system as a contrast to the grouse system.

Rinderpest in African ungulates

Rinderpest is a disease of ungulates caused by a morbillivirus and an equivalent of a 'buffalo measles' transmitted through an aerosol of virus during coughing and sneezing. This is an infection of domestic animals that invaded Africa in about 1889 and then spread at a frightening pace across the continent to reach the Cape within just 8 years (more details in Chapter 8). The impact on wild ungulate populations was dramatic, Buffalo and Wildebeest were decimated by about 95% and local populations of greater kudu, bongo, and eland were totally wiped out. Here was a huge 'experimental' perturbation to the ecosystem of the African savannah that is probably still influencing the functioning of several ecosystems today, more than 100 years later.

We probably know most about the situation in the Serengeti than elsewhere since the disease became endemic in the park (Dobson 1995a). In the 1960s there was a heavy vaccination programme that eventually ringed the whole park and by 1968 the disease was eradicated. At that time there was a series of detailed and fascinating scientific studies that have since followed changes in the population of ungulates and the ecology of the area and so recorded the recovery of the ungulates and changes in the ecosystem since rinderpest. Over this period, the wildebeest and buffalo exhibited what the Serengeti workers describe as 'an eruption': wildebeest and buffalo increased dramatically, at a rate of about 10% per annum, so wildebeest numbers were up 7-fold in 17 years and buffalo showed a parallel increase. There is a circumstantial evidence that this increase was a consequence of rinderpest since other grazing species like the zebra (not an ungulate) were unaffected. Furthermore, subsequent epidemics all provide good evidence that rinderpest was a significant factor limiting the size of several ungulate species. The interesting feature is the effect this must have had on the rest of the ecosystem. Sinclair (1979a) examined the effect that the eruption of wildebeest had on the Serengeti ecosystem; the increase in the wildebeest changed the seasonal grazing pattern and the abundance of the grasses and the herbs, reduced the combustible material and reduced fires that in turn allowed tree regeneration and through these processes influenced the whole community of herbivores. The wildebeest compete with the buffalo so while buffaloes increased after the removal of rinderpest, the population subsequently levelled off, probably because of competition for food from the wildebeest. They probably influenced the other grazers such as zebra (−ve: direct competition) Grants gazelle (+ve: herbs increased) and giraffe (+ve: trees increased). While there was some evidence that the predator populations responded to the increased prey base (Spinage 1962) the evidence is not clear, probably because wildebeest are seasonal migrants and the predator populations are regulated by the prey base during the intervening periods but also because a subsequent increase in predators led to the outbreak of other diseases (see Dobson 1995a; and Chapter 8).

While these data are limited and the story is pieced together from anecdotal evidence, Sinclair's deep understanding of the natural history of the Serengeti and Dobson's analysis, illustrate well how the removal of a pathogen can shape the processes within an ecosystem. Essentially the pathogen acted to reduce the abundance of the primary consumers and so influenced competition with other grazers, the vegetation structure and no doubt the flow of energy through the ecosystem.

Parasitic worms in British grouse

Red grouse inhabit the open, semi-natural moor lands characteristic of the British uplands where the vegetation is dominated by heather (*Calluna* *vulgaris*), the primary food plant of the red grouse. While the red grouse is the only species that relies solely on the heather, the habitat is home to a number of other species including the mountain hare, red deer, roe deer, and an avifauna that is predominantly migratory. These large tracts of heather moorland are managed by keepers to produce a harvestable surplus of grouse each year and provide grazing for the sheep farmers. The grouse are preyed on by foxes, golden eagles, hen harriers, short eared owls, and peregrines, but the keepers legally control the foxes and frequently interfere or kill the protected raptors (Thirgood *et al.* 2000).

The grouse exhibit unstable population dynamics with oscillations in abundance and a period usually between 5 and 12 years. The maximum growth rate the grouse is determined by the quality of the main food plant but subsequent changes in abundance are a tension between the natural enemies of the grouse (Hudson *et al.* 2002). Grouse are infected with a caecal nematode which reduces their condition and breeding production and these demographic effects coupled with the low degree of parasite aggregation in the host population generates instability in the population that can account for the cyclic fluctuations in red grouse abundance recorded in harvesting records (Dobson and Hudson 1992; Hudson *et al.* 1992b). Large-scale experiments that have removed parasites at the population level effectively stop the periodic crashes in abundance indicating that parasites play an important role in the cyclic nature of this species (Hudson *et al.* 1998). So while the direct interaction between parasite and host plays a major role in generating the cyclic fluctuations in abundance, we should now ask how this parasite–host interaction shapes the effects of other natural enemies in the ecosystem.

The parasites major influence on the host is to reduce body condition and make the host morbid, thus less able to produce young (Hudson 1986a), defend territories (Fox and Hudson 2001), and become more vulnerable to predation (Hudson *et al.* 1992a). The grouse emit a characteristic scent which trained pointing dogs can smell at remarkable distances, when the hens commence incubation they close off their caeca and no longer produce caecal

faeces and at the same time stop emitting scent the trained dogs can locate. It is here, in the caeca, that the parasitic worm lives and interferes with the workings of the caeca (Watson *et al.* 1987). Highly infected grouse have difficulties controlling their scent emission and the dogs, searching by scent, can locate these grouse significantly more frequently than individual who have had their worms experimentally removed (Hudson *et al.* 1992*a*). Incorporating this selective predation into a model of parasite–host interaction predicts an increase in the host population (Hudson *et al.* 1992*a*; Packer *et al.* 2003); initially a counterintuitive finding since we would not expect the addition of predation mortality to increase prey abundance (Fig. 0.1). However, since the predation is selective, the predator is removing the heavily infected individuals from the population, thus reducing the regulatory

role of the parasite, dampening the oscillatory behaviour of the population, and leading to an overall increase in the population. Interestingly, one may predict that harvesting by humans is another form of predation, and should also remove parasites and lead to dampening of the cycles but an examination of the time series data shows this is clearly not the case (Hudson and Dobson 2001). In fact what seems to be happening is that much of the infection process has already taken place prior to the start of harvesting. The infective stages are on the ground before the harvesting commences so the 'dye is cast' and the infection process will continue irrespective of any removal of grouse by the hunters although reduced density over an extended period will of course lead to reduced infection levels.

Grouse are also infected with a tick borne virus that causes the disease louping-ill, a significant

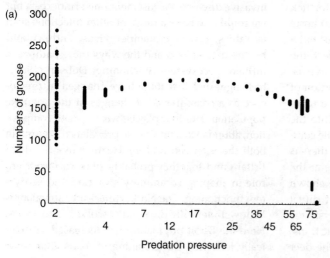

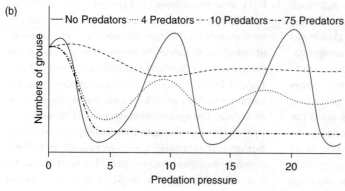

Figure 0.1 Consequences of selective predation on the dynamics of red grouse as predicted from modelling (a) Bifurcation figure showing an Increase in host density with increasing predation pressure that reduces the effects of parasite regulation and dampens oscillations (b) simulation runs showing reduction in oscillations with increasing predation.

Source: After Hudson et al. (2002).

cause of mortality in populations where it is prevalent. The ticks themselves cause little mortality unless numbers are high, but in areas like the North Yorkshire Moors the grouse pick up the ticks from bracken dominated ground, a habitat that provides high humidity and assists tick survival (Hudson 1986*b*). In the past, bracken was cut for livestock bedding and these activities restricted the bracken beds to the steep slopes but poor heather burning practices coupled with heavy grazing has allowed the bracken to escape from these slopes and invade the heather moorland, thus bringing the ticks into the habitat used by the grouse and exposing them to infection. The tick life cycle is dependent on the presence of a large mammalian host and the removal of these hosts should lead to the eradication of the tick and the louping-ill. In areas where the sole host is the sheep, they can be treated with acaricides and the ticks eradicated. In other areas, deer may be important hosts for tick but (unlike the sheep) they are not competent hosts for the virus, thus deer act as a 'sink' for virus but a 'source' for ticks (Gilbert *et al.* 2001). In effect, the deer act as 'dilution hosts' for the virus. Theoretically, if the deer host a large proportion of the ticks then they can reach a point where more virus is lost through these 'wasted bites' into the dead-end deer host than is generated from the competent hosts the virus levels fall. However, there is another interesting player in this ecosystem: the mountain hare. Experimental studies have shown that while mountain hares do not permit direct viral amplification through the normal systemic route (like deer) they do permit non-systemic transmission between co-feeding ticks, unlike the deer (Jones *et al.* 1997). Laurenson *et al.* (2003) undertook a large-scale experiment where they removed hares from a large area of moorland habitat and showed a significant decrease in the louping-ill seroprevalence in the grouse. Interestingly they were able to show that much of this decline in infection was because they removed the effects of co-feeding transmission. In other words, the hares were the key players that kept the virus persistent in this ecosystem and while the hares were a source of ticks the important role they played was in providing a suitable habitat that permitted transmission

between ticks. The final outcome of the role played by the virus is in the interplay between hares, deer, and grouse as hosts for both ticks and louping-ill. The whole vertebrate community plays a role in determining louping-ill dynamics but the specialist grouse–nematode interaction has far-reaching repercussions to the predators and the balance between these mortality factors and then the quality of the vegetation moulds the grouse dynamics.

Both of these studies are interesting since they illustrate the role of parasitism in the ecosystem. An understanding of the simple parasite–host relationship at the individual level allowed an understanding of population dynamics, the interaction with other natural enemies and how changes in the community structure influenced the ecology within the ecosystem. The Serengeti is a unique and rich ecosystem of international significance that rightly provides a wildlife spectacular but is vulnerable to invasive diseases. We just mentioned rinderpest but we could also have a range of other infections such as rabies, canine distemper virus, anthrax and bovine tuberculosis and the ways these pathogens influence ecosystem functioning. Dobson (1995*a*) has suggested that these have changed in prevalence as a consequence of changes in the ungulate population. This interplay between species composition, abundance, and disease prevalence is found in both the Serengeti and the Heather moorlands of Britain and together probably play an important role in shaping community structure and ecosystem functioning. The North American equivalent is Yellowstone with Bison, elk, wolves, and bears. Here the Bison population has increased dramatically following the grooming of roads after snow fall and so brucellosis has become a major concern, not directly to the Bison population but indirectly by the perceived threat wildlife pose as reservoirs of infection to cattle on neighbouring ranches. In the same vane, humans also interact with the semi-natural heather moorland where grouse management is a dominant form of land use in the uplands of Britain. The grouse management favours a multiple land use system that incorporates sheep farming, conservation, and recreation but if the economics of this collapses then this land is sold for commercial forestry, a single land use system of

little benefit to wildlife. But there is an irony here, another one of the dualities of parasitism if you like. The significance of the disease to the grouse has arisen primarily because the keepers reduce the predation pressure from the foxes so allowing grouse numbers to rise and parasitism to become a problem. This parasitism leads to highly unstable dynamics so that every few years the disease forces the grouse down to low levels. At this point if the predators are allowed back then the grouse will be held at this low level, the grouse no longer become a viable crop to harvest and the land can be sold to a single land use such as commercial forestry. The parasites are quite capable of knocking the host population down to a low density that inverse or simple density-dependent effects prevent them from rising.

These two examples show us that a specialist parasite like the nematode worm in the grouse can have important implications for the population dynamics of the host and consequently other natural enemies. For a generalist pathogen like rinderpest or even louping-ill (a specialist pathogen but with a generalist vector), there are also important repercussions to the whole community. We have evidence to suppose parasites should be important to the ecosystem.

Does biodiversity affect parasitism?

There is increasing evidence that species composition and diversity influence ecosystem functioning (Tillman *et al.* 1997*a*) and productivity (Tillman *et al.* 2001). These studies and others are showing us that biodiversity really matters to the quality of our environment and since we are losing biodiversity fast we are changing ecosystem functioning. But what are the processes involved in these effects? There are two dominant explanations in the literature, the first is that with low diversity, the probability of having species with the key traits present in the community is reduced and so productivity falls (the sampling effect). An alternative explanation is that at diversity falls, so fewer species utilize resources less completely (niche complementarity). Both of these hypotheses are focused on resource acquisition and use, but nearly 50 years ago, Elton

(1958) proposed that reduced biodiversity would increase the severity of diseases. An important hypothesis and one very relevant to the objective of this book. Does parasitism increase in severity with reduced biodiversity? Can the influence of disease be an important mechanism that reduces plant productivity? These are not trivial questions but questions that help us identify the role of parasites in the ecosystem. Two recent studies have shown us that that biodiversity really does influence disease severity. The first examines the pathogen load of grass swards and shows that biodiversity does indeed reduce disease severity and influences productivity from an ecosystem. The second examines the tick borne zoonotic Lyme disease and shows that reduced wildlife biodiversity increases the risk of infection to humans.

Fungal pathogens and plant biodiversity

Foliar fungal pathogen spores are spread by wind and rain and are common among grass species; they darken leaves, reduce leaf life span through the loss of nutrients, and photosynthate and so depress productivity. Mitchell (2003) demonstrated experimentally that excluding these pathogens from intact grassland with fungicide dramatically increased root production and biomass by increasing leaf longevity and photosynthetic capability. The interesting aspect of this work was that he also undertook a factorial experiment that included insect herbivores along with the pathogens and showed that herbivory had relatively little impact compared with pathogens and thus demonstrated that pathogens were potentially the important regulators of ecosystem processes.

The hypothesis that the severity of disease is greater when diversity is reduced is explained in plants because reduced biodiversity results in increased abundance and in turn this facilitates specialist pathogen transmission. This is well illustrated by studies in agriculture where the impact of a pathogen in a cereal system is reduced when mixtures of multiple genotypes of one crop species are grown together (e.g. Zhu *et al.* 2000). In another excellent study Mitchell *et al.* (2002) tested the diversity–disease hypothesis by using experimental communities of perennial grassland plants where diversity was controlled directly by hand weeding. They showed that decreased diversity increased pathogen load to an extent that pathogen load was three times greater in a monoculture than in a plot with 24 grass species, the equivalent to a natural system (Fig. 0.2). They showed that increased pathogen load and the severity of the attack was essentially the consequence of changes in host abundance although they also noted that changes in the relative species composition played an important role.

These studies are interesting and important since they show us that as species diversity increases so the impact of the pathogens falls and plant productivity increases. At the current time it is not clear how much these pathogens account for the total changes in productivity observed and as Mitchell points out the overall impact depends on species composition and the presence of basal species. Nevertheless, one point is clear, parasites are an important component that have a dominating influence on ecosystem functioning and their presence should not be ignored.

Lyme disease and biodiversity

The previous section examined how pathogen load changed with the diversity of grassland species and showed that as diversity fell, so relative abundance

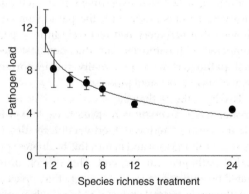

Figure 0.2 Decrease in pathogen load with increasing species richness of grassland plants as determined by experimental manipulation.

Notes: $p < 0.001$; $r^2 = 0.181$; and $n = 147$.
Source: After Mitchell *et al.* (2002).

of the few remaining species increased and this resulted in increased transmission of the specialist pathogens. However, one of the interesting findings was that while the overall pattern was clear, there was high variation between trials since the species composition varied and certain species influenced the pathogen load more than others. When species richness falls, the order of loss is often not random, we can predict that some species are more vulnerable to disturbance than others and these will be lost first. If these species vulnerable to disturbance are also less important in disease transmission, then this will result in a relative increase in the pathogen load over and above changes in relative density.

What happens when the less competent hosts are lost first? LoGiudice *et al.* (2003) examined this question in the zoonotic Lyme disease system of North America, a disease caused by the spirochete, *Borrelia burgdorferi* that is transmitted between hosts by the tick, *Ixodes scapularis*. The immature stages of the ticks are generalists and feed on a wide range of mammalian, avian, and reptilian hosts but while these hosts may provide a blood meal for the ticks they are not all competent hosts for the transmission of the Lyme disease pathogen. In other words, when an infected tick bites some hosts the pathogen is introduced into a 'dead end' host and lost from the system so that when naïve ticks bite the host they do not become infected. The dominant competent host in the system is the white footed mouse, they can infect up to 90% of larval ticks depending on how many of the larvae are feeding on the non-competent host. When woodland habitat is degraded from pristine woodland to wood lots, the charismatic larger species (and those not competent for Lyme disease transmission) are often the first to go while the mice are invariably the last species. As species richness falls so proportionately more ticks feed on the white footed mouse and since the mouse is a competent host, and many of the other hosts are not competent, the overall level of prevalence in the ticks should increase and the risk of infection to humans increase. LoGiudice *et al.* (2003) tested this hypothesis by showing that the non-mouse hosts are relatively poor reservoirs for Lyme disease and dilute the disease by feeding ticks but not infecting them with the spirochete.

They captured the 10 main groups of hosts, estimated their relative density, counted the immature stages of ticks on each, the proportion that engorged and then moulted, and also estimated the relative abundance of the host to transmit the pathogen. These unique data provided a community level insight into the relative role of each vertebrate host species in the transmission of the disease. Since they also knew the approximate order of species loss (mice are lost last) from woodland areas as they become degraded they were able to show clearly that as species diversity increases so the infection prevalence in the nymphs, and so the risk of human infection falls. The mouse is so dominant that the effects depend strongly on what the mouse density is within the study area. Even so, squirrels had the highest dilution effect reducing prevalence by about 58% whereas shrews provided a rescue effect; they acted to dilute the effects of the most competent mouse hosts but could also maintain the spirochete in the community when mouse density was low.

This study shows that the buffering of disease prevalence is an interesting and important function provided by high biodiversity. The finding is important since it shows us that while the presence of the ticks is important, we must consider the biological role of the different host species. While changes in climate may influence the development and survival of vectors when they are not on the host, it may also influence the distribution and abundance of host species that may have a large effect on disease prevalence and risk of infection to humans.

How does parasitism affect biodiversity? Parasite mediated competition

I now want to turn the biodiversity and parasitism question on its head and ask the inverse of the previous question: how does parasitism affect biodiversity. There are well-defined hypotheses that a major driving force behind the evolution of species diversity is parasitism (e.g. Janzen 1970; Connel 1978). Rather than consider this in detail I wanted to introduce an important threat of parasitism to biodiversity and conservation: the effects of a reservoir host on the abundance and existence of a

more vulnerable species through the process of apparent competition, sometimes referred to as parasite mediated competition (Price 1980; Hudson and Greenman 1998). The evidence that parasites may drive some populations to extinction is frail. There is the clear example of a microsporidian parasite killing the last-known individual snail *Partula turgida* (Cunningham and Daszak 1998) but there are few documented cases where parasites alone have driven a species to extinction.

General theory assumes, logically, that the transmission of most parasites can be considered density dependent; so if a virulent pathogen is introduced into a susceptible host population it will reduce density but once density is reduced, transmission will fall and the population will not be extirpated. However, parasites could lead to local extinction when transmission is frequency dependent, independent of density but dependent on the contact rate between conspecifics. So, for example, when wild dog density fell in the Serengeti, they still lived in packs and held their social structure with daily contact rates and the reduced overall density did not mean the few remaining individuals were spread independent of each other. Sexually transmitted diseases depend on the frequency of partner exchange and not host density so HIV or syphilis increases with the number of partners each infected individual has a sexual relationship with and not the total density of hosts in the population. HIV would fade out in a dense community of strictly monogamous couples unless maintained through other forms of transmission such as blood transfusion. Similarly vector borne disease are frequency dependent since transmission depends on being bitten by a vector then the more often a host is bitten the more likely they are of being infected, irrespective of host density. Indeed vector borne diseases often exhibit inverse density dependence since as host density decreases, due to parasite induced mortality, so the remaining vectors focus on a smaller and smaller number of hosts thus increasing the likelihood of them being exposed and dying from the vector borne disease. In this instance disease could drive species to extinction.

Another way in which parasites can reduce biodiversity is in shared parasitism where two or more species share a parasite, in one the parasite causes little mortality but this species sustains the infection and there is between-species transmission such that a second, more vulnerable species receives the infection and suffers significant mortality and eventually becomes wiped out. This is parasite mediated competition. A preliminary glance at the literature and some of the reviews indicates that parasite mediated competition may indeed be rife. However few workers have examined the effects of parasites in detail and clearly separated the effects of direct from parasite mediated competition in the wild. One of the classic experiments sometimes referred to as parasite mediated competition is the laboratory study of *Tribolium* beetles by Park (1948) where he showed that a competitive interaction between two species was reversed when an *Adelina* parasite was added to the system but this appears to be a special case where the competitive ability of one species is reduced by the parasite rather than indirect competition via a shared parasite. The clearest example is the beautifully designed laboratory study by Bonsall and Hassell (1997) summarized in Chapter 8, but here I wish to highlight two field studies that have examined parasite mediated competition in the wild.

Squirrel invasion and parapox virus

Since its first introduction into Britain, the grey squirrel has spread and replaced the red squirrel. While the dominant theory was that resource competition was the underlying cause of the replacement, simulation modelling indicates that this alone can not account for the rate and pattern of red squirrel decline (Rushton *et al.* 1997). When the grey squirrel was introduced, it brought with it a parapox virus that may have had a big impact on red squirrels but not grey squirrels and may have had at least a helping hand in the demise of the red squirrel. A critical piece of evidence comes from the study by Tompkins *et al.* (2002) who showed that the virus caused a severe and deleterious disease in the red squirrels but had very little effect on the grey squirrels. Detailed modelling of the system shows that the parapox virus was likely to have

had played a critical role in the demise of the red squirrel even though the prevalence of infection was low and may have led to previous workers dismissing its role (Tompkins *et al.* 2003). These studies show that parasite mediated competition could be taking place but at this stage it is not clear whether this acts alone or the pathogen may be interfering with aspects of competitive ability. Disentangling the effects of parasites on direct competition from parasite mediated competition is not simple but the introduced squirrel appears to be having an effect on the endemic species through the effects of parasitism.

Game birds and gastrointestinal nematodes

One of the clearest examples of parasite mediated competition are some detailed studies on the shared nematode parasite *Heterakis gallinarum* that infects both the ring necked pheasant and grey partridges (Tompkins *et al.* 2000*a*). In this system it would appear that there is little direct competition between the two hosts, but the impact of the parasite on the grey partridge is enough to drive it to localized extinction. Intensification of agriculture in the British countryside has led to the loss of weeds and the invertebrates associated with them leading to the demise of the grey partridge (Potts 1986). Land owners and farmers faced with the loss of a quarry species replaced the partridge by rearing and releasing large numbers of pheasants. A number of farms and other areas have extensified their land management practices, introduced conservation headlands, beetle banks, and encouraged the habitat so partridges should recover. Unfortunately many of these have failed and there appears to be some other process acting that could be preventing recovery and this may be parasite mediated competition.

Tompkins *et al.* (2000*b*) infected pheasants and partridges in captivity with 500 eggs of the gastrointestinal nematode *H. gallinarum* and then followed changes in body condition and worm egg production. Both host species exhibited the classic self-cure and after 100 days had cleared infection but the fitness of the worms was roughly two orders of magnitude greater in the pheasants in that

each worm introduced into the pheasant produced 100 times more eggs than the in the partridge. What was more, the impact of the worms was such that the partridges lost condition while the pheasants did not, in other words worms in pheasants produce more infective stages but only partridges suffered from the infection indicating that the pheasants are potentially an important reservoir of infection. They then released partridges into five different areas where they had monitored infections in wild pheasants and the uptake of infective stages in the partridges was directly dependent on the intensities in the pheasants (Tompkins *et al.* 2000*b*). This demonstrated that the partridges were being infected through a common pool of infective stages and the larger this was the more this increased infection in the partridges and the bigger the impact on partridge condition. By modelling the multiple host system they were then able to demonstrate that these effects were sufficient to lead to the local demise of the partridges by the pheasants through the process of parasite mediated competition (Tompkins *et al.* 2000*c*). This is one of the few field based studies to show clearly that parasite mediated competition operates in the wild and that parasites have an effect on biodiversity.

The role of parasites in ecosystems

In the first section of this chapter I looked at how a specialist parasite (*T. tenuis* in red grouse) can influence host dynamics and so other natural enemies, the complexities of a specialist pathogen with a generalist vector (louping-ill), and then the far-reaching repercussions of a generalist pathogen in a relatively complex food web (rinderpest in the Serengeti). I think that detailed studies on the impacts and transmission of parasites provides us with an understanding of how parasites influence host dynamics and thus allows us to examine the consequences of parasitism in an ecosystem. These first order interactions are examined in more detail in Chapter 3.

In the second section I touched on two of the central questions to this book: how does biodiversity affect parasitism? And how does parasitism affect biodiversity? I have but scratched the surface and I

hope by doing so wetted your appetite for the main meal of the tome that follows. The following chapter (Chapter 1) takes this approach much further by applying our understanding of ecosystem science to the world of parasitism. One thing becomes abundantly clear, parasitism is a dominant part of many ecosystems and while there maybe few measures of the relative biomass of parasites in a system we should appreciate that parasites often have a very large turnover rate, so parasites may have a relatively large effect on energy flow. The whole idea that parasites redirect energy away from other trophic levels and how they operate within a spatially heterogeneous environment is explored fully in Chapter 4. These discussions are never far from the dominating concerns that many environmentalists, as well as biologists, have about our ecosystems, the loss of biodiversity and what we can do about it from a conservation strategy, as examined in Chapter 8. Perhaps we could take a few lessons from the few but beautifully detailed studies that have been undertaken in disturbed (Chapter 7) and hostile environments (Chapter 6).

One key issue I was keen to address, but felt I should leave for others more qualified, is the fact that parasites and hosts are in a wonderful evolutionary tension where the host develops increased resistance and the parasite generates the means of avoiding the host's response. The immune system is a complex and highly adaptive system that must outstrip the within host growth of the pathogen and at the same time carry a memory. This memory can have important implications to the way the infracommunity of the parasites develops and is observed (e.g. Lello *et al.* 2004) and current levels of infection within any mature vertebrate are the ghosts of infection past. The whole manner in which parasite communities are structured in ecosystems is examined in detail in Chapter 2 and then examined in relation to the structure of the food web in Chapter 4. The evolutionary tension I refer to, of course, not only influences the host's response but also the response of the pathogen, this includes features such as parasite induced susceptibility to infection (explored in Chapter 9) and the important effects of anthropogenic impacts on the environment, examined in Chapter 10. The integration of

evolutionary principles with those of epidemiology and ecosystems is extremely exciting and I feel the need to point you at a recent exciting paper by Grenfell *et al.* (2004) that brings together an evolutionary, phylogenetic approach with our understanding of epidemiology into a new field of investigation that has become known as phylodynamics. I suspect many of the dualities we observe in parasite–host systems arise because of the tensions between the evolutionary and epidemiological characteristics of parasites.

There is little doubt that the whole concept of how parasites fit into the ecosystem, how they interact with the community of hosts and climatic changes is rapidly becoming an exciting new field. One group of workers calls this integrated discipline conservation medicine (Aguirre *et al.* 2002; Hudson 2004) and there is now a new journal of *EcoHealth* which publishes papers on disease and ecosystem sustainability. This journal and study area examines in detail how the emergence and effects of pathogens and parasites interact with anthropogenic pollutants and the environment to influence disease dynamics. The vertical integration of the subject is essential; at the subcellular level we need to understand the very binding processes that allow a virus to enter a cell of a specific host and in the case of zoonotic diseases a second host that is phylogenetically distant (e.g. Hanta virus in mice and humans, West Nile virus in birds, humans, and horses). We need to embrace the machinations that results in relatively high rates of mutation in the RNA viruses, the genetic processes of recombination that can lead to the evolution of new strains, and how these interact with the complex arms of the hosts immune system. At the cellular level we need to examine how parasites locate the organ they inhabit, the complexities and shifting sands of antigen expression and then the consequences this can have for such subjects as assessing optimal mates for sexual reproduction. At the level of the individual host, we need to include the patterns of the parasite communities within and between hosts and how species interact and can influence the within host selective forces that

occur during the course of infection. One major development over the past 25 years has been the production of logical, generic models that integrate the findings at the individual level and then predict the epidemiological consequences at the population level (Anderson and May 1991; Hudson *et al.* 2002). This approach has served to identify the means of controlling infections in a number of important epidemics. We now need to extend these to the community level to understand how and when species may act as reservoirs, the means by which to control them (vaccinate reservoir or target host?), and the ecosystem consequences of our actions. The final frontier is to combine the fundamental aspects of ecology with immunology and evolutionary biology to understand the role of parasites in a dynamic ecosystem.

I want to finish with one final and ironic duality; the perceived importance of parasitism in natural systems is a simple reflection of the importance of disease in the human population. Before the 1950s ecological workers often considered parasitism of major importance in wild host systems. Later, the development of penicillin and vaccination programmes reduced the threat of infectious disease to humans and in the same manner the ecological texts of the time played down and usually ignored the role of parasites in population dynamics or community structure. Since then we have seen increased concern over emerging diseases, a series of important epidemics (HIV, SARS, Foot & Mouth, BSE, Avian Influenza, etc.), the development of resistance to anthelmintic drugs, and the threats of bioterrorism all of which have highlighted the role of parasites. In parallel we have seen increased concern and funding for disease related issues and modern ecology texts include parasitism in natural systems in at least one chapter. Modern parasitologists and the ecological parasitologists that are the authors of this book now see that the parasite–host relationship is full of nonlinearities that need investigating if we are to control the threats of infectious diseases, but at the same time encourages us to appreciate the pivotal role parasites can play in an ecosystem.

CHAPTER 1

Linking ecosystem and parasite ecology

Michel Loreau,[1] Jacques Roy,[2] and David Tilman[3]

Parasites are rarely considered in ecosystem studies. The current interest in the relationship between biodiversity and ecosystem functioning, however, has stimulated the emergence of new synthetic approaches across the traditional divide between population and ecosystem ecology. Here we provide a brief introduction to ecosystem ecology, an overview of current trends in the field of biodiversity and ecosystem functioning, and ideas about how parasites should and could be brought into ecosystem ecology.

1.1 Introduction

Host–parasite interactions have traditionally been approached from the viewpoint of population dynamics and epidemiology. In contrast, ecosystem ecology has traditionally focused on the 'big picture' of stocks and flows of mass and energy at the whole system level, in which parasites at first sight seem irrelevant because they account for such a low biomass. Parasites are rarely considered in ecosystem studies. For example, since its launch in 1998, the journal *Ecosystems* has not published a single paper containing the words *parasite, parasitism,* or *parasitoid* in its title, key words or even abstract! This nearly complete separation between parasites and ecosystems in modern ecology is an expression of the broader separation between population/community and ecosystem ecological approaches. The current interest in the relationship between biodiversity and ecosystem functioning, however, has stimulated the emergence of new synthetic approaches across the traditional divide

between community and ecosystem ecology (Jones and Lawton 1995; Kinzig *et al.* 2002; Loreau *et al.* 2002). In this chapter we provide a brief introduction to ecosystem ecology, an overview of current trends in the field of biodiversity and ecosystem functioning, and ideas about how parasites should and could be brought into the 'big picture' of ecosystem ecology.

1.2 Ecosystem ecology, an integrative science in need of further integration

Because of its central role in ecological thinking, the ecosystem concept has been extensively analysed by ecologists, historians, philosophers, and linguists (e.g. Hagen 1992; Golley 1993; Dury 1999; Pickett and Cadenasso 2002). A historical overview of this concept helps to grasp the fundamentals of ecosystem science, its progress in half a century of existence and its current challenges.

1.2.1 The ecosystem concept in a historical perspective

Since it was first introduced by Tansley (1935), the ecosystem concept has designated not only the sum of the organisms and their abiotic environment, but also the 'constant interchange of the most various

[1] Laboratoire d'Ecologie, UMR 7625, Ecole Normale Supérieure, 46 rue d'Ulm, F-75230 Paris Cedex 05, France.
[2] Centre d'Ecologie Fonctionnelle et Evolutive, UMR 5175, CNRS F-34293 Montpellier Cedex 5, France.
[3] Department of Ecology, Evolution and Behavior, University of Minnesota, St. Paul, MN 55108 USA.

kinds within each system, not only between the organisms but between the organic and the inorganic' (Tansley 1935). Lindeman (1942) and Odum (1959, 1969) also stressed the exchange of energy and materials between the living and non-living parts in their definitions of the ecosystem. Odum (1959) recognized 'four constituents as comprising the ecosystem: (1) abiotic substances, basic inorganic and organic compounds of the environment; (2) producers, autotrophic organisms, largely green plants [...]; (3) consumers (or macro-consumers), heterotrophic organisms, chiefly animals [...]; (4) decomposers (micro-consumers, saprobes or saprophytes), heterotrophic organisms, chiefly bacteria and fungi, which [...] release simple substances usable by the producers.' With these constituents, an ecosystem is a 'life-support system [...] functioning within whatever space we chose to consider whether it be a culture vessel, a space capsule, a crop field, a pond, or the Earth's biosphere' (Odum 1964). These initial views are still prevailing among current ecosystem ecologists. Thus, for Chapin *et al.* (2002), 'ecosystem ecology addresses the interactions between organisms and their environment as an integrated system. [...] The flow of energy and materials through organisms and the physical environment provides a framework for understanding the diversity of form and functioning of Earth's physical and biological processes'.

Some authors, however, gave extended, more abstract definitions of the ecosystem. Evans (1956) suggested that the ecosystem concept could be used at any organizational level of life. On this view, any organism with its micro-environment could potentially constitute an ecosystem. Higashi and Burns (1991) distinguished two ecosystem concepts: 'the ecosystem as a physical entity' following Tansley (1935), and 'the ecosystem as a paradigm for science: an entity–environment unit'. With such extended definitions, a host and its parasites could be viewed as an ecosystem (as in Thomas *et al.* 1999a, see also Chapter 8). Pickett and Cadenasso (2002) emphasized the flexibility of the definition, the ecosystem concept being scale independent and free of narrow assumptions such as equilibrium. This general concept can then be applied to an array of models whose characteristics depend on the issue being addressed and on the nature of the system under study. Depending on the model, energy, nutrient, biodiversity, or economics can be the focus of the study (Pickett and Cadenasso 2002).

For a large majority of ecologists, and in particular for those who bridge fundamental research and applied problem solving, an ecosystem is clearly a 'spatially explicit unit of the Earth' (Likens 1992). As such it comprises abiotic substances, autotrophic and heterotrophic organisms, and their interactions. The nature and consequences of these interactions, however, has fuelled a recurrent debate: do these interactions lead to emergent properties and integration of the ecosystem into a self-regulated functional unit? A number of scientists working on subsets of ecosystems, such as physiologists Engelberg and Boyarsky (1979) and community ecologist Simberloff (1980) questioned this idea, whereas ecosystem ecologists generally supported the cybernetic nature of ecosystems (McNaughton and Coughenour 1981; Patten and Odum 1981). The strongest, and most controversial, form of this view is probably Lovelock's (1995) Gaia theory of the total Earth system as a single self-regulating unit. The debate on these issues, however, has often been led astray to one-sided positions. The theories of complex adaptive systems (Levin 1999) and multilevel natural selection (Wilson 1980), for instance, provide frameworks to understand the ecological and evolutionary emergence of properties at higher levels of organization without invoking strong top–down, integrated regulation.

Chapin *et al.* (1996, 2002) provide a useful synthetic view of what controls ecosystem structure and functioning. According to them; five external factors set the bounds for ecosystem properties: parent rock material, topography, climate, time, and potential biota. Within these bounds, actual ecosystem properties are set by a suite of interactive controls: (1) resources (soil, water, air); (2) physical and chemical modulators (such as local temperature and pH, which affect organisms without being consumed by them); (3) disturbance regime; (4) the biotic community, and; (5) human activities which affect all the other controls. The biotic community

influences ecosystem functioning through the trophic levels present, the number of species within each trophic level, their relative abundances, and their identity. Dominant species (in term of biomass) and species with particular functional attributes (like mycorrhizal fungi) are the species with *a priori* the largest role. Populations of these species are regulated by a set of negative and positive interactions among species, and parasitism is one of them. With only an indirect role on one of the five interactive controls of ecosystem processes, parasites understandably are not of first concern to most ecosystem ecologists.

1.2.2 Ecosystem science, its achievements and frontiers

Ecosystem science is characterized by the processes it addresses rather than by the type of system it deals with, although it is more often conducted at high levels of organization (several trophic levels) and large spatial scales (from a plot to the whole Earth). It is concerned mainly with the pools and the fluxes of energy and materials among ecosystem components (in contrast to population and community ecology which are concerned with the demography, diversity, and interactions of the organisms living in ecosystems). Its aim is usually to understand how these pools and fluxes are regulated by the interactive controls mentioned above, but also how they set constraints on the structure of ecosystems (community types and diversity). Temporal and spatial patterns of ecosystem processes and ecosystem management are also of primary concern. The increasing impact of humans on all these aspects and its consequences are often at the forefront of ecosystem ecology (Vitousek *et al.* 1997).

Accomplishments of ecosystem science have been numerous (Pace and Groffman 1998), and include understanding the flow of water and chemical elements and compounds in watersheds, rivers, lakes, estuaries, and oceans; analysing feedbacks between plants and animals and their biophysical environment; understanding the causes of, and remedies to, eutrophication; understanding the biophysical basis of production and its coupling to climate; assessing the importance of below-ground

processes in terrestrial ecosystems; and recognizing the scale dependence of most ecosystem processes (Carpenter and Turner 1998). These accomplishments have been mainly achieved through (1) comparative studies of natural ecosystems (e.g. Matson and Vitousek 1987; Turner *et al.* 2001); (2) long-term field studies (Gosz 1996; Hobbie *et al.* 2003); (3) experimental manipulation of ecosystems from model laboratory systems to large-scale field experiments (Beyers and Odum 1993; Lawton 1995; Schindler 1998); and (4) theory and mathematical modelling (Tilman 1988; DeAngelis 1992; Ågren and Bosatta 1996; Loreau 1995, 1998*a*).

The ecosystem approach is fundamental to managing the Earth's resources. Ecosystem ecology often bridges fundamental research and applied problem solving. When environmental concerns moved from the local scale in the 1960s to the regional and now global scales, so did ecosystem science. These three scales cannot replace each other, however, and basic research is still needed at all scales. For example, the knowledge of a basic ecosystem process such as primary productivity, whose study was fostered from the 1960s by the International Biological Programme, is still developing fast, integrating new techniques, control factors, and scales (Canadell *et al.* 2000; Roy *et al.* 2001). But the main challenge ahead is getting more strongly involved in solving the ever-increasing environmental problems and working towards a more sustainable future (Lubchenco 1998; Gosz 1999). Integrating across scales is a prominent task (Levin 1992; Carpenter and Turner 2000*b*), as is integrating the various controls of ecosystem processes. Taking into account the role of biodiversity in ecosystem functioning is a critical, fast-developing area, which we develop in the next section. Integrating the socio-economic aspects of human activities from local to global scales is a novel dimension which will be crucial for achieving a sustainable management of ecosystems (Carpenter and Turner 2000*a*; Costanza 2000; Di Castri 2000). Efforts are also needed to develop stronger communication and cooperation among the research, policy and public spheres (Baron and Galvin 1990; Rykiel 1997). The Millennium Ecosystem Assessment is an example of such efforts (Ayensu *et al.* 1999; Samper 2003).

Despite its achievements in basic and applied science, ecosystem ecology has developed until recently in growing isolation from other fast-moving ecological subdisciplines such as population ecology, community ecology, and evolutionary ecology. The level of integration that it promotes has stimulated links with other scientific disciplines such as chemistry and geology, but has also tended to diminish links with other biological disciplines. Reciprocally, population ecology, commun-ity ecol-ogy, and evolutionary ecology have until recently largely ignored the higher level of integration offered by ecosystem ecology. This separation between subdisciplines that provide different per-spectives on the same ecological reality is a funda-mental limitation which needs to be overcome if we are to understand the predominantly biological basis of ecosystems, the reciprocal constraints that individual species and ecosystems exert on each other on ecological and evolutionary time-scales, the role of biodiversity in ecosystem functioning, and more particularly the role of parasites and of their diversity in ecological systems.

1.3 Biodiversity and ecosystem functioning, a new area that synthesizes population, community, and ecosystem ecology

The relationship between biodiversity and ecosys-tem functioning has emerged as a new research area at the interface between community ecology and ecosystem ecology which has expanded dra-matically during the last few years (see syntheses in Loreau *et al.* 2001, 2002; Kinzig *et al.* 2002). This new area finds its origin in a questioning that started only a decade ago on the potential con-sequences of biodiversity loss which results from the increasing human domination of natural ecosystems, a domi-nation that is likely to further develop considerably during the twenty-first century (Schulze and Mooney 1993).

Three types of reasons have been put forward to justify current concerns about threats to biodivers-ity. First, biodiversity is the source of natural resources that lead to the direct production of goods that are of economic value, such as food, wood fibre, new pharmaceuticals, genes that improve crops, or organisms that are used for bio-logical control of pests. Second, biodiversity is viewed as linked to human well-being for aesthetic, ethical, and cultural reasons. Third, biodiversity may contribute to the provision of ecosystem services that are of value to society, but are generally not given an economic value, such as primary and sec-ondary production, plant pollination, climate regu-lation, carbon sequestration, the maintenance of water quality, and the generation and maintenance of soil fertility. It is this third possibility that gave rise to the interest in biodiversity and ecosystem functioning: could biodiversity loss alter the func-tioning of ecosystems, and thereby the ecological services they provide to humans?

When this question was posed in the early 1990s, scientific ecology had a number of theories and empirical data that clearly showed the importance of 'vertical' diversity, that is, functional diversity across trophic levels along the food chain, in ecosystems. An eloquent example of the dramatic impacts that changes in vertical diversity can have is provided by the kelp–sea urchin–sea otter food chain in the Pacific. Removal of sea otters by Russian fur traders allowed a population explosion of sea urchins that overgrazed kelp (Estes and Palmisano 1974). Reduction in kelp cover in turn leads to extinction of other species living in kelp, as well as increased wave action, coastal erosion, and storm damage (Mork 1996). More intense herbivory in the absence of sea otters has also been shown to trigger evolution of chemical defences in kelp (Steinberg *et al.* 1995). Thus, removal of a single top predator generates a cascade of population dynam-ical, physical, and even evolutionary effects within ecosystems.

In contrast, little was known on the ecological significance of 'horizontal' diversity, that is, genetic, taxonomic, and functional diversity within trophic levels. Different theories of coexistence among competing species have vastly different implications for the relationship between species diversity and ecosystem processes. To take two extreme examples, neutral theory assumes that all species in a commun-ity are equivalent (Hubbell 2001). This implies functional redundancy among species, and hence an

absence of any effect of changes in diversity on aggregate community or ecosystem properties. At the other extreme, niche theory postulates that all species differ to some extent in the resources they use. This implies functional complementarity among species, and hence increased productivity and other ecosystem processes with diversity (Tilman *et al.* 1997*a*; Loreau 1998*b*).

To investigate the effects of 'horizontal' diversity on ecosystem processes, a new wave of experimental studies was developed using synthesized model ecosystems. Many of these studies were focused on effects of plant taxonomic and functional-group diversity on primary production and nutrient retention in grassland ecosystems. Because plants, as primary producers, represent the basal component of most ecosystems, they represented the logical place to begin detailed studies. Several, though not all, experiments using randomly assembled communities found that plant species and functional-group richness has a positive effect on primary production and nutrient retention (e.g. Tilman *et al.* 1996, 1997*b*; Hector *et al.* 1999; Fig. 1.1). Although the interpretation of these experiments was hotly debated (e.g. Huston 1997; Huston *et al.* 2000; Hector *et al.* 2000), this controversy has been largely resolved by a combination of a consensus agreement on a common conceptual framework (Loreau *et al.* 2001), the development of a new methodology to partition selection and complementarity effects (Loreau and Hector 2001), and new experimental data (Tilman *et al.* 2001; van Ruijven and Berendse 2003). These new studies have all shown that plant diversity influences primary production through a complementarity effect generated by niche differentiation (which enhances resource exploitation by the community as a whole) and facilitation. Thus, there is little doubt that species diversity does affect at least some ecosystem processes, even at the small spatial and temporal scales considered in recent experiments, although it is still difficult to assess how many species are important to generate functional complementarity.

Even if high diversity were not critical for maintaining ecosystem processes under constant or benign environmental conditions, it might nevertheless be important for maintaining them under changing conditions. The 'insurance' and 'portfolio' hypotheses propose that biodiversity provides a buffer against environmental fluctuations because different species respond differently to these fluctuations, leading to functional compensations

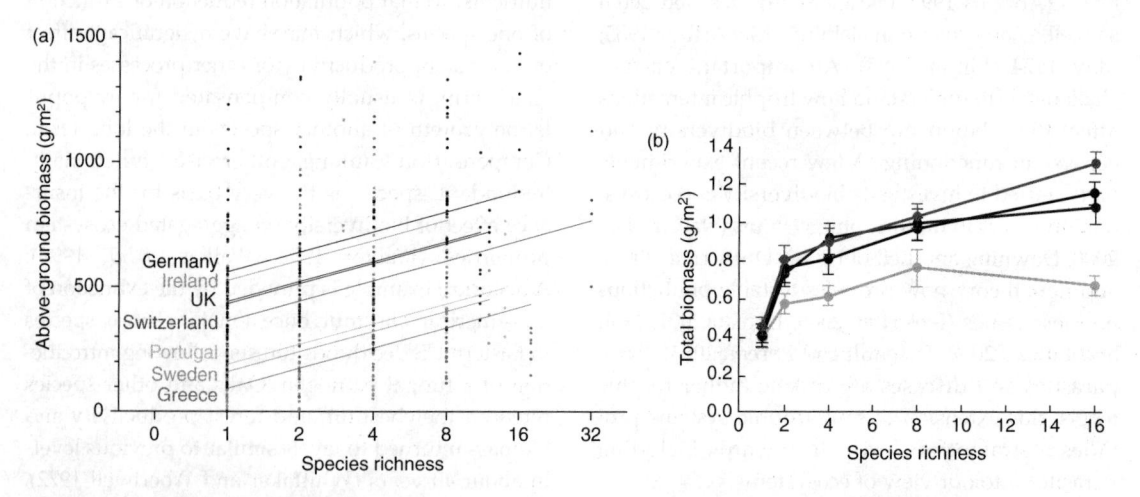

Figure 1.1 Responses of plant biomass to experimental manipulations of plant species richness in grassland ecosystems: (a) above-ground plant biomass in eight sites across Europe; (b) total plant biomass (mean ± standard error) during several years in Minnesota.

Note: Points in (a) are data for individual plots.

Sources: (a) Modified from Hector *et al.* 1999; and (b) Modified from Tilman *et al.* 2001.

between species and hence more predictable aggregate community or ecosystem properties. A number of studies have recently provided theoretical foundations for these hypotheses (e.g. McNaughton 1977; Doak *et al.* 1998; Tilman *et al.* 1998; Yachi and Loreau 1999; Lehman and Tilman 2000). Several empirical studies have found decreased variability of ecosystem processes as diversity increases, despite sometimes increased variability of individual populations, in agreement with the insurance hypothesis (e.g. Tilman 1996; McGrady–Steed *et al.* 1997). The interpretation of these patterns, however, is complicated by the correlation of additional factors with species richness in these experiments, which does not fully preclude alternative interpretations (e.g. Huston 1997).

An important limitation of virtually all recent theoretical and experimental studies on the effects on biodiversity on ecosystem functioning and stability is that they have concerned single trophic levels—primary producers for the most part. Although they have contributed to merging community and ecosystem ecology, they have unintentionally disconnected 'vertical' and 'horizontal' diversity and processes. Yet it is well known that trophic interactions can have important effects on the biomass and productivity of the various trophic levels (Abrams 1995; Oksanen and Oksanen 2000) as well as on ecosystem stability (MacArthur 1955; May 1974; Pimm 1984). An important current challenge is to understand how trophic interactions affect the relationship between biodiversity and ecosystem functioning. A few recent experiments have started to investigate biodiversity and ecosystem processes in multitrophic systems (Naeem *et al.* 2000; Downing and Leibold 2002; Duffy *et al.* 2003), and new theory now provides testable predictions on these issues (Ives *et al.* 2000; Loreau 2001; Holt and Loreau 2002; Thébault and Loreau 2003). Since parasites and diseases are cryptic higher trophic levels, this extension to multitrophic systems provides a straightforward path towards including parasites into our view of ecosystems.

1.4 Parasites in ecosystems

Parasites are typically small-sized organisms that exploit their host both as a food resource and as a habitat. They affect their host negatively either because they alter specific physiological functions or because they multiply and develop large populations within their host; individually their effect is often very small. Even collectively, their biomass and the amount of material and energy they process is often much smaller than the biomass and the material and energy flows of their host. This explains why parasites have traditionally been ignored by ecosystem ecology: they are hidden within their host, and their direct ecosystem impact is seemingly negligible.

Yet their indirect impact on ecosystem processes can be substantial through their effect on their host. Here we explore some of the ways in which they exert strong indirect influences on the biodiversity and functioning of ecosystems.

First, parasitism and disease are probably one of the most significant causes of population regulation in many species under natural conditions (see Chapter 3). By regulating populations of dominant species they can have significant effects on ecosystem processes (see Chapter 8). Massive mortality or fertility reduction in individual species, however, may be of little long-term significance for ecosystem properties under natural conditions, especially in plants. Plants compete strongly for space, light, and nutrients, so that population reduction or extinction of one species, which may have a significant effect on ecosystem productivity or other processes in the short term, is usually compensated for by population growth of another species in the long term. Compensation among otherwise functionally 'redundant' species is the very basis for the insurance effect of biodiversity on aggregated ecosystem properties (Walker 1992; Walker *et al.* 1999). A historical example is provided by the extinction of the American chestnut, once a major canopy species in Eastern US deciduous forests, following introduction of a fungal pathogen. Oaks and other species replaced the chestnut, and forest productivity and biomass returned to levels similar to previous levels in about 40 years (Whittaker and Woodwell 1972). Effects of parasites on individual animal populations might be more significant for ecosystem processes and services—at least as we perceive them from our anthropocentric perspective—because animals often have more specific roles in the complex interaction

networks of natural ecosystems. For instance, in a successful attempt to control the proliferation of the European rabbit, introduction of the myxoma virus in Australia led to decimation of rabbit populations (Fenner and Ratcliffe 1965). Rabbit mortality helped restore the vegetation which supported sheep populations utilized for wool production in range and pasture lands. Little is known on the net effect of myxomatosis on total primary and secondary production, but wool production at least was strongly influenced by the presence of the myxoma virus.

Second, by exerting top-down control on populations from lower trophic levels, parasites may substantially alter the diversity of their host species and its effects on ecosystem processes. Higher trophic levels can generate hump-shaped or other complex nonlinear relationships between species diversity and ecosystem processes (Thébault and Loreau 2003). These nonlinear relationships are critically dependent on where and how top-down or bottom-up controls occur in the food web. For instance, when all plant species are controlled from the top down by specialized herbivores, there is a monotonic increase in total plant biomass as diversity increases. By contrast, when some plant species escape top-down control or when herbivores are generalists, a unimodal relationship can emerge between total plant biomass and diversity (Fig. 1.2). Whether the agents of top-down control are herbivores or parasites is immaterial to these theoretical results. Therefore these should apply to parasites as well. Application of insecticide to a biodiversity experiment revealed major effects of insect herbivores on the relationship between plant diversity and primary productivity: there was a strong positive response of above-ground plant biomass production to plant diversity when insect herbivores were reduced, which was not apparent when herbivores were unchecked (Mulder et al. 1999). The reason for this difference lies again in the top-down control exerted by insects on plants, which diverts part of primary production to the herbivore trophic level. There is no reason why this should not apply to parasites too. Seed predators and pathogens are hypothesized to be one of the main factors maintaining tropical tree diversity (Janzen 1970; Connell 1971; Wright 2002, see also Chapter 8). If

this is the case, they may have a major influence on ecosystem processes in tropical forests despite their very low biomass. Similarly, viruses are arguably one of the major factors that maintain (through selective exploitation), and even create (through gene transfer), microbial diversity (Weinbauer and Rassoulzadegan 2004). Their indirect impact on microbially driven ecosystem processes, in particular nutrient cycling, should accordingly be considerable, although it is still poorly known.

Third, a well-established body of theory and empirical evidence shows that there is a gradual transition from parasitism to mutualism on both ecological and evolutionary time scales (Maynard Smith and Szathmary 1995; Johnson et al. 1997). In particular, the nature and intensity of symbiotic interactions can be highly variable, and change from mutualistic to parasitic, and vice versa, depending on local environmental conditions. For instance, mycorrhizal fungi are usually mutualists for their associated plant partners because they help them to better capture soil nutrients. In fertile soils with high nutrient concentrations, however, they become parasitic because plants no longer need them to gain access to soil nutrients while they still incur the cost of providing them with carbon resources (Johnson et al. 1997). As a consequence of this high variability in the benefits and costs derived by the two partners, mycorrhizal fungi have highly species-specific effects on plants, and may strongly affect the diversity, species composition, and relative abundances of plant communities (van der Heijden et al. 1998). Mycorrhizal diversity thereby contributes to maintaining plant diversity and primary productivity under nutrient-limited conditions (van der Heijden et al. 1998; Klironomos et al. 2000). Under nutrient-rich conditions, however, mycorrhizal fungi may behave as plant parasites. Their impacts on plant diversity and productivity are then expected to be more complex as discussed above. Similar shifts in impacts on plant-based ecosystem processes are likely to occur for other plant parasites as environmental conditions change and alter the physiological status of the two partners.

Lastly, nutrient cycling is a key process that determines the productivity of all trophic levels in nutrient-limited ecosystems (DeAngelis 1992;

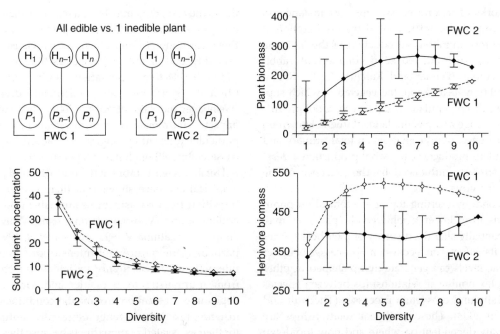

Figure 1.2 Expected soil nutrient concentration, total plant biomass and total herbivore biomass (mean ± standard deviation across all species combinations) as functions of species richness for two food web configurations: a food web in which each plant species is controlled by a specialized herbivore (dotted lines), and a food web in which one plant species is inedible and lacks a specialized herbivore (solid lines). Herbivore species richness varies parallel to plant species richness to keep the same food web configuration along the diversity gradient.

Source: Modified from Thébault and Loreau (2003).

Loreau 1995, 1998*a*; de Mazancourt *et al.* 1998). Heterotrophic consumers such as herbivores, carnivores, and parasites can substantially influence primary production through nutrient cycling. They can even increase primary production if they channel limiting nutrients towards more efficient recycling pathways, that is, to recycling pathways that keep a greater proportion of nutrients within the system (de Mazancourt *et al.* 1998). Although this theory has been mainly applied to the debated grazing optimization hypothesis, that is, the hypothesis that herbivores maximize plant production at a moderate grazing intensity, this theory should apply to parasites as well. By altering the timing and spatial location of their host's death, parasites may contribute to release nutrients locked in their host's biomass at times and places that are favourable for the conservation of these nutrients within the ecosystem or, conversely, for their loss from the ecosystem by such processes as leaching, volatilization, and sedimentation. In the former case

they will tend to enhance local productivity; in the latter case they will tend to depress local productivity. Stoichiometric constraints also come into play. For instance, bacterial decomposers often immobilize a substantial amount of limiting nutrients such as nitrogen and phosphorus because their carbon : nutrient ratio is lower than the carbon : nutrient ratio of plant dead organic matter which is their main resource (Tezuka 1989; Ågren and Bosatta 1996). Parasitic viruses are likely to enhance nutrient cycling, and hence primary production, by killing bacteria and making nutrients available again to plants.

1.5 Concluding remarks

Ecosystem ecology has provided an integrative perspective of the interactions between biological organisms and their abiotic environment, especially at relatively large spatial scales. However, it would be strengthened by better ties to, and synthesis of,

the insights and approaches of population ecology, community ecology, and evolutionary biology. After all, organisms simultaneously experience all the forces of nature, including those that are the foci of evolutionary, population, community, and ecosystem ecology. Each of these perspectives has been, and will continue to be, useful simplifications. Their synthesis, we assert, is likely to provide novel and important insights into all branches of ecology.

Recent theoretical and experimental work provides evidence that biodiversity dynamics can have profound impacts on functioning of natural and managed ecosystems and their ability to deliver ecological services to human societies. Work on simplified ecosystems in which the diversity of a single trophic level—mostly plants—is manipulated shows that taxonomic and functional diversity can enhance ecosystem processes such as primary productivity and nutrient retention. Theory also strongly suggests that biodiversity can act as biological insurance against potential disruptions caused by environmental changes. One of the major challenges, however, is to extend this new knowledge to multitrophic systems that more closely mimic complex natural ecosystems.

The role of parasites in ecosystem functioning has usually been underestimated and poorly investigated because of their low biomass, low visibility, and small direct contribution to energy and material flows in natural ecosystems. We have provided several arguments why they may nevertheless have significant indirect impacts on ecosystem properties, by controlling numerically dominant host species, by exerting top-down control and maintaining the diversity of lower trophic levels, by shifting from parasitic to mutualistic interactions with their hosts, and by channelling limiting nutrients to more or less efficient recycling pathways.

Despite recent progress towards greater convergence and dialogue between population, community, and ecosystem ecology, much remains to be done to achieve full integration of these subdisciplines. In particular, the potential importance of parasites and disease emphasize the need to take into account both direct and indirect effects in our view of ecosystems. Although indirect effects have received increasing attention in community ecology recently (Wootton 1994; Abrams 1995), their importance for ecosystem functioning has seldom been considered. Parasites, just as microbes, remind us that small causes can have large effects. Unless we better develop our understanding of the ecological significance of the whole of biodiversity, including that of parasites, we have an insufficient understanding of the functioning of natural and managed ecosystems.

CHAPTER 2

Are there general laws in parasite community ecology? The emergence of spatial parasitology and epidemiology

J.-F. Guégan,[1] S. Morand,[2] and R. Poulin[3]

Recent insights into both population and community ecology of host-parasite relationships have shown the importance of spatial processes in influencing the structure of local parasite and microbe communities. This now requires from parasitologists, epidemiologists and evolutionary biologists working on those interactions that they place their analyses into a broader spatial perspective. Because local species richness and composition in parasites and pathogens depend on large-scale species pools, a greater consideration of epidemiological processes will favour the emergence of spatial parasitology and epidemiology devoted to understanding population dynamics and community structure.

2.1 Introduction

There is an increasing interest in parasite and infectious disease population (Grenfell and Dobson 1995; Hudson *et al.* 2001) and community (Esch *et al.* 1990; Poulin 1998*a*; Rohde 2001) ecology, and interestingly this has developed at a time when mainstream ecologists have shown increasing interest in metapopulation theory and habitat fragmentation (Hanski and Gilpin 1997; Hanski 1999), population dynamics in fragmented landscape (Hassell and Wilson 1997; Ferguson *et al.* 1997; Grenfell and Harwood 1997; Rohani *et al.* 1999) and macroecology

[1] GEMI, UMR IRD-CNRS 2724, Centre IRD de Montpellier, 911 Avenue Agropolis BP 64501, 34394 Montpellier Cedex 5, France.
[2] CBGP, UMR INRA-IRD-CIRAD-Agro.M., Campus International de Baillarguet, CS 30016, 34988 Montferrier sur lez cedex, France.
[3] Department of Zoology, University of Otago, PO Box 56, Dunedin, New Zealand.

(Brown 1995; Rosenzweig 1995; Maurer 1999; Lawton 2000; Gaston and Blackburn 2000). The development of what is now called spatial ecology is one of the great triumphs of modern population and community ecology (Tilman and Kareiva 1997), which has showed the critical importance of space and spatial characteristics for understanding a wide range of ecological phenomena (Holt 1993, 1999). There are clear analogies between modern spatial ecology and parasite–infectious disease population and community ecology, and this chapter will be devoted to a review of the recent development in parasite–infectious disease population and community ecology within this fruitful cross-fertilizing arena.

There are considerably more studies available on parasitic systems today than 10 years ago (see Poulin 1997, 1998*a*; Poulin *et al.* 2000 for a review), and many of these investigations have clearly showed the role of dynamical processes in a spatial context (Ferguson *et al.* 1997; Grenfell and Harwood 1997; Morand and

Guégan 2000*a*; Poulin and Guégan 2000; Morand *et al.* 2002). Most if not all parasites live as populations that are divided into metapopulations on several spatial scales, and each host operates as a patch. Many of the topics in modern spatial ecology have also their parallels in within-host infection processes with the individual host body forming an heterogeneous environment (see Holt 1999). From the perspective of a parasitic larval form or microbe, an individual host is an extraordinary landscape to invade with heterogeneity in resource availability and colonization-extinction risks. An infrapopulation is thus defined as all the members of a given parasite species within a single host individual, and an infracommunity includes all of the infrapopulations within an individual host. The next hierarchical level includes all the infrapopulations sampled from a given host species within an ecosystem, and which forms the metapopulation. Parallel to the metapopulation is the component parasite community which represents all of the infracommunities within a given host population. Then the highest level of parasite organization is the suprapopulation which represents all individuals of a given parasite species within an ecosystem. Next, the parasite compound community consists of all the parasite communities within an ecosystem (see Esch *et al.* 1990). This creates at least a third-order scaling of habitat fragmentation for the parasites which has a significant impact on the development of theory regarding the evolution of populations and communities of parasites and pathogens. Infrapopulations and infracommunities may form many replicates from one host to another, thus providing a remarkable opportunity for comparative analyses of the variability of organizational patterns at several hierarchical levels, so much more difficult to explore for free-living organisms. Furthermore, this hierarchical organization means that larger-scale processes may have a strong influence on local community structure (see Poulin 1998*a*; Poulin *et al.* 2000) and dynamics of parasites and microbes (see Grenfell and Harwood 1997; Rohani *et al.* 1999), indicating that these larger-scale phenomena cannot be ignored anymore. For instance, recent advances in epidemiology of childhood diseases have clearly shown the influence of spatial fluxes on local disease dynamics, recognizing structural similarities between the processes of

metapopulation biology and infection dynamics (Grenfell and Harwood 1997; Rohani *et al.* 1999).

Price (1990) in his contribution to the seminal book by Esch, Bush, and Aho (1990) argued that parasite community ecologists should take a leading role in advancing areas of ecology with many parasite studies being attractive complements for investigating some of the major questions in population and community ecology. Nearly one decade and a half after, the intent of the present chapter is to synthesize the more recent developments in parasite and microbial community ecology, and to assess current perspectives regarding our knowledge of these communities.

To see where we are heading, consider a few simple questions one could ask about a parasite or microbial community. What determines the number of parasite species one host individual can harbour? Why are some parasite species extremely rare when others are very common? What is the local population abundance of a widely distributed macroparasite when compared to that of a rare species? These very basic questions of (parasite and infectious disease) community ecology have extraordinary little to do with small-scale processes, but on the contrary need that we explore larger-scale phenomena. Having described the different macroscopic patterns, we then explore the consequences of this research framework in population and community ecology of parasites and microbes. We opted in the present chapter to use examples from both the microbial and parasitological community literature to illustrate the various concepts. We conclude by highlighting what these findings mean for further study of population dynamics and community ecology of parasitic and infectious disease in wildlife and humans.

2.2 Parasite community organisation and species coexistence

2.2.1 The emergence of spatial ecology in infectious and parasitic diseases population and community dynamics

As for mainstream population and community ecology (see Putman 1994; Begon *et al.* 1996; Weiher and Keddy 1999), parasite population and community ecology has concentrated on local processes with an emphasis on local interactions between parasite species, and between these species and their host environment (see Kennedy 1975; Cheng 1986; Esch and Fernandez 1993; Combes 1995; Bush *et al.* 2001). Another aspect is the strong research effort made over the past decades on very untidy small-scale studies, which do not take a strongly quantitative approach to issues such as spatial patterns in species richness at very large scales or patterns in species distributions (but see Price 1980; Rohde 1982; Poulin *et al.* 2000; this chapter). For instance, what determines the structure of the local community of human infectious diseases in a given place of Western Africa has certainly as much to do with large-scale, biogeographical processes as it has to do with local conditions. This view is shared by an increasing number of researchers interested in the influence of climate variability on regional disease dispersion and diffusion, for instance (Dobson and Carper 1993; Rogers and Williams 1993; Hay *et al.* 1996; Patz *et al.* 1996; McMichael and Haines 1997;

Rapport *et al.* 1998; Epstein 1999; Rogers and Randolph 2000; Aron and Patz 2001).

2.2.2 Some definitions and basic conceptual framework

Many processes studied by parasite and microbial community ecologists have clear linkages with larger-scale, regional phenomena, but since studies have focused too largely on small spatial scales they are obviously unable to put the communities into perspectives (see Lawton 2000; Poulin *et al.* 2000; this chapter). The assembly of parasites or microbial communities as for free-living ones is a multistage, multi-layered process, and this forms a conceptually important framework on which to base further research investigations. First, it starts at the top of Fig. 2.1 with the largest-scale pool of species. The existence of a global-scale pool of parasite and microbe species is entirely relevant for many organisms like in the case of crop pests, viruses, bacteria, and fungi, or human infectious and parasitic diseases (see Rapport

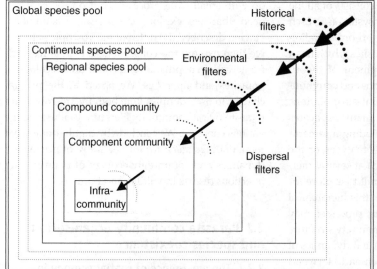 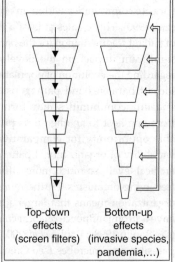

Figure 2.1 Schematic illustration of the main determinants of parasite or pathogen species richness in hosts. On the left: processes influencing species diversity are arranged into hierarchies in which different temporal and spatial factors may act. Any given level includes all lower levels, and it is included within all higher levels. On the right: the hierarchical scheme indicates that parasite or pathogen community assemblages at lower-spatial scales, for example an individual host, are strongly dependent on upper scales (top-down effect on community richness), but the opposite situation where lower-spatial scales may influence higher levels (bottom-up effect) is also possible.

Notes: Invasive species like crops or pandemia like HIV are illustrative of global impact of parasites and pathogens. Further research should reveal the respective roles played by 'top-down' and 'bottom-up' effects on community assemblages of parasites and microbes.

et al. 2002). This notion is at present made highly relevant by the development of transcontinental transports and economical exchanges between two distant biogeographical regions, making the Earth today a global village for many pathogens and diseases (see Poulin 2003*a*, and the later Section 2.3.2.4 'Latitudinal gradient in species richness'). The regional pool of parasite and microbe species, or metacommunity, is a more conventional and accepted notion in community ecology, and it exists within a biogeographic region like a continent or a subcontinent. Then, understanding the origin of the parasite or microbe pool requires a knowledge of the evolutionary history of host–parasite associations, of the geographic isolation of the continent, of the linkages between pathogens and the biological diversity present within the area, and so on (see Brooks and McLennan 1991; Combes 1995; Poulin 1998*a*). Local communities like those of macroparasites in fish populations or microbes in human populations assemble themselves from this regional pool through a series of filters (see Fig. 2.1). Differences in host population size and density or the spatial arrangement of host population habitat patches (high to low connectivity), for instance, may be responsible for the persistence or the extinction of parasite or pathogen populations moulding local communities (see Grenfell and Dobson 1995; Grenfell and Harwood 1997; Keeling and Grenfell 1997, 2002; Rohani *et al.* 1999). At the individual host level, infracommunities of parasites and pathogens again assemble themselves from the local pool of available species (see Fig. 2.1). If species can reach a host they may still find the environment unsuitable, species interactions may also operate or may be constitutive, and induced defences against invasion can intervene. Many processes described here have also clear parallels in within-host infection since each host individual is composed of many different sites more or less connected with one another, and available, or not, for parasite or pathogen establishment (see Holt 1999). This framework shows that different environmental filters work on all these communities representing important steps in community assembly and constitution (see Murray *et al.* 2002; Rapport *et al.* 2002).

Mainstream community ecologists have long debated on the important steps in community assembly working down from the larger scales largely dominated by regional, not local processes (see Ricklefs and Schluter 1993; Brown 1995; Rosenzweig 1995; Maurer 1999; Lawton 2000; Gaston and Blackburn 2000). This can be contrasted with a more traditional approach in both parasitology and epidemiology through the study of local phenomena for understanding the structure and dynamics of parasitic or microbe assemblages. However the 'top-down' and 'bottom-up' paths are clearly complementary, and the recognition of the importance of a regional or even global perspective in parasite and microbe population dynamics and community assembly theory would clearly benefit from more detailed attention than would be possible from either approach alone. Recent studies on the impacts of global environmental changes on disease population and communities dynamics (see Harvell *et al.* 1999, 2002; McMichael 2001; Martens and McMichael 2002) provide several good examples of how largest-scale studies are of particular relevance to both parasitology in wildlife and human epidemiology. It is also obvious that the 'context' of the beginning of an infectious disease outbreak transmitted from wildlife is clearly local (see for instance the cases of HIV), some having dramatically increased in incidence and expanded in geographic range panglobally (see Hahn *et al.* 2000; Daszak and Cunningham 2002).

As we know today from the study of complex hierarchical systems inspired by physics, both processes from the top to the bottom and from the bottom to the top of Fig. 2.1 (see also Allen and Hoekstra 1992; Allen *et al.* 1993) are certainly acting as forces controlling parasitic and infectious disease community assembly. Recognizing that often the determinants of both host animal (or plant) and individual human health may occur at levels higher within the ecosystem hierarchy is thus one of the major tasks of modern parasitology and epidemiology.

2.3 Emergent properties of parasite and infectious disease communities

2.3.1 On the search for regularities in parasite and infectious disease community structure and processes

One of the principal advantages of this two-way viewpoint of community assembly organisation is that it takes a sufficiently distant view of parasitological and epidemiological systems that the idiosyncratic details disappear, and only the important generalities remain (see Brown 1995; Rosenzweig 1995; Gaston and Blackburn 2000 for application in mainstream ecology; Poulin et al. 2000 and this volume for parasitological–epidemiological investigations). This may reveal general patterns, or regularities, that would otherwise have been entirely neglected (see Morand and Poulin 1998; Morand 2000; Morand and Guégan 2000a). Any scientific discipline must pass through a phase where the phenomena of interest are clearly and quantitatively identified (this chapter), and then the mechanisms underlying the observed patterns are explored and challenged with rigourous theoretical and empirical testing. Specifically, the recent developments in population dynamics of infectious diseases have clearly shown how useful generalizations, but not at the level of local human communities, might be indicative of significant regulation in spatial dynamics of those diseases (Ferguson et al. 2003). Nevertheless, the impossibility of using manipulative experiments in natural systems (case of wildlife diseases) and anthropogenic systems (case of epidemiology) means that it is often difficult to retain a single hypothesis among competing alternative solutions to understand the mechanisms that underlie the patterns (see Morand and Guégan 2000a for an illustration). Consequently, this inability to exploit manipulative experimentation over large-scales have forced parasitologists, plant-associated insect ecologists, and epidemiologists to use comparative approaches (see Aho and Bush 1993; Cornell 1993; Lawton et al. 1993; Poulin 1995a; Cornell and Karlson 1997; Morand and Poulin 1998; Choudhury and Dick 2000; Morand and Guégan 2000a; Guégan et al. 2001; Brändle and Brandl 2003; Nunn et al. 2003; Guernier et al. 2004) as macroecologists did before (Brown 1995; Rosenzweig 1995; Maurer 1999). The following sections attempt to illustrate this using different examples from the literature, and to discuss the consistency of the results with the various hypotheses used to explain the observed patterns in the light of recent advances in macroecology (see Gaston and Blackburn 2000), and comparative analysis in parasitology-epidemiology (see Poulin et al. 2000).

2.3.2 Parasite and infectious disease species richness

Undoubtedly, parasites and other kinds of microbes and associated organisms like phytophagous insects may represent more than half of the living organisms (Price 1980; de Meeüs et al. 1998; Morand 2000; Poulin and Morand 2000; Curtis et al. 2002; Brändle and Brandl 2001, 2003; Nee 2003), even if very few attempts have been made to rigorously quantify and delineate the differences in richness between free-living organisms on the one hand and their associated organisms on the other hand (but see Strong and Levin 1975; Strong et al. 1985; Hillebrand et al. 2001; Guernier et al. 2004 and hereafter). Most, if not all, organisms are hosts for parasites, comprising helminths, arthropods, fungi, or microbes, and if the pioneer work by Guernier and colleagues is representative of other (host) species, the overall biodiversity on Earth may be currently underestimated by more than an order of magnitude due to the unsuspected species diversity of parasites and other kinds of microorganisms (see Guernier et al. 2004 and hereafter). Investigations attempting to identify determinants of species richness of parasites and other kinds of associated organisms are now well represented in the recent parasitological literature (Poulin 1995a; Morand 2000; Poulin and Morand 2000; Brändle and Brandl 2001; Guernier et al. 2004) much more than a decade ago, and more research is needed on different symbiotic systems to examine the extent to which the observed patterns are supported or undermined. Here we draw attention to the four most popular and striking 'macroecological' patterns in the species richness of assemblages of parasitic and other kinds of associated organisms, and we promulgate how the search for consistency in common patterns, or not, from other symbiotic systems will contribute to the emergence of a more rigorous research agenda in parasitology and epidemiology.

2.3.2.1 Species–area relationship
The variation in species richness of parasites and other associated organisms is generally not random, but shows one regular pattern which is the species–area relationship (Simberloff and Moore

1997; Poulin 1998*a,b*; Morand 2000), a now-classical factor explaining the number of species likely to be found at any site in mainstream ecology (Brown 1995; Rosenzweig 1995; Gaston and Blackburn 2000). Usually, widespread hosts tend to have more parasite or infectious disease species than hosts with a more restricted geographical range because the increase of host range may allow the host to encounter more species (Gregory 1990). This pattern seems to hold both within host species (see Freeland 1979; Marcogliese and Cone 1991; Goüy de Bellocq *et al.* 2003; Calvete *et al.* 2004) and across host species (see Poulin 1998*a*; Morand 2000; Brändle and Brandl 2003 for a review). Alternatively, host body size may be taken to represent area size for parasites (see Guégan *et al.* 1992; Guégan and Hugueny 1994). Indeed, in interspecific comparisons among host species, many positive relationships have been reported between host body size and parasite species richness; however, the relationship is not universal, and there are many exceptions (see reviews in Poulin 1997; Morand 2000).

Still, the species–area relationship is generally a strong pattern in parasite diversity studies. For instance, the species–area relationship explains more than 50% of the total variation in species richness of phytophagous insects (Kennedy and Southwood 1984; Brändle and Brandl 2001), and around 14–30% of the variance in species richness of parasitic fungi (Strong and Levin 1975; Brändle and Brandl 2003). More interesting than the existence of a species–area relationship for symbiotic systems is what does this pattern mean? Or, put in other words, which processes may be responsible for the relationship?

One problem with this approach is that the search for correlates of species richness across populations of host species characterized by different geographical ranges (or size area for isolated systems) is often not independent of sampling effort (see Gregory 1990), and most studies have thus controlled for the effects of differential sampling of host species (Gregory 1990; Poulin 1998*a*). Whether it is correct still remains to be determined (Guégan and Kennedy 1996). Sampling effort may be strongly correlated with host geographical range and other biological factors which may covary with host range, for example, host size, host niche

breadth, extinction/colonization, and temporal dynamics (see Gaston and Blackburn 2000), which then disappear after such a statistical control (see Guégan and Kennedy 1996).

The paper by Goüy de Bellocq *et al.* (2003) shows that the parasite species richness in the woodmouse, *Apodemus sylvaticus*, on western Mediterranean islands depends on the surface areas of the different surveyed localities (but see Dobson *et al.* 1992*a,b* for contradictory results). Other factors, that is, parasite species life-history traits and host species diversity, were also important as determinants of helminth richness and composition across isolated rodent populations. Fig. 2.2 illustrates the relationship between parasite species richness and surface area for the eight Mediterranean islands and three continental regions used in the study.

More interestingly, the authors demonstrate that the positive relationship observed between parasite species richness and surface area across woodmouse populations is not the result of a random process, but, on the contrary, it shows the existence of order, that is, nestedness (see corresponding Section, 2.3.2.4) in the presence/absence matrix of parasite species across different localities. Helminth parasites are organized according to a hierarchy of species, also called nested species pattern (see Guégan and Hugueny 1994), across surveyed areas: some parasite species are widespread across localities, and some others are

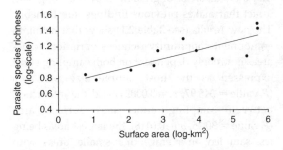

Figure 2.2 The relationship between log-species number and log-surface area for parasites of the woodmouse, *Apodemus sylvaticus*, on different islands and regions from Europe.

Notes: The line of best fit is a linear function of the form $y = 0.11x + 0.72$, $r^2 = 0.68$, $p = 0.0017$. The three points in the upper right side of the diagram are for Continental Europe, Spain, and Italy. Other points are for western Mediterranean islands, that is, Mallorca, Menorca, Formentera, Ibiza, Corsica, Port-Cros, Porquerolles, and Sicila. Source: Redrawn from Goüy de Bellocq *et al.* (2003).

uncommon and found in only fewer areas (see Goüy de Bellocq *et al.* 2003). The richness and distribution of the different helminth species across areas depends not only on area size as previously shown but also on the type of parasite life cycle and host mammal species diversity. The explanations given by the authors are that larger areas may sustain larger host populations, an important parameter in epidemiology which determines the host resource needed for a parasitic or infectious disease agent to persist (see Grenfell and Harwood 1997; Keeling 1997), and thus may favor the existence of a higher parasite diversity (see Morand and Guégan 2000b). Then the difference observed in both the parasite species richness and composition across areas may be due to the fact that a parasite with a direct life cycle may have more chances to succeed in the colonization of a new host population compared with a parasite with an indirect life cycle which absolutely needs to find a suitable intermediate or definitive hosts to complete its cycle. Finally a locality with high host species diversity may be more favourable for parasitic or disease persistence, for example parasites with a complex life cycle finding more definitive host species to achieve their development.

A reconsideration of the study by Goüy de Bellocq *et al.* (2003) using Generalised Linear Models (see Wilson and Grenfell 1997; Venables and Ripley 1999) instead of simple regressions indicates that the data are strongly flawed by sampling bias, a fact that makes previous findings questionable. The new results (see Table 2.1) show that helminth component community richness variation across areas is strongly dependent on both sampling effort expressed as the host sample size per area (F value = 345.97, p = 0.00034) and the interaction between sampling effort and surface area (F value = 30.35, p = 0.0118), the largest areas being less sampled, on average, than smaller areas, with all other parameters kept constant in the model. Mammal host species diversity is then just marginally significant in the multivariate analysis (see Table 2.1). Surface area does not appear to be significant anymore. A stepwise elimination procedure using the Akaike criterion yielded similar results.

This new result indicates that sampling effort may exert strong bias in estimation of parasite

Table 2.1 Summary of Generalized Linear Model with a gaussian error structure for explaining the parasite species richness variation in woodmouse across 11 different areas

	Deviance	Resid. *df*	Resid. dev	*F* value	*p* (*F*)
Null	678.73				
Sampling size	538.77	9	139.96	345.97	0.00034
Host diversity	22.12	8	117.84	14.21	0.03269
Surface area	3.33	7	114.51	2.14	0.23978
Sampling × Host	6.33	6	108.18	4.06	0.13721
Sampling × Area	47.26	5	60.92	30.35	0.01178
Host × Area	40.53	4	20.39	26.02	0.01457
Sampling × Host ×Area	15.72	3	4.67	10.09	0.05021

Notes: Resid. *df* and Resid. dev are the residual degree of freedom and the residual deviance at each step of the procedure, respectively; *p(F)* is the probability statistics associated with the *F* test; Host diversity is the host mammal species diversity per unit area; Sampling × Host is the two-way interaction term between sample size and surface area, and so on (see text for explanation). The rank of introduction of terms in the successive models did not alter the main results as illustrated here.

species richness variation across different areas, notably in the case of the largest areas where rare parasite species may be missed during parasitological investigations. This point has been made before (Gregory 1990; Poulin 1998b), and sampling effort must be taken into account in any investigations of species–area relationships using parasite data. We also strongly recommend the use of multivariate analyses to take into account perverse effects exerted by sampling bias on statistics instead of *a priori* regressing species richness data against sampling and the use of residuals since sampling effort may also covary with other independent variables under study.

2.3.2.2 Species richness–isolation relationship

More generally, species diversity is dependent on the fragmentation and isolation of habitats (Whittaker 1998). Fragmentation and isolation have promoted organism speciation and the build-up of endemic faunas on Earth (Brown 1995). Isolation and fragmentation are, of course, one part of the many factors promoting species diversity and composition (see Brown 1995; Rosenzweig 1995; Gaston and Blackburn 2000). Most organisms, including parasitic and infectious diseases in hosts (this chapter), exhibit patterns of similarity in composition and richness depending on geographic distance and isolation (see Poulin and Morand 1999; Morand and Guégan 2000b).

Perhaps one of the most important lesson to be learned from recent studies in host–parasite systems is the demonstration by Poulin and Morand (1999) that the geographical distance between component communities of parasites in freshwater fish is often the best, most general explanation of similarity in parasite species composition and to a lesser extent, of species richness across localities. Using multivariate analysis based on permutation methods (see Legendre et al. 1994), these authors conclude that patterns of parasite species composition across distinct isolated areas (i.e. distinct lakes in their study) strongly depend on the distances that separate the different localities, shorter geographical distances between isolated areas being associated with a greater similarity in parasite composition between them, and nearby lakes harbouring numbers of parasite species more similar than those of distant localities. Put in other words, there is in these parasitological data a tendency for species composition and richness to be autocorrelated over space. As mentioned by Poulin and Morand (1999), it might be more accurate to say that it is the isolation of a given locality within the network of patch areas that here matters instead of geographical distances among sites that roughly approximate this isolation. It is possible to have exceptions to this pattern, and many other factors may promote similarity between close localities in the composition and richness of parasite communities (see Kennedy et al. 1991; Hartvigsen and Kennedy 1993; Poulin 1998a). Nevertheless, the authors strongly suggest to consider the effect of geographical distance as a good index of isolation in further comparative analyses of parasite communities, and it should therefore become a basic requirement to control for the confounding and often important effect of geographical distance on the determinants of species composition and richness in parasite component communities.

Recently, Poulin (2003b) has shown that the influence of geographical distance on the similarity between parasite communities may follow a regular pattern. In the majority of parasite communities of fish and mammal hosts, the similarity in the species composition of communities decays exponentially with increasing geographical distance between localities (Poulin 2003b). Exponential rates of decay in similarity have also been reported for plant communities (Nekola and White 1999), and further emphasize the importance of geographical isolation.

One illustrative example of the effect of geographical distance and isolation on parasitic and infectious disease communities is that of oceanic islands. Typically islands have fewer species per unit area than the mainland, and this distinction is more marked the smaller the island and the farthest it is from a continental source (Rosenzweig 1995; Whittaker 1998). Very few studies have investigated parasite species community richness and composition on islands (see Kennedy 1978; Mas-Coma and Feliu 1984; Kennedy et al. 1986a; Dobson 1988a; Dobson et al. 1992a; Miquel et al. 1996; Goüy de Bellocq et al. 2002, 2003). The main conclusions reached by Dobson et al. (1992a) concerning parasite species richness and composition in Anolis lizards from northern Lesser Antilles islands are that they show a relatively depauperate parasite community when compared with lizards sampled on the larger Carribean islands, for example, Cuba, or on continental areas, and that these differences are associated with the life history attributes of the different parasite species in the assemblages. On the whole, Goüy de Bellocq et al. (2003) reach the same conclusions (see above), but they were unable to more formally characterize an effect of geographical distance and isolation in-between Mediterranean islands on parasite species richness and composition in the woodmouse. Examples of studies of pathogen communities and species composition in animals and humans on islands are even more scarce. Collares-Pereira et al. (1997) provide the first epidemiological data on pathogenic leptospires serovars diversity in insectivore and rodent species in the Azores archipelago, and they conclude to a low serovars diversity of three within this group of islands out of a total of nineteen serogroups more largely represented over the world. Surprisingly, there are only few studies that have quantified pathogen species richness and composition on islands for human communities in a way similar to what is traditionally done in community ecology. Based on unpublished data from one of us

(Guégan and Guernier, unpubl. data), we show here that on a total set of 197 different countries all over the world, the species richness of pathogens including viruses, bacteria, fungi, protozoa, and helminths is lowest on the 73 islands compared with the mainland countries (see, Table 2.2). Even after controlling for confounding effects that may be exerted by factors like the socio-economical power, the population size in number of inhabitants or the latitudinal position of the country, the island factor still explains 24.1% of the total variation in pathogen species diversity across areas. This result on pathogen species assemblages in humans is indicative that isolation and/or distance from a continent may be highly responsible for lower species richness in those localities. As suggested by the work of Goüy de Bellocq *et al.* (2003), islands sample only from the dispersive portion of the mainland pool. This effect must, of course, be distinguished from area size since the human pathogen study (see Table 2.2) has kept its effect constant in a multivariate analysis. Present-day distances between isolated islands and a regional biogeographic pool, even if they do not strictly reflect distances at the time of disease colonization, may however provide a rough index of isolation.

From the present human diseases data set, information concerning the Caribbean islands and the surrounding continental countries from Northern, Central, and Latin America representing the mainland, it is informative to see that remote islands from the Gulf of Mexico are poorest in pathogen species when compared to what is observed on the continent and even on close islands like Trinidad & Tobago, for instance (see Fig. 2.3). When considering the confounding effects exerted by covariate factors (notably the economic power of a nation) on disease species richness across localities, the influence of distance still remains (statistical data not illustrated): distance from a continental regional pool (49.1% of the total variation explained; $p = 0.00001$), the total area size of islands (28.9% of the total variation explained, $p = 0.00001$), and to a lesser extent human community size (5.9% of the total variation explained, $p = 0.028$) are the best predictors of pathogen species richness in human populations among the nineteen Caribbean islands and the fifteen surrounding continental countries. These findings based on a study of community assemblage of human diseases are in accordance with studies on population dynamics of infectious diseases on islands (see Black 1966). According to

Table 2.2 Summary of Generalized Linear Model with a poissonian error structure and a log-link for explaining the species richness variation in human infectious diseases across a set of 197 different continental (124) or isolated (73) countries

	Deviance	Resid. *df*	Resid. dev	*p* (χ^2)
Null	265.56			
Island	64.07	196	201.49	0.00000
GNP	2.63	195	198.86	0.10466
Population size	40.32	194	158.54	0.00000
Surface area	19.60	193	138.94	0.00001
Latitude	8.80	192	130.14	0.00301
Island × Surface area	2.98	191	127.16	0.08404

Notes: Dispersion parameter in the model is 0.67. Island is coded 1 and continental area is coded 0; GNP is the Gross National Product per country (in US $) to control for its effect on final statistics; Population size is the number of inhabitants; Latitude is the geographical position of each country in degrees and minutes; see also Table 2.1; *p* (χ^2) is the probability statistics associated to χ^2 test. Models were built with independent variables and their two-way and three-way interaction terms. The rank of introduction of terms in the successive models did not alter the main results as illustrated here.

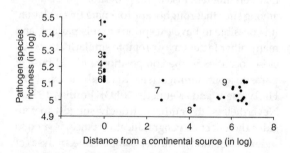

Figure 2.3 The relationship between log-species number and log-distance (in km^2) from the continent for human infectious diseases in different Caribbean islands ($n = 19$) and surrounding continental countries ($n = 15$) from Americas.

Notes: Codes are 1 : USA; 2 : Brazil; 3 : Venezuela and Mexico; 4 : Colombia and Panama; 5 : Surinam, Guyana, Honduras, Guatemala, and Costa Rica; 6 : French Guiana, San Salvador, Trinidad & Tobago, Puerto Rico, Nicaragua, Haiti, Dominican Republic, Belize; 7 : Guadeloupe, Cuba, Martinique, Jamaïque; 8 : Dutch Antilla, 9 : Aruba, Caïman Islands, Montserrat, Grenada, St Kitts & Neville, Antigua & Barbuda, Barbados, St Vincent and Dominica.

Price (1990), as local host abundance increases, so the effective population size for maintaining parasite populations increases, resulting in more parasite species being maintained in larger communities. Thus species dynamics of infectious diseases might be in many ways analogous to population dynamics (e.g. critical community size threshold), (see Grenfell and Harwood 1997; Keeling 1997; Broutin et al. 2004). There are obviously similarities between species and population maintenance and dynamics (see Brown 1995), and all these points need to be developed in further research on infectious diseases in wildlife and humans as we learn more about the patterns of variation in infectious disease species richness with respect to island size, isolation, and host community size. The above findings do not stipulate that modern events, and more particularly transcontinental exchanges, do not influence infectious disease dispersal and maintenance in heterogeneous environments, but they strongly suggest that we need to explore the impact of spatial heterogeneity on the course of infectious disease species dynamics, and the importance of these variables, for example, area size, isolation, community size, to better grasp the abundance, distribution, and identity of pathogen species within local habitats.

A major outcome of fauna (or flora) isolation is endemism; thus old and/or remote islands tend to generally have a large degree of endemism (see Whittaker 1998). The term endemism refers to the restricted ranges of taxa in biogeography, and it is used differently in epidemiology–parasitology. We refer here to the former definition. Little is known about parasite or infectious disease community assemblages in endemic hosts, and Morand and Guégan (2000b) have concluded with some predictions based on both empirical studies and mathematical modelling that hot spots of (host) endemism are also the foci for a large diversity of endemic parasites and pathogens, and that restricted areas and/or low host community sizes are associated with a decrease in parasite or pathogen species numbers (see above). Regarding the existence of endemic parasites or pathogens, we can only speculate on their existence on Earth, and they are probably legions. Many parasitological investigations have focused their efforts on the impacts of exotic pathogens and pests on native host species, but an important advance in future research should be the recognition of endemic parasites and pathogens, and the role they may play in maintaining and regulating biodiversity and ecosystem dynamics. Recent research on emergent pathogens might shed light on their importance in nature (see Aguirre et al. 2002).

2.3.2.3 Local–regional richness relationship

Much parasitological literature on species diversity patterns has been devoted to local mechanisms whereas in recent years large-scale processes have been regarded as important determinants of the species richness of local communities in free-living organisms (see Lawton 1999; Gaston and Blackburn 2000). Implicit in many ecological studies is the important recognition that regional and historical processes may profoundly affect local community structure (Brown 1995; Rosenzweig 1995; Lawton 1999, 2000). Questions of spatial scale have been addressed only very recently in parasite community ecology (Price 1980; Aho 1990; Aho and Bush 1993; Kennedy and Bush 1994; Kennedy and Guégan 1994; Barker et al. 1996) probably because traditional parasitology has been too medically orientated over a long time with a major focus on very fine-scale studies.

Simply because every parasitic, parasitoid, or even microbe species cannot be present everywhere, we do not expect every species occurring within the regional pool to affect the composition of every local community. A central method used for the recognition of the importance of regional and local processes is the regression of regional species richness against local species richness plots (see Lawton 1999; Srivastava 1999; Hillebrand and Blenckner 2002), and several contributions have tried to disentangle the regional and local constraints in free-living communities (see Srivastava 1999; Shurin and Allen 2001; Hillebrand and Blenckner 2002, for recent reviews). The test requires estimates of species richness for a given group of organisms at both local and regional spatial scales and a statistical evaluation of their relationship. When local richness is regressed against regional richness and the relationship is linear, the communities are

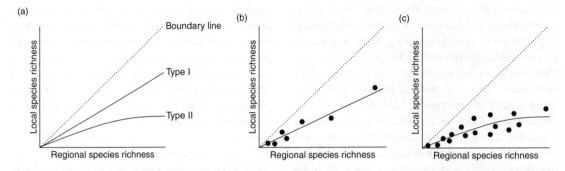

Figure 2.4 The relationship between regional and local species richness for parasite, pest and pathogen organisms. (a) Because of the hierarchical processes illustrated on Fig. 2.1, every parasite, pest or pathogen species present within the regional pool is unlikely to occur everywhere, which thus determines a boundary line never reached for which y equals x. Local species richness both at the infracommunity and component community scales are usually less than regional richness, and two kinds of relationships may then exist. A type I curve indicates proportional sampling in which local species richness increases linearly with regional richness. A type II curve saturates with local species richness above a threshold for higher regional richness. In nature, parasite, pest or pathogen community assemblages lie anywhere between type I and type II systems (see Cornell and Karlson 1997). (b) Example where the richest regional (component) communities are associated with the richest infracommunities and vice versa, suggestive of no saturation in parasite species. Helminth parasites of introduced freshwater fish in the British isles (Guégan and Kennedy 1993) or of natural populations of partridges in Spain (Calvete et al. 2003) are clear examples of unsaturated systems. (c) Examples where local species richness may be fixed by internal constraints indicative of saturation in local species communities. Intestinal helminth infracommunities (see Aho 1990; Kennedy and Guégan 1994; Calvete et al. 2004) generally are examples where saturation in parasite species may occur (see also text for contradictory results).

unsaturated and are said to exhibit 'proportional sampling' of the regional species pools. If the relationship is a somewhat curvilinear function, the possibility of saturation may then arise (see Fig. 2.4). Although this type of analysis appears to be straightforward, several pitfalls have been discussed elsewhere (see Creswell et al. 1995; Srivastava 1999; Shurin et al. 2000). A major question is the definition of what exactly regional and local richnesses mean for parasite communities, and the way their measurements are best comprehended (Kennedy and Guégan 1994; Barker et al. 1996). Usually, regional (Kennedy and Guégan 1994) or continental (Aho and Bush 1993) parasite species richnesses, that is to say spatial scales reflecting a naturally occurring hierarchy from which local parasite communities may be drawn, have been used to represent regional pools. Morand et al. (1999) used another estimate of regional species richness defined as the component community richness, but the authors were faced to the problem of defining exactly the spatial hierarchy at which processes may

operate in open marine systems. The measure of local parasite species richness adopted sometimes is the mean or maximum number of parasite species in the parasite component communities known to the authors (see Aho and Bush 1993; Kennedy and Guégan 1994), but a more correct definition of local species richness seems to be the mean (Kennedy and Guégan 1994; Morand et al. 1999) or even maximum (Poulin 1996a, 1997; Calvete et al., 2004) infracommunity parasite species richness.

Generally, all the studies agree on an important influence of both regional and local factors, but their relative importance may differ between categories of organisms. Concerning parasite, parasitoid, and microbe species communities, the diversity of species in local (at the host population level) assemblages is intuitively regulated both by local (e.g. interspecific competition, habitat heterogeneity) and by regional factors (e.g. evolution, migration, history). The studies published so far on free-living organisms have stressed the prevalence of type I communities (see Fig. 2.4) interpreted as an indication of unsaturation of local assemblages with generally weak or no effects of local interactions on species richness. Analyses of local to regional species richness performed on host-associated organisms like parasites or parasitoids (Cornell 1985; Aho 1990; Bush 1990; Hawkins and Compton 1992; Aho and Bush 1993; Lawton *et al.* 1993; Dawah *et al.* 1995; Kennedy and Guégan 1994, 1996; Poulin 1996a, 1997; Morand *et al.* 1999; Frenzel and Brandl 2000; Calvete *et al.* 2004) have, on the contrary, shown the existence of both type I and II communities (see Fig. 2.4). Particularly, the commonness of type II communities in host–parasite systems, and more specifically for helminths, may be illustrative of limiting factors shaping these local communities. In addition to the occurrence of saturated assemblages shown in parasite systems compared to other organisms, the three clearest results that emerge from the published parasitological literature available today may be summarized as follows.

First, most studies on herbivorous insects (see Dawah *et al.* 1995; Frenzel and Brandl 2000) and fish ectoparasites, that is, helminths and copepods (see Morand *et al.* 1999), have identified that empty niches are common and that local communities are unsaturated. As such, many natural enemy communities are subject to strong regional influences then providing opportunities for new invasive species to become established. The opposite can be observed for many studies on internal parasites like intestinal helminths of fish (Kennedy and Guégan 1994), of amphibians and reptiles (Aho 1990), and of birds (Bush 1990; Calvete *et al.* 2004) where local forces may contribute to parasite community structure. Poulin (1996a, 1997) showed, on the contrary, that for 31 intestinal helminth communities in bird hosts and 37 in mammal hosts the relationship between the maximum infracommunity richness and component community species richness was linear, indicating the absence of species saturation and the availability of vacant niches in organisms accepted to generally have species-rich helminth communities (see Bush and Holmes 1986a,b; Stock and Holmes 1988). The contrasting results may in part be due to the fact that

Poulin's (1996a, 1997) analysis included different host species, whereas many of the studies that found a curvilinear relationship between infracommunity richness and component community richness included only different host populations of the same host species (e.g. Kennedy and Guégan 1996; Calvete *et al.* 2004).

Then, second, the only intestinal helminth communities which exhibit unsaturated assemblages are those occurring in introduced fish species in the British Isles, that consist of non native fishes not having had enough time to accumulate sufficient helminth species from the native pool (Guégan and Kennedy 1993) to develop ecologically interactive communities (Kennedy and Guégan 1994). A recent study by Torchin *et al.* (2003) which compared the parasite species richness between introduced and native populations for 26 host species of molluscs, crustaceans, fishes, birds, mammals, amphibians, and reptiles also confirmed the reduced parasitization of introduced organisms suggestive of an absence of saturation in those parasite communities (see Figs. 2(c) and (d) in Torchin *et al.* 2003).

Third, major advances in our understanding of saturation versus non saturation of local parasite communities have been made in recent years. Cornell and Karlson (1997) and Srivastava (1999) drew attention to the necessity of the demonstration of other lines of evidences of niche and habitat relationships combined with information on local versus regional relationships. In particular, Rohde (1998) using randomization procedures highlighted the many scenarios in which a curvilinear local to regional relationship might be generated without requiring the necessity of species saturation. Using a comprehensive survey of marine fish ectoparasite communities, Morand *et al.* (1999) examined the effects of interspecific aggregation on the level of intraspecific aggregation in infracommunities, and they demonstrated that interspecific interactions were reduced relative to intraspecific interactions thus facilitating species coexistence in rich communities (see Tokeshi 1999 for further details on species coexistence). This pattern was highly coincidental with a positive linear relationship between infracommunity species richness and total parasite species richness obtained after controlling for the confounding effect exerted by phylogeny, indicative of no saturation in ectoparasite communities of marine fish (see Rohde 1991, 1998). In a recent study, Calvete *et al.* (2004) showed the existence of a curvilinear relationship between local and regional species richnesses of intestinal helminth infracommunities for eight populations of the red-legged partridge in Spain, even after checking for the confounding effect of geographical distance among localities on species richness calculations. Interestingly, this finding was confirmed by a demonstration of negative interspecific associations for the helminth species community, especially between cestodes and other helminths parasitizing the bird intestines.

All these results illustrate a number of important issues about the understanding of local–regional richness relationships in parasite or microbe community assemblages. First, demonstrating the effects of saturation, or not, in infracommunity assemblages requires that we simultaneously use additional investigations of interspecific interactions, or that published examples of the types of interactions exist, to test for the possible existence of interspecific competition. Neither of the two patterns for community assemblage organization, that is, the local to regional richness relationship and the demonstration of interspecific competition, is conclusive on its own. Interestingly, conclusions about the degree of concordance in the saturation of local communities, or not, between two or more methods may yield generality, but most of studies to date have only considered one option to test for the shape of interspecific relationships (but see Calvete *et al.* 2004). Second, the study of interspecific competition in local communities of parasites and pathogens has shaped the development of our understanding of species interactions, that is, importance of local processes. Thus, consideration of local to regional richness relationships in parasite or pathogen communities should lead to more attention being paid to the importance of large-scale patterns in parasitology

and epidemiology. Third, one area in which the combination of these two methods should be fruitful is in the connection that might exist with the density of parasites or microbes. Indeed, if competition is important within parasite or microbe communities, one would expect to see density compensation in those communities with few species. If complete compensation occurs, there should be no relation between parasite density or biomass and local species richness, while if there was no density compensation a linear trend would thus be expected (see Oberdorff *et al.* 1998; Griffiths 1999 for taxonomic groups other than parasites). Furthermore, the linkages between interspecific competition, (un)saturation and density compensation in parasite or microorganism community assemblages will require more research from community ecologists, parasitologists and epidemiologists. Notably, these issues should be highly relevant in the field of veterinary and medical sciences since any alteration of local habitats (from the point of view of one parasite species, for example, one intestine) and other disturbances exerted by humans (e.g. the use of drugs like helminthicides or antibiotics) should reduce parasite or pathogen populations from time to time, making ways for more resistant species or aliens to increase or to invade. The idea of saturation predicts that an invasive species (like a crop or an emerging virus) should not invade an infracommunity in individual hosts, or should do so only with the consequence of excluding a resident member species, that is, density compensation by new individual invaders. The study of interconnectedness between these patterns and our efforts at understanding the processes behind will require judicious choices of both host and parasite or pathogen taxa, and at different levels of spatio-temporal organization. As discussed before, patterns of within-host microbial species richness will also likely profit from a greater consideration of dynamical processes in patchy and discontinuously distributed environments, microbial persistence, and abundance in a particular tissue-habitat being influenced in several ways by biogeographical-like processes within host individuals.

2.3.2.4 Latitudinal gradient in species richness

Latitudinal gradients in species richness of free-living organisms are one of the most consistent large-scale trends that we can observe in nature, and one of the best documented patterns in the ecological literature (see Hawkins *et al.* 2003 for a recent review), but there are still exceptions (Brown 1995; Rosenzweig 1995; Gaston and Blackburn 2000). This pattern does hold not only for the hosts as a whole, but also for some parasitic and infectious disease organisms (Rohde 1992; Poulin and Rohde 1997; Rohde and Heap 1998; Calvete *et al.* 2003; Guernier *et al.* 2004; but see Poulin 2001; Poulin and Mouritsen 2003), which means that it could be a simple consequence of the observed latitudinal cline in host species diversity. Parasitological or epidemiological investigations on large spatial scales are rare, and there is undoubtedly a need for more comparative studies. Expectations of rich low-latitude parasite communities have been suggested by some authors (Kennedy 1995; Salgado-Maldonado and Kennedy 1997) while it is intuitive from other studies (Poulin and Guégan 2000; Guégan *et al.* 2001) that a latitudinal richness gradient exists for fish ectoparasite and human infectious disease communities, respectively.

Across 80 localities from 16 Spanish provinces, there is a marked cline in helminth species richness and composition in the red-legged partridge (*Alectoris rufa* L.) (Calvete *et al.* 2003). The highest levels of helminth richness are encountered in southern provinces of Spain (e.g. Badajoz, Huelva) and the lowest in northern ones (e.g. Alava Burgos Santander, Alava Navarra) (Calvete *et al.* 2003). Statistical analyses controlled for the effect of host age, sex, body condition, and time at which the study was carried out, and thus it is unlikely that these variables contributed to the spatial variation in helminth distribution and species richness across the study area. The cline in parasite species richness and composition across Spain was mirrored by the measures obtained for both infra- and component communities levels, indicating that the poorest component communities in the north correlated well with the poorest infracommunities. In the north, parasite communities were characterized by having one, that is, *Dicrocoelium* sp., or a few dominant widespread species. In contrast, in southern provinces, helminth communities were more species-rich with several codominant helminth species, *Raillietina tetragona*, *Subulura suctoria*, *Cheilospirura gruweli* (see Calvete *et al.* 2003 for further details). The answer to why there are more helminth species in the red-legged partridge in southern Spanish provinces than in the north is that the pattern may be related to variation in definitive host densities and to the distribution and diversity of intermediate hosts. According to Calvete *et al.* (2003), red-legged partridge populations are usually denser in the centre and south of Spain which represent the core-area of their range. This finding would tend to suggest that high densities of definitive hosts might be associated with a greater abundance of helminths and greater helminth species richness as a whole. In addition, and this second hypothesis is not mutually exclusive of the definitive-host density hypothesis, a greater abundance or diversity of arthropod intermediate hosts in the centre and south of Spain might result in the exposure of partridges to a wider variety of potential parasite species (see Poulin 1995*a*). Calvete *et al.* (2003) argued, based on correlation statistics using factorial scores from multivariate analysis, that the north–south abundance and richness variation of helminths in partridges might be to a large extent determined by higher temperature in southern regions of Spain, environmental conditions causing an increase in the survival or activity of intermediate forms of parasites with life-history stages outside their definitive hosts. These results suggest that variation in the distribution of helminths in partridge hosts are probably associated with variations in the distribution of their definitive and intermediate hosts and the local ecological conditions that may act on these host–parasite relationships.

One recent study (see Guernier *et al.* 2004) goes one step further in the explanation of the existence of a latitudinal gradient of species richness for pathogen species in human populations. Compiling data on parasitic and infectious diseases for a total set of 229 different species of pathogens in human hosts, including bacteria, viruses, fungi, protozoa, and helminths, Guernier and colleagues (2004) showed that after correcting for cofactors, that is, area, socio-demographic variables, physical and environmental parameters, that could exert a strong influence on the relationship between latitude and parasitic and infectious diseases species richness, one still observes that the species richness in human pathogens is strongly correlated with latitude (Fig. 2.5) with, on average, tropical areas harbouring a higher pathogen diversity than more temperate areas. This new result shows that

pathogen species richness is not distributed homogeneously across the planet but there is a marked cline with the highest levels of parasitic and infectious diseases species diversity near the Equator, and the lowest in northern areas. The cline in pathogen species richness occurs for the vast majority of groups of pathogens the authors analysed (7 times out of 10, exceptions being bacteria, viruses with direct transmission, and fungi). Interestingly, this similarity between many free-living organisms and pathogens is useful, because several common extrinsic and/or intrinsic factors might cause these common sets of pattern. Other variables are indeed

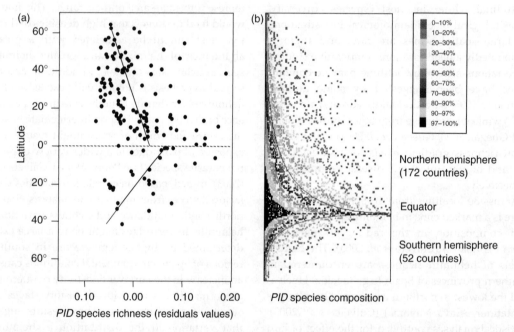

Figure 2.5 (a) Relationship between parasitic and infectious disease species richness and latitude in human populations across the two hemispheres. Linear relationships between species richness and latitude (dotted lines) are highly significant ($F = 12.29$, $df = 29$, $p = 0.0015$ and $F = 18.01$, $df = 130$, $p < 0.0001$ for Southern and Northern hemispheres, respectively). Residuals of species richness on the x-axis were extracted from minimal GLIM models controlling for the effects of confounding factors on disease species diversity estimates. Latitudes are expressed in degrees. (b) Presence/absence matrix for the 229 distinct parasitic and infectious disease species across the two hemispheres. The spatial distribution of pathogen species were organized according to the procedure adopted by the *'Nestedness Temperature Calculator'* (see Atmar and Patterson 1995). One hundred and seven ubiquitous pathogen species where eliminated from the entire data base since information they contained where entirely redundant with the most ubiquitous species already present into the matrix. Figure 2.5 (b) was generated after 1000 randomized permutations. This distribution is nonsymmetrical because of the 224 studied countries, 172 countries are found in the Northern hemisphere versus only 52 in the Southern one. Figure 2.5 (b) indicates that species diversity decreases as we move northwards or southwards from the equator ($F = 28.2307$, $df = 161$, $p < 0.001$). The occurrence boundary lines (black exponential curves) were fitted by non-linear regression ($y = 1.51 + 20.01e^{-0.29x}$ and $y = 1.65 + 35.87e^{-0.36x}$ for Northern and Southern hemispheres, respectively). See Guernier *et al.* (2004) for further details.

Source: Courtesy by *PloS* (Biology).

important in explaining global-scale patterns of human pathogens (e.g. modernization, urbanization, or impoverishment, especially in developing countries), but the fact that the authors considered such effects in their multivariate analyses tends to indicate that biogeographical forces are also important indeed in shaping the distribution and abundance of pathogen species. These results thus challenge the conventional wisdom that socioeconomic conditions are of preponderant importance in controlling or eradicating diseases.

Over the last three decades, the number of hypotheses advanced to explain the latitudinal gradient has increased from six (Pianka 1966) to nearly thirty, proffered by Rohde (1992). Some of these include spatial heterogeneity and patchiness, competition, predation, parasitism, mutualism, area, environmental stability, productivity, seasonality, solar energy (see Rohde 1992 for a general description). In reality, several factors may be acting in concert or in series, and among the numerous explanations given it is possible to substantially narrow the list of the most plausible explanations since many factors may be entirely redundant or untestable (Hawkins *et al.* 2003). For instance, Gaston and Blackburn (2000) listed only three plausible explanations for latitudinal richness gradients: area, energy, and time. They discarded the possibility that random location of species might be responsible for the latitudinal species gradient based on the absence of formal evidence. Area has often been cited as a simple hypothesis (see section on species–area relationships) to explain that the tropics, which really harbour the highest richness for many groups of organisms, also have the largest terrestrial surface area, that is, the geographical area hypothesis (see Rosenzweig 1995). This explanation may not be plausible when considering the latitudinal gradient of pathogen species in human populations as observed by Guernier and colleagues since both the surface area and continental mass effects have been taken into account in multivariate analysis.

Furthermore, and probably more important than the existence of a latitudinal gradient in species richness is the demonstration of an overall pattern of spatial distribution of parasitic and infectious

diseases species in human populations on Earth that conforms to a nested species subset hierarchy (see Guégan and Hugueny 1994; Guégan *et al.* 2001; Guernier *et al.* 2004). Nestedness structure indicates that species (here pathogens) that compose a depauperate community (here temperate conditions) statistically constitute a proper subset of those occurring in richer communities (here warmer conditions in tropical areas), but the converse situation, that is, pathogen species solely occurring in depauperate communities but not in the richest ones, is either not found or not properly substantiated. This pattern, although not considered in the ecological literature (but see Gaston and Blackburn 2000 who suspected the existence of a connection between the two patterns; see later), was strongly associated with latitude, indicating that the progression of pathogen species richness is from species-poor countries in more temperate areas to species-rich ones when reaching tropical zones (see Fig. 2.5).

Using Monte-Carlo simulations (see Manly 1991; Guégan and Hugueny 1994) to test the hypothesis of parasitic and infectious diseases spatial organisation on the largest scale, Guernier *et al.* (2004) assessed the degree of nestedness of the system using two different but complementary programmes: (*i*) 'Nestedness' (Guégan and Hugueny 1994) and (*ii*) 'Nestedness Temperature Calculator' (Atmar and Patterson 1995). In the former programme, pathogen species were either selected with uniform probability (R_0) or with a probability proportional to their incidence (R_1) (Guégan and Hugueny 1994) whereas in the latter one only a R_{00} procedure was retained (Atmar and Patterson 1995); (see also Wright *et al*, 1998; Cook and Quinn 1998; Gaston and Blackburn 2000 for further details). Results from Monte Carlo simulations showed that the global distribution of human pathogens was strongly nested ($N_s = 2481.4$, R_0 and R_1 procedures, $p < 0.0001$), with some slight differences that were found across the different groups of aetiological agents (all groups, $p < 0.0001$, but except for vector-borne viruses, with the R_1 procedure ($N_s = 1787$, $p = 0.0015$). When considering the Northern and Southern hemispheres separately, both were highly nested (R_0 and R_1

procedures, $N_s = 6602$, $p < 0.0001$ and $N_s = 1230$, $p < 0.0001$, respectively). This was confirmed by the R_{00} procedure used by the 'Nestedness Temperature Calculator' program (Atmar and Patterson 1995), which provides a useful graphic representation of the results (Fig. 2.5), showing that parasitic and infectious diseases species diversity decreases as we move northwards or southwards from the equator ($F = 28.2307$, $df = 161$, $p < 0.001$), (see Guernier et al. 2004). Results from all three nestedness models (see Wright et al. 1998) explained reasonable amounts of the nested pattern in human pathogen species across latitudes.

Wright et al. (1998) have suggested a cogent explanation for nestedness as a series of probabilistic filters, screening species with particular characteristics: local habitat suitabilities, differential colonization capacities of species, and sustainability of viable populations within their environment. Additionally, as pointed out by Gaston and Blackburn (2000), nestedness might be an inevitable second-order consequence of the same factors that cause variation in species richness and range size along latitudinal gradients. It is exactly the view with which Guernier et al. (2004) totally agree with the example of pathogen species diversity in human on a broad-scale.

Searching for the common causes explaining the existence of both patterns, i.e. gradient in species richness and nested structure, for parasitic and infectious diseases diversity and composition in human communities, Guernier et al. (2004) retained the energy hypothesis as a likely candidate for explanation. The energy hypothesis is a climate-based hypothesis that claims that energy availability generates and maintains species richness gradients (see Hawkins et al. 2003 for a recent review). Many studies have successfully correlated gradients in species diversity with variation in the climatic environment, a relationship thought to shape large-scale biogeographic patterns (Hill et al. 1999). The authors decomposed the potential effect of climate on pathogen diversity into Pearson's correlations to more deeply analyse the kinds of relationships between each of the four climatic variables and disease richness under study. The results show significant positive correlations

between pathogen species richness and the maximum range of precipitation after Bonferonni multiple correction for all six of the parasite or infectious disease taxa considered: bacteria ($r = 0.3545$, $df = 213$, $p < 0.0001$), viruses directly transmitted from person-to-person ($r = 0.2350$, $df = 215$, $p < 0.0001$), viruses indirectly transmitted via a vector ($r = 0.3575$, $df = 215$, $p < 0.0001$), fungi ($r = 0.3554$, $df = 216$, $p < 0.0001$), protozoa ($r = 0.3744$, $df = 216$, $p < 0.0001$), and helminths ($r = 0.4270$, $df = 215$, $p < 0.0001$). On the other hand, the relationship between pathogen species richness and monthly temperature range was only significant for three groups of pathogens: bacteria ($r = 0.3016$, $df = 213$, $p < 0.0001$), viruses directly transmitted ($r = 0.2142$, $df = 214$, $p = 0.0015$), and helminths ($r = 0.2590$, $df = 213$, $p = 0.0001$). No relationship between parasite species richness and mean annual temperature appeared to be significant after Bonferroni corrections. Finally, only the relationship between bacterial species richness and mean annual precipitation was significant ($r = -0.1987$, $df = 213$, $p = 0.0034$). No or very slight differences between total and categories of pathogen species richnesses and some climatic factors were observed between Northern and Southern hemispheres.

Many factors are involved in the determination of the climate in an area, particularly latitude, altitude, and the position of the area relative to oceans and land masses. In turn, the climate largely determines the species of plants and animals that live in those areas. According to the results of Guernier et al. (2004), the maximum range of precipitation is highly correlated with latitudinal gradient of pathogen species, the parasitic species diversity significantly increasing with this climate-based factor. Interestingly, the variation of precipitation around the mean was overall a better predictor of pathogen species distribution than its average value, thus indicating that pathogen species, their vector and host populations might best function over only a wide range of precipitation, which is actually found in many tropical regions of the world, those regions having more or less distinct wet and dry seasons during the year. Many parasites obviously require water as the basic medium of their existence, and many others strongly need wet conditions to complete their life cycle, for example vector-borne diseases. Often, many microorganisms are also constrained by the humidity of the atmosphere. Undoubtedly, the physical factor of precipitation variation may affect parasitic and infectious microorganisms, vectors and/or hosts over a range from low precipitation variation at one extreme, for example, deserts, to high precipitation variation such like in the tropics. This relationship might be related to biological cycles and a variety of features in parasitic and infectious stages that have evolved so that they are specifically well adapted to the variability of precipitation. So, prolonged drought should not be fatal for some well-adapted microbes if wet conditions are encountered once to complete their life cycle. Curiously, average precipitation was not retained as a good candidate for explaining the latitudinal gradient of pathogen species diversity except for bacteria (see Guernier *et al.* 2004). If we consider the Earth as a simple body with an environmental gradient, such as the annual precipitation range, which runs from wet and hot equatorial regions northward and southward to Arctic and Antarctic areas with harsh conditions, then distance and isolation from pathogen species-rich regions in the tropics may screen pathogen species by their extinction and colonization tendencies. Moreover, habitat suitability, for example presence of new hosts and reservoirs, and passive sampling may screen them by their habitat preference and availability, and abundance, respectively. Guernier *et al.* (2004) reached the same conclusions as Calvete *et al.* (2003) indicating that parasite species richness, their spatial distribution and organization on very different scales, climate-based forces, and the interplay between habitat conditions and host–parasite interactions might be intimately connected to generate the observed patterns of parasite species diversity.

The overall conclusion that can be drawn from this section on the latitudinal gradient in species richness for parasite and pathogen organisms is that a better understanding of parasitic and infectious diseases species diversity and community dynamics over wide ranges of spatial scales is now clearly needed. The similarity in the patterns of some

parasitic or pathogen taxonomic groups and free-living organisms suggests that common mechanisms are at work. Regardless of whether the richness of parasitic and infectious diseases simply tracks host diversity or, rather, is determined to a greater extent by exogenous factors, for example, climate-forced variables, is now a challenge that needs to be pursued in parasitology and epidemiology. In addition, the significant findings illustrated in this and other sections confirm that investigations on parasite and microorganism community assemblages should also be performed at greater scales than the scale at which local variation in species richness and composition is too often examined. Integration of systematics, biogeography, population and species dynamics, community ecology, and evolutionary biology is now essential for a complete understanding of the many scaling processes affecting parasitic and infectious diseases.

2.4 Linking parasite and microorganism communities and ecosystems. Directions for further work

The results summarized above underscore the important role of large-scale determinants on local parasite and pathogen community organization and assembly processes. They suggest that the success of invasive parasite species may vary depending on how physical conditions, for example, size of the area, geographical distance from a continental source, affect the rates of colonization. Additionally, the regional effects on local parasite or microorganism species richness also tend to suggest that research should be directed at regional processes and their effect on local diversity, regional processes being just as important as local ones in setting levels of richness in parasite or microbe communities. Then, the existence of a latitudinal gradient of parasite and microbe species richness both at subregional (gut helminths in Spanish red partridges) and global (infectious diseases in human populations) scales confirms that investigations on parasites and microorganisms should expand the scale at which local variation is traditionally examined. More interestingly, the demonstration of the existence of a nested species subset pattern for human infectious diseases (also true for gut helminths in Spanish red partridges) suggests that the interplay between well-differentiated species of pathogen species within assemblages and distinctive requirements for resources, including micro- and macro-habitats, contribute to the overall nestedness pattern and species diversity we observed.

It is important to recognize that similar patterns, that is, species area, species isolation, local to global relationships, and latitudinal gradient in species diversity, have been observed for many free-living taxonomic groups (see Gaston and Blackburn 2000; Hawkins *et al.* 2003), suggesting that common mechanisms might be at work in generating the observed patterns of species diversity for parasitic and infectious agents (see previous sections). But what kinds of common properties and characteristics between, say, a virus species and a bird species, may produce similar patterns?

So far, the concern of parasitology and epidemiology has been widely aimed at defining the characteristics in which pathogens may differ from other groups, focusing on details instead of searching for similarities. One pattern that appears to be pervasive across many communities, parasites, microbes, and free organisms alike, and at different spatial hierarchical scales, is the nested species subset pattern. Nested patterns were observed in some parasitological–epidemiological studies (Guégan and Hugueny 1994; Hugueny and Guégan 1997; Guégan and Kennedy 1996; Guernier *et al.* 2004; Poulin and Valtonen 2001, 2002; Vidal-Martinez and Poulin 2003), but in many other cases communities were observed to form random, unstructured assemblages (Rohde *et al.* 1994, 1995, 1998; Poulin 1996*a,b*; Worthen and Rohde 1996). The previous sections make clear something that should be obvious, that is, that the four patterns documented above are not independent of each other. Morand *et al.* (2002) have provided a cogent explanation for the generation of nested patterns in parasite communities, thus reinforcing the importance of both spatial and demographic stochasticity in parasite species distribution and composition. Fig. 2.6 illustrates the way Morand and collaborators synthesize the hypothetical distributions of parasite species among local communities and the many

factors that may be implied in shaping these communities. Considering that nested versus non-nested species patterns are strongly linked to unimodal versus bimodal species frequency distributions, respectively (Morand *et al.* 2002), the authors concluded that observed patterns should be the result of differential colonization/extinction processes acting at the level of each parasite (or pathogen) species. Differential colonization/extinction processes, also called epidemiological processes in the field of epidemiology (see Anderson and May 1991), are attributes of parasite or microbe species, and are related to birth and death processes in population dynamics (see Morand *et al.* 2002). Assemblages of microbes and parasites displaying nestedness on different hierarchical scales, ranging from within individuals of the same host population to the global scale, as shown in the present chapter, provide a strong indication that the same forces are at work across different spatial and temporal scales.

So what emerges from this concept is that both the recruitment of parasite (or pathogen) species from local (host) patches and the differential capacities of parasite (or pathogen) larval or adult forms to colonize those patches may have strong impacts on the composition and richness of those communities from the local to the global scale, and vice versa. This is an approach which is roughly similar to that adopted by metapopulation biologists (see Hanski 1999; Hanski and Gaggiotti 2004) and ecologists working on the diffusion of infectious diseases (see Grenfell and Dobson 1995; Diekmann and Heesterbeek 2000), and in which invasibility is a key factor for explaining local presence and distribution of a given species.

Invasibility is dependent on the availability of local resources and potential competition with local species, for example, presence of empty niches. In mainstream ecology (see Elton 1958), it has been hypothesized among other predictions that greater diversity should increase resistance to invasions because the levels of limiting resources are generally lower in more diverse habitats (Tilman *et al.* 1996; Knops *et al.* 1999), thus giving rise to the so-called diversity–invasibility hypothesis. A decrease in host species diversity allows remaining

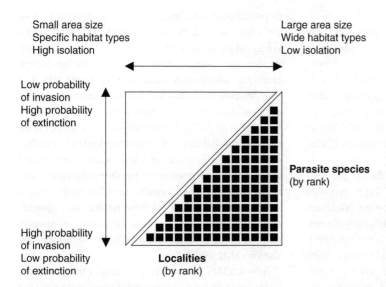

Figure 2.6 Nested (versus non-nested) species subset patterns are strongly associated with an interplay of specific characteristics of parasite (or microbe) species themselves and local habitat conditions to sustain or not a given parasite (or microbe) species population and its life-cycle associates, for example, vectors, reservoirs. Parasite (or microbe) community ecology would benefit greatly by considering the importance of epidemiological patterns in generating the processes observed.

species to increase in abundance as a result of decreased competition for limiting ressources, a phenomenon called density compensation by ecologists (see Begon *et al.* 1996). As a result, the increase in remaining host abundances should facilitate the invasibility of parasite or microbe species (Anderson and May 1978; Burdon and Chilvers 1982; Antonovics *et al.* 1995). Indeed, host living in high densities may harbour a high diversity of parasite species (Morand *et al.* 2000; Poulin and Morand 2000; Stanko *et al.* 2002), which in turn often achieve high abundances (Arneberg *et al.* 1998*a*). Not all available results agree with this, however (Stanko *et al.* 2002). Hence, both host diversity and disease invasibility are related to host species abundances within communities (Burdon and Chilvers 1982; Mitchell *et al.* 2002). As already stated, this hypothesis has received some support in general ecology (Tilman 1997), but further work needs to be done in both parasitology and epidemiology to determine whether higher parasite (or microbe) species diversity at different hierarchical scales, that is, within-host, within-population, and within-metapopulation community dynamics protects parasite or pathogen communities from invaders (but see Torchin *et al.* 2003).

Another prediction available from the general literature is that greater species diversity of microbes or parasites should decrease the severity of each of

the component diseases, a theorem which is called the species composition–disease hypothesis (see Elton 1958). This hypothesis predicts that any changes in parasite or microbe community composition within a host community should impact on the severity and virulence of one, or a group of, diseases, which then proliferate to the detriment of other parasite or microbe species. Both these hypotheses, that is, the diversity–disease hypothesis and the species composition–disease hypothesis, have received some empirical support from the plant disease literature (Knops *et al.* 1999; Mitchell and Power 2003), but they have never been tested to our knowledge in the case of both animal or human pathogens. Research on community ecology of parasite or microbe species, at the different hierarchical scales discussed in the present paper, should clearly benefit from developing both experimental designs and comparative studies to explore in greater detail the potential linkages between species diversity and composition, invasibility, and other life-history traits such as virulence.

In addition, theoretical studies on parasite (or microbe) invasibility rest upon epidemiological principles, in which the local abundance of hosts is the key determinant of whether a parasite can establish into a naïve host population (Anderson and May 1978; Burdon and Chilvers 1982; May and Anderson 1978). Within the study of human infectious diseases, recent investigations have clearly illustrated the importance of local community size for the maintenance and diffusion of, in particular, childhood diseases (see Grenfell and Harwood 1997; Rohani *et al.* 1999; Dieckmann and Heesterbeek 2000; Grenfell *et al.* 2001). Unfortunately, very few studies have gone a step further in demonstrating that higher local abundance in hosts should yield higher parasite or microbe species diversity (but see Poulin and Morand 2000).

For macroparasites, and especially for contact-transmitted microparasites, both high parasite abundance and high parasite species richness should be attained more easily at high host abundance or density. This is expected from epidemiological theory (see Grenfell and Dobson 1995) and supported by empirical studies on the determinants of parasite species richness (see Poulin and Morand 2000, and the present chapter). Indeed, host abundance/density is positively correlated with parasite (or microbe) species richness in several groups of vertebrates (Poulin and Morand 2000). This trend has yet to be demonstrated for human populations (and plant pathogens as well) in which large communities as resources should harbour and sustain a greater diversity of pathogens over space and time. Some, though very few, studies have even shown that parasite species richness covaries positively with host species richness across localities, an interesting finding that is yet to be more seriously quantified across different host taxonomic groups (Krasnov *et al.* 2004).

Any loss of host biological diversity should intuitively have a direct influence on the diversity and abundances of parasites or pathogens, and in turn on their virulence and pathogenicity within host populations. In addition, changes in host community composition should also influence the component community of parasites or microbes living there. Host species diversity should not randomly affect parasite species richness and composition as some of them will be more affected than others

depending on the exact causes of diversity loss. This loss may depend on habitat fragmentation, which should impact on rare host species living at low abundance, or may be due to landscape changes, which may influence the more adapted host species, and thus will favour either the spread of rare species or invasion by exotic species (Daszak *et al.* 2001). Unfortunately, we cannot easily refer to parasitological or epidemiological studies on animals or humans that have specifically tested these different hypotheses. Further empirical studies are thus needed, which should contribute to a better understanding of how native host species and their associated parasite or pathogen species may prevent or limit the attacks of invaders (see Torchin *et al.* 2003).

In summary, there are, of course, other interesting parasitological or epidemiological patterns that are not discussed in the present chapter. However, our aim here was to argue that both parasitology and epidemiology would first largely gain from developing both a much broader perspective and a more quantitative approach, as advocated recently by both macroecologists and community ecologists. Parasitology and epidemiology are two scientific fields that have evolved with an individual-centred research perspective, and which have then concentrated their research efforts on only local phenomena over the past decades. One important message from the present work is clearly that we need to adopt a much broader research perspective in parasite (or microbe) community ecology. Second, we are also intimately convinced that a better consideration of the linkages between host species diversity and composition, and parasite or microbe species diversity and composition within communities on the one hand, and the risk of altering any (host) species and composition on host invasiveness, host defences and increase of virulence by parasites or pathogens on the other hand. We are definitely advocating here a community ecology perspective for host–parasite systems.

CHAPTER 3

Parasitism and the regulation of host populations

Anders Pape Møller

Parasites can regulate hosts when parasites affect density-dependent mortality or fecundity, particularly among early life stages. Evidence consistent with parasite regulation of host populations include experiments that manipulated parasites; analyses of area–abundance relationships showing that host populations strongly affected by parasites are relatively small; and analyses of introductions of hosts to novel environments where they are released from parasite regulation.

3.1 Introduction

Factors that determine the size of populations of animals and plants have been studied for over two centuries. Understanding such factors is important because of applied aspects of ecology such as harvesting and conservation, but also for purely scientific reasons. Population regulation depends on density-dependence since only reduced fecundity or survival under high densities can regulate populations. Only biotic agents such as conspecifics, predators, or parasites can play this role. Density-dependent regulation is more likely during earlier stages of the life cycle because density-dependence at later life stages occurs 'too late' to allow regulation. Our knowledge of population regulation is still relatively poor despite more than a century of research (see reviews in Begon *et al.* 1996; Newton 1998). This opens up the possibility that even major factors contributing to population regulation may have been neglected. Parasites may be an example of such a neglected factor.

I will start this chapter by reviewing studies that show that parasites locally at least can regulate host populations. This aspect is important because such

Laboratoire de Parasitologie Evolutive, CNRS UMR 7103, Université Pierre et Marie Curie, Bât. A, 7ème étage, 7 quai St. Bernard, Case 237, F-75252 Paris Cedex 05, France.

cases are the basis for any potential of population regulation at larger scales. Second, I will review the literature on population introductions, parasitism, and success of introduction. Introduced species in novel environments provide examples of some of the most dramatic examples of population growth, the reverse of strict population regulation. A careful study of such examples might therefore provide important insight into mechanisms that could be of importance for population regulation. Third, I will investigate the possibility that indices of parasite impact on hosts, such as host investment in immune function, may predict population size of hosts. I do that by investigating the literature on abundance–area relationships and how that relates to parasitism. Finally, I briefly describe parasite-mediated competition and predation and how these mechanisms may play an important role in population regulation of hosts.

3.2 Parasites and density-dependent effects on host fecundity and survival

Like most natural enemies, parasites are quite capable of regulating their host population when they reduce the fecundity or the survival of their host population in a density dependent fashion. The fundamental understanding of how macro-parasites can regulate their host population arises from a series of papers by Anderson and May (Anderson and May 1978; May and Anderson 1978) that have since been reviewed by Tompkins *et al.* (2001). In these models, Anderson and May not only identified the conditions when parasites would be capable of regulating their host population but also the processes that could destabilize or stabilize the host–parasite relationship. In essence regulation will occur when the growth rate of the parasite exceeds that of the host population assuming that the parasite has a harmful affect on the survival or fecundity of the host population so the growth rate of the host population is reduced through the parasite induced effects. In essence the parasites 'outstrip' the host population such that when the host population size gets high the parasites reduce it and when the host population is low, so levels of parasite infection tend to be low allowing the levels of infection to rise. Demonstrating that regulation by parasites occurs in free-ranging wildlife is not so easy, particularly when the host population is at equilibrium and requires careful, large-scale experimental treatment in which levels of parasite infection are either reduced or host numbers increased. In this section I review the features of the Anderson and May model and some of its assumptions and then examine some experimental data that test the hypothesis that parasites can play an important role in regulating host populations.

The basic Anderson and May model for macro-parasites is based on two linked differential equations that describe changes in the size of the host and the parasite population, respectively. The model encompasses two important elements. First and foremost it captures the aggregated distribution of the parasites within the host population by assuming the parasites distribution conforms to the negative binomial distribution and uses the inverse estimate of aggregation described by the aggregation parameter, k. When k is low, aggregation is high but when k is above 5, the distribution approaches a random distribution as described by a Poisson distribution. By far and away the majority of macro-parasites exhibit an aggregated distribution; Shaw and Dobson (1995) examined 269 data sets of parasite distribution within wild animal

species and found a tight relationship between log-variance and log-mean that followed Taylor's power law with the variance rising faster than the mean, indicating an aggregated distribution (e.g. Fig. 3.1). The second feature is that the model incorporates both the effects that the parasites have on the survival of the host and on their ability to produce young. Macro-parasites tend to generate morbidity rather than mortality in the host, reducing the condition and the ability of the host to search for resources, defend a good territory or provide food for their offspring so heavily infected individuals tend to have lower reproductive success and increased vulnerability to secondary causes of mortality such as predation or secondary infections.

The Anderson and May model demonstrates that it is the parasites of moderate virulence that will regulate a host population and not those of high virulence. Those parasites that are effectively benign have no impact on the growth rate of the host population so naturally they are unlikely to control the host growth rate while those of high virulence tend to cause mortality in the host before the parasites have had time to infect many other hosts, and thus the host tends to escape the regulating role of the parasite. The regulation is more likely when the parasites have a random distribution as opposed to an aggregated distribution. When the parasites are highly aggregated, most of the parasites are concentrated in a small proportion of the host population and since they influence just a small proportion of the host population, the host can escape the regulatory role. There will in fact be a threshold of aggregation, above which the host escapes regulation by the parasite but below which the parasite can still regulate. Aggregation will tend to stabilize the host–parasite dynamics, but as the distribution becomes more random, the population will become increasingly unstable. Two other features will tend to stabilize the parasite–host relationship: density-dependent constraints on parasite growth rate such as acquired immunity or competition for space or resources within the host and nonlinear increases in parasite-induced host mortality.

The features of the parasite–host system that will tend to destabilize the dynamics include a random

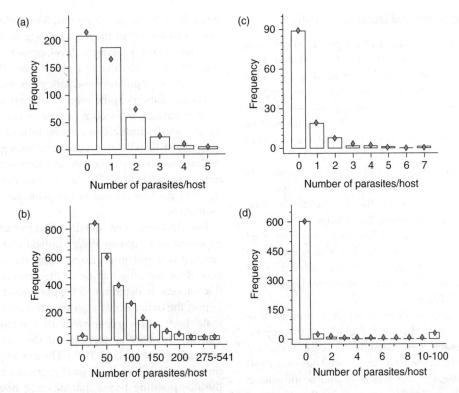

Figure 3.1 Examples of aggregated distributions of parasites within their host population. The bars represent observed frequencies and the points the fit of the negative binomial distribution for (a) perch *Perca fluviatilis* infected with the parasitic tapeworm *Triaenophorus nodulosus*, (b) reindeer *Rangifer tarandus* infected with the warble fly *Hypoderma tarandi*, (c) starling *Sturnus vulgaris* infected with the nematode *Porrocaecum vulgaris*, and (d) the pond frog with the nematode *Spiroxys japonica*.

Note: More details can be found in Shaw *et al.* (1998).

Source: After Shaw *et al.* (1998).

distribution of parasites within the host population, time delays in the development of the parasite, direct parasite multiplication within the host, and relatively high impact of parasites on fecundity compared with the impact of the parasites on survival. Time delays in parasite development are an integral part of most parasite systems, since the parasites usually leave the host in an egg form and then there is either a period of further development through several stages, or at least a time period where the infective stage is lying dormant outside the host or even a period of diapause within the host (known as arrested development). Parasite induced reduction in fecundity is likely to be prevalent in parasite–host systems in wild animal populations and has been recorded in a number of parasite–host systems. An experimental approach to this with red grouse *Lagopus lagopus scoticus* found that grouse with experimentally reduced levels of infection had greater clutches, higher hatching success, and chick survival than controls, but the important effect is to note that this was not because chicks were being infected but because these effects were operating through the condition of the female (Hudson 1986*a*; Hudson *et al.* 1992*b*). A common misunderstanding of the Anderson and May model is that the oscillations in abundance are caused by a delayed density dependent feedback, this feedback is to some implicit within the system, but there again the tendency to cycle is an emergent property and usually a consequence of the parasite induced effect on host fecundity.

3.2.1 Regulation and laboratory studies

While a number of studies have provided evidence that parasites can reduce host fecundity and/or survival (Tompkins *et al.* 2001: table 3.1), few studies have provided sufficient evidence to suppose that these effects are acting in a density-dependent manner and are capable of leading to regulation. One of the problems is to show that the effects of the parasites on host fitness are acting in an additive or compensatory manner. To demonstrate regulation we need to perturbate the system experimentally by changing the host–parasite relationship and predicting the outcome (Scott and Dobson 1989). This is not that simple, since we need both the right scale, replication, and importantly ensure that the mechanism is truly altered at the right scale and in the correct manner (Hudson and Bjørnstad 2003).

Marilyn Scott undertook two free-running laboratory studies and provided evidence to suppose that parasites were capable of regulating host populations. In the first (Scott and Anderson 1984) she investigated the effects of the monogenean *Gyrodactylus bullaturdus* on guppies *Poecilia reticulata* in fish tanks. They showed that parasites had a dramatic effect on reducing host abundance and reduced abundance below the level it would have been without infection although after about 95 days the parasite went extinct. In another free-running experiment Scott (1987) introduced the nematode *Heligmosomoides polygyrus* into large mouse enclosures and showed that with a wet peaty substrate (that encourages transmission) the parasite regulated the host population for a period of 16 weeks after which anthelmintic treatment resulted in an increase in the population.

3.2.2 Regulation and a case study: red grouse and *Trichostrongylus tenuis*

Dobson and Hudson (1992) explored the dynamics of the red grouse–*T. tenuis* system by extending the basic Anderson and May model and incorporating a third equation to describe changes in the size of the free-living stages. They also included terms to evaluate the effects of time delays introduced through arrested development. Simulations from the model show that the degree of aggregation and the time delays in parasite development identified through arrested development are not the main cause of the population oscillations in red grouse. Cyclic oscillations in the model occurred when the ratio of parasite induced reduction in host fecundity to parasite induced reduction in host survival was greater than the degree of parasite aggregation within the host population. In other words, the impact on fecundity is the feature that destabilizes the host population and is the principal cause of oscillation.

The findings from detailed field studies and experiments (Hudson 1992; Hudson *et al.* 1992*b*) coupled with the model do not show that the parasites were the actual cause of the observed cyclic fluctuations. To determine if the parasites do indeed impact the dynamics of their host required a large-scale population experiment that manipulated parasite burdens and predicted the subsequent dynamics (Hudson *et al.* 1998). The model provides the means by which these predictions can be made. By incorporating terms that describe how worm mortality is affected by anthelmintic treatment they predicted the proportion of grouse that must be treated to have a significant impact on the population and the timing of this application. Treatment of just a small proportion of the population just at the peak grouse density would have an impact on the dynamics of the host population but worm eradication would require almost total treatment of the whole population.

Six study areas were used for the experiment. The first predicted crash was 1989 and in the winter before the crash, large numbers of grouse were caught at night on four of the six populations and treated orally with an anthelmintic to reduce intensities of worm infection. The remaining two were left untreated as control areas. On three of the treated populations about 20% of the population were caught, treated, and tagged although on the remaining one about 15% were treated. A second population crash was predicted in 1993 and in this year two of the areas treated in 1989 were treated again while two of those treated in 1989 were left as controls to provide comparisons within and between estates. The treatment of grouse reduced the variance of the bag records and in effect reduced the extent of the population crash following treatment in all the years and sites where treatment was applied (Fig. 3.2). Note that all treated sites showed some level of decline as predicted from the model and one site exhibited a decline that was similar to the extent of a decline when no treatment was applied, this was the site where just 15% of the population was treated. Nevertheless, the conclusion is clear, the removal of parasites reduced the amplitude of the oscillations in agreement with the model predictions implying that in these sites at these

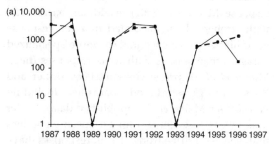

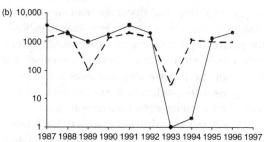

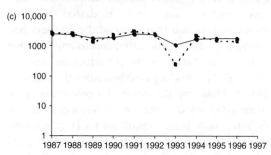

Figure 3.2 Population changes in red grouse, as presented through bag records, in (a) the two control plots, (b) the two plots with a single anthelminthic treatment, and (c) the two plots with two anthelminthic treatments.

Note: Further details can be found in Hudson *et al.* (1998).

times the parasites were necessary for the cycles observed. This statement does not refute the possibility that other factors may influence the dynamics of the grouse at other sites or at other times (e.g. Mougeot *et al.* 2003), simply because when we consider all the long-term monitoring and experimental data together it seems clear that parasites play an important role in the regulation of red grouse. Lambin *et al.* (1999) suggested that the use of bag records rather than traditional population density estimates may have affected the conclusions. However, Hudson *et al.* (1999) suggested that this was unlikely for the population densities considered.

In summary, it is clear that parasites are quite capable of regulating populations. Demonstrating this in free-ranging wild animal populations is not so simple and it is interesting to note that the clearest case comes from an unstable host population. The models indicate that parasite–host relations are intrinsically unstable so maybe this is an area where studies of host–parasite dynamics should focus.

3.3 Population introductions, parasitism, and success of introduction

Population introductions now constitute model systems that are used to study colonisations, the fate of small populations and conservation. Introductions of hosts into novel environments are interesting because they provide large-scale natural experiments that are replicated at a sufficient level to assess the ecological and evolutionary determinants of success. Successful colonizations also comprise some of the most dramatic population increases known, with the populations of several birds and mammals reaching hundreds of millions of individuals in their novel environments. Three studies of parasites are relevant for the problem of population regulation of hosts by parasites. First, Mitchell and Power (2003) and Torchin *et al.* (2003) investigated success of colonizations of a range of different organisms in relation to reduction in parasites in the newly colonized environment (see also Chapter 7). The parasites that were left behind were an important predictor of successful colonization (Mitchell and Power 2003; Torchin *et al.* 2003). Second, an analysis of introduction success of birds released in novel environments revealed a difference in T-cell mediated immune response between successful and unsuccessful colonizers (Møller and Cassey 2004). I will briefly review these studies and address their importance for determining the importance of parasites for population regulation of hosts.

First, Mitchell and Power (2003) and Torchin *et al.* (2003) investigated success of colonisations of a range of different organisms in relation to reduction in the parasite fauna in the newly colonized environment. The parasites that were left behind were an important predictor of successful colonization in plants (Mitchell and Power 2003) and animals (Torchin *et al.* 2003). Introduced populations of animal hosts on average had more than 40% fewer species of parasites than the native populations, and the average prevalence was also more than 40% reduced (Torchin *et al.* 2003). The parasites that were left behind were those that were less prevalent, perhaps because such species are less likely to be present in a small sample of hosts. In addition, less prevalent parasites are more likely to be more virulent since virulent parasites by definition have a higher probability of killing their hosts, thereby reducing their prevalence in the host population. However, prevalence of parasites that did invade the novel environment with their host had as large prevalence as in the native environment, and native parasites that subsequently invaded introduced host populations had as large prevalences as those that were introduced with the invading host (Torchin *et al.* 2003).

Second, an analysis of introduction success of birds released in novel environments, mainly New Zealand, revealed a difference in T-cell mediated immune response between successful and unsuccessful colonizers (Møller and Cassey 2004). Deliberate human introductions of birds to novel environments were a passion for many immigrants of European origin in the eighteenth and nineteenth centuries. The number of introductions, the number of individuals introduced, and the outcome of many of these introductions has been meticulously documented, allowing statistical analysis of factors that potentially contribute to colonization. Introductions of birds to New Zealand and other

islands have been particularly well documented, and a number of key features of successful introductions such as number of introductions, number of individuals, habitat generalism, and sexual dichromatism have been identified as significant predictors of success (Cassey 2002). Møller and Cassey (2004) suggested that the stressful conditions associated with introductions would render individuals that suffered from chronic parasitic infections less likely to survive than less infected individuals. Therefore, individuals of species with strong immune responses would have a reduced likelihood of succeeding because bird species with stronger nestling mortality directly caused by parasites have stronger T-cell mediated immune response (Martin *et al.* 2001). In addition, bird species with stronger T-cell mediated immune response have higher prevalence and/or more specialized parasites than species with weak responses (fleas: Møller *et al.* in press; chewing lice: Møller and Rozsa in press; blood parasites including *Plasmodium*: Møller *et al.* unpublished data). Møller *et al.* (2004*a*) have shown for European passerines that species with strong T-cell responses have longer natal dispersal distances than species with weak responses, even when controlling statistically for potentially confounding variables. Therefore, individuals with strong immune responses are better able to disperse long distances successfully, giving rise to the prediction that successful colonisers should have stronger immune responses.

A comparative analysis of species of birds introduced to New Zealand, and for which information was available on T-cell mediated immune response, revealed that successful colonisers had stronger T-cell responses after controlling for other factors associated with introduction success (Fig. 3.3). This finding clearly refutes the first prediction. However, the effect of T-cell response on success depended on the size of the inoculum, with the effect only being significant for large releases. Introductions usually failed immediately after release before the birds had a chance to reproduce, making it unlikely that inbreeding played a role in success. Since introduced populations typically are derived from small subsets of native populations, they generally have fewer species and

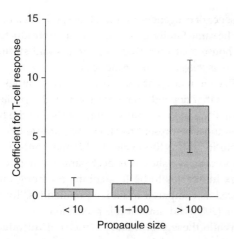

Figure 3.3 Introduction success of birds to New Zealand and other islands in relation to T-cell mediated immunity among nestlings. The values are partial regression coefficients (± SE) after controlling for habitat generalism, other potentially confounding variables and phylogenetic relationships among taxa.

Source: Adapted from Møller and Cassey (2004).

lower prevalences of parasites than the ancestral population, as described by Torchin *et al.* (2003) and Mitchell and Power (2003). A strong immune response would not be important in such a situation. In contrast, large introductions with more than 100 individuals are much more likely to harbour a diverse and abundant parasite fauna, where a strong immune response might play a key role between life and death.

Which are the interpretations of these findings of introductions, parasitism, and immunity? I suggest that at least four not necessarily mutually exclusive explanations are possible. (1) The results could be an effect of parasite absence from the inoculum of introduced hosts; (2) the findings could be an effect of host physiology during stressful introductions, which only secondarily is associated with parasitism; (3) the results could arise as an effect of the local host and parasite community in the novel environment; and (4) many parasites have complex life cycles that depend on more than one species of host. I will discuss each of these explanations in detail.

First, missing parasites from introduced populations may arise as a consequence of such populations often representing a small subset of the native population. Both Torchin *et al.* (2003) and Mitchell and Power (2003) found that diversity and prevalence of parasites among introduced species were dramatically reduced compared to the native populations. While this may make a difference in terms of probability of successful colonisation, this conclusion rests upon the untested assumption that the introduced population is a random sample derived from the native population. Small populations are by definition likely to suffer from strong sampling bias giving rise to demographic stochasticity that could affect host–parasites interactions as well. This is most likely the case for bird species introduced to New Zealand (Legendre *et al.* 1999). Drake (2003) has shown in a simple model that the relationship between the number of individuals introduced and the probability of establishment depends upon the relationship between virulence and the fraction of the population infected at introduction. The finding that T-cell mediated immune response did not predict introduction success for small inocula, but did so for large inocula (Cassey and Møller 2004; Fig. 3.3) is in agreement with this suggestion.

Second, the ability of hosts to cope with parasites depends strongly upon their current physiological state. Stressful conditions may especially provide an impediment to efficient host defences. The birds that were introduced to New Zealand and other islands went through extraordinary ordeals before reaching their final destination for release. First, they were captured in the United Kingdom and placed in cages that subsequently were transported to the harbour for shipment. The sea voyage took several months, often under very different weather conditions than commonly experienced by the species in question. These conditions must have resulted in strong physiological stress responses including a strong corticosterone response and induction of heat-shock proteins. Corticosterone is notorious for depressing immunity including T-cell mediated immunity (von Holst 1999), giving the parasites an upper hand during this long-lasting part of the introduction process. Such responses will only be important in vertebrates with a corticosterone-based stress physiology, although other mechanisms such as induction of stress proteins occur across almost all organisms. It seems very likely that natural selection will have played an important role in shaping the phenotype of the individuals that eventually arrived in New Zealand and other islands, since particularly individuals that were able to cope with the stressful conditions by not developing strong stress responses should be able to succeed. An important parallel situation similar to that experienced during introductions can be seen in the process of domestication (Kohane and Parsons 1988). Individuals with weak stress responses are strongly favoured by initial selection, and such individuals are characterized by low levels of testosterone, low levels of aggression, and a life-history characteristic of r-selected strategies. Since domestication can result in dramatic micro-evolutionary changes in just a few generations (see examples in Kohane and Parsons 1988), it seems likely that introductions might likewise have resulted in intense natural selection and changes in genetic composition in just a single or a few generations.

Third, the local host and parasite community in the novel environment may differ dramatically from those of the native environment. Parasites are often disproportionately common on islands (review in Hochberg and Møller 2001). Models of the coevolution of parasite virulence and host resistance on islands have shown that costly host resistance genes tend to be lost from populations inhabiting oceanic islands (Hochberg and Møller 2001). As a consequence, island populations are very susceptible to introduced parasites from continents that tend to be more virulent than their island counterparts (Hochberg and Møller 2001). These theoretical arguments would suggest that species of birds introduced to oceanic islands in general would run low risks of parasite-induced mortality due to local island populations of parasites. If anything, the absence of virulent parasites on islands should provide introduced species with a considerable fitness advantage. The local parasite community in the area of introduction will often differ considerably from that of the native populations. Since horizontal transfer of parasites is more common among sister taxa than among more distantly related taxa, introduced hosts would encounter relatively few novel parasites due to the

absence of congeners or even other species belonging to the same family, as was the case for introductions of house sparrows *Passer domesticus* and starlings *Sturnus vulgaris* to the Americas.

Fourth, many parasite species have complex life cycles that depend on more than a single species of host. Each of these intermediate hosts and the final host must be present to allow the parasite to complete its entire life cycle. If this mechanism was important, we should expect parasites with complex life cycles to be less common among introduced as compared to native populations. Torchin *et al.* (2003) did not test this prediction.

While these preliminary studies of introduced species, successful colonization, and missing parasites may appear to be clear-cut, there is clearly scope for much more work to be done before the exact mechanisms that led to this situation can be resolved.

3.4 Abundance–area relationships and parasitism

Abundance–area relationships constitute one of the most robust findings in ecology, with locally abundant species being widely distributed, while those with low local densities tend to have restricted distributions (Gaston 1994; Brown 1995). While this 'law' seems to be general, accounting for 20–30% of the variance across related species (Gaston 1996), the factors contributing to this relationship remain largely unknown. For example, Gaston and Blackburn (2003) failed to find general support for the hypothesis that greater dispersal propensity among birds could account for some of the variance in abundance–occupancy relationships.

If parasites are able to reduce the abundance of hosts compared to the level of abundance expected in the absence of virulent parasites, we should expect that indices of parasite impact on hosts would be able to partly explain some of the variance in the abundance–occupancy relationship. In particular, given that population regulation is more likely to be due to density-dependent effects at early rather than late life stages, we can make the prediction that estimates of the impact of parasites on hosts during early life stages should be better predictors of abundance after controlling for occupancy than estimates of parasite impact during later life stages. Recently, Møller *et al.* (2004*c*) tested this prediction using a measure of T-cell mediated immune response to a challenge with a novel mitogen as an estimate of parasite impact on hosts. This assumption is supported by the fact that bird species with stronger nestling mortality directly caused by parasites have stronger T-cell mediated immune response (Martin *et al.* 2001). In addition, bird species with stronger T-cell mediated immune response have higher prevalence and/or more specialized parasites than species with weak responses (fleas: Møller *et al.* in press; chewing lice: Møller and Rozsa in press; blood parasites including *Plasmodium*: Møller *et al.* unpublished data). Thus, there is empirical evidence supporting the assumption that species with stronger T-cell mediated immune response indeed suffer from greater mortality due to parasitism, and that such species are exploited by more specialized and virulent parasites.

In a comparative analysis of 73 European bird species for which information on continental population size, occupancy, and T-cell mediated immune response was available, Møller *et al.* (2004*c*) found the commonly reported positive association between abundance and occupancy, accounting for more than 30% of the variance. In a stepwise regression analysis T-cell mediated immune response of nestlings was a significant predictor of abundance with a negative impact as expected (Fig. 3.4). The relationship accounted for 12% of the variance, suggesting that parasites can have dramatic impact on host population size on a continental scale. In contrast, there was no significant relationship due to T-cell mediated immunity in adult birds. Likewise, a number of potentially confounding variables that has been found to be related to T-cell mediated immunity (e.g. hole nesting (Møller and Erritzøe 1996) and coloniality (Møller *et al.* 2001)) did not account for the relationship between abundance and immunity in nestlings.

These findings are suggestive and certainly raise a number of questions about future studies of abundance–occupancy relationships in general and about the impact of parasites on population size of hosts in particular.

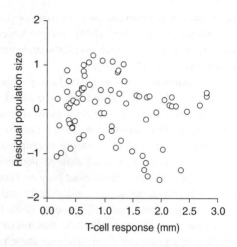

Figure 3.4 Relative population size of European passerine birds in relation to T-cell mediated immune response (mm) of nestlings. Population size was estimated as the residuals from a regression of log (population size) on log (distribution area).

Source: Adapted from Møller *et al.* (2004*c*).

3.5 Parasite-mediated competition and predation

There is clear evidence of intraspecific competition causing population regulation (e.g. Begon *et al.* 1996; Newton 1998). Likewise, there is clear evidence consistent with the suggestion that predators regulate prey populations (e.g. Newton 1998; Murdoch *et al.* 2003). Most population ecologists are happy to accept the notion that predators can regulate prey populations (e.g. Hudson and Bjørnstad 2003; Murdoch *et al.* 2003). This is fair judgment based on available evidence. However, we do *not* know what was the case of successful predation in the first place. Therefore, a more appropriate conclusion would be that predation, or any factor associated with successful predation, has been shown to result in population regulation. This distinction is crucial since most studies are only based on correlations, thereby preventing firm conclusions about exact cause and effect. What is must less appreciated is that these effects of competition and predation may be parasite-mediated, thereby changing the important role of population regulation from conspecifics and predators to parasites. There is firm empirical and experimental evidence showing that both intraspecific competition and predation can be parasite mediated. Begon *et al.* (1996) review several studies showing that parasites can play an important role in intraspecific competition.

A few studies have investigated how parasites may mediate which host individuals are killed by predators and which survive. In a now classical study Temple (1986) used a tame red-tailed hawk *Buteo jamaicensis* to collect a sample of chipmunk *Tamias striatus*, cottontail rabbit *Sylvilagus floridanus*, and grey squirrel *Sciurus carolinensis* prey and compared the health status of such individuals with a random sample that he simultaneously collected with a gun. There was a dramatic difference in intensity and prevalence of parasitism of these two categories of potential mammalian prey, with mammals killed by the predator suffering significantly more from parasitism by trematodes, nematodes, and ectoparasites (Temple 1986). In a second study Møller and Erritzøe (2000) investigated 535 individuals of 18 species of common passerine birds that had either been killed by domestic cats *Felis catus* or died accidentally by crashing with a window. All individuals were sexed and aged and a large number of characters were recorded for each specimen, including the size of the spleen. Bird species with larger spleens suffer more from parasite-induced mortality (Møller and Erritzøe 2002, 2003), and prey to cats were therefore expected to have smaller spleens than non-prey reflecting their poor ability to defend themselves against parasites. Prey had consistently smaller spleens than non-prey in a paired comparison across the 18 species, implying that they had weak immune systems. The data set did not indicate that sex or age, month of death, body mass, body condition, liver mass, wing length, or tarsus length differed significantly between prey and non-prey. These studies suggest that parasites can play an important role in determining which individuals are eaten and which survive.

The mechanisms behind differential predation on heavily parasitized hosts could simply be that such individuals are weakened and therefore easier to catch. Alternatively, predators may interact directly with the immune system of the host, thereby increasing the probability of prey being heavily parasitized (Møller *et al.* 2004*b*). Predation risk may affect the allocation priorities of limiting resources by potential prey, reducing investment in immune function when hosts are exposed to predators because of the costs of immune function. Møller *et al.* (2004*b*) tested this hypothesis by randomly exposing adult house sparrows to either a cat or a rabbit *Oryctolagus cuniculus* for six hours while assessing their ability to raise a T-cell mediated immune response. Sparrows exposed to a cat had a significant reduction of on average 18% and 36%, respectively, in T-cell response in two different experiments, as compared to sparrows that were exposed to a rabbit. In a field experiment in which sparrows were exposed to a barn owl *Tyto alba* or a rock dove *Columba livia* placed next to a nest box during laying Møller *et al.* (2004*b*) found a mean reduction in T-cell mediated immune response of 20%. In a third experiment conducted during spring, when blood parasite infections relapse, house sparrows were either exposed to a barn owl or a rock dove, while development of malarial infections was recorded during the subsequent 6 weeks. Individual sparrows exposed to a predator had a significantly higher prevalence and intensity of *Haemoproteus* malarial infection than control individuals. Therefore, exposure to predators reduced that ability of hosts to cope with parasitism mediated through effects on immune function. This was the case in aviaries and under field conditions. In conclusion, predators may have direct effects on the ability of hosts to defend themselves against parasites. This is likely to increase the mortality rate since house sparrows infected with *Haemoproteus* are more likely to die than uninfected individuals (González *et al.* 1999). Reductions in immune defence will also result in increased levels of parasitism that subsequently will affect risk of predation. The importance of this mechanism will increase with the density of predators, but potentially also with the density of prey, if availability of shelter from predators is limiting.

As these few examples suggest, parasites may play a much more indirect role in predator–prey interactions than previously thought. Such mechanisms must be taken into account when evaluating studies of population regulation of hosts by predators.

3.6 Concluding remarks

This chapter has provided one important insight that was clear at the outset: we have a very incomplete knowledge of population regulation of hosts by parasites. This may stem from the fact that a large part of parasitology and immunology is restricted to the laboratory. Another reason is that most population ecologists are not interested in parasites, perhaps stemming from the influence of David Lack who wrote that 'While further evidence is needed, it seems unlikely that disease is an important factor regulating the numbers of most wild birds' (Lack 1954: p. 169). This clearly has to change and there are signs that parasitism finally has started to claim its rightful position in population ecology. Here, I suggest four areas for which more research is needed.

First, more long-term field studies of hosts and parasites. Studies of population regulation have traditionally been based on long-term studies such as those on mammals in Canada, rodents in many parts of the northern hemisphere and hole nesting birds in Europe. Almost all of these ongoing studies have completely neglected the role of parasites. This clearly has to change if we want to understand if parasites play a role in many population ecology processes.

Second, more large-scale field experiments. I am aware of very few large-scale field experiments that have manipulated the level of parasitism and subsequently investigated the consequences for host populations. Clearly, many more such studies are needed.

Third, we need more comparative studies of population regulation. We may be able to gain insights into the role of parasites in population regulation of hosts by studying the explanatory power of estimates of parasite impact on hosts as predictors of host population sizes.

Finally, we need more studies of mechanisms. As emphasized above, we need to include parasitism in studies of predator regulation of prey populations to be able to partition the variance into predator and parasite components. We also need to know much more about the effects of parasites as mediators of competition and predation under field conditions. Finally, we need to know more about the microevolutionary changes that have happened during introductions. I believe that such efforts will provide a much better basis for judging the role of parasites in population regulations of their hosts.

CHAPTER 4

Food web patterns and the parasite's perspective

Michael V. K. Sukhdeo and Alexander D. Hernandez

The way that parasites fit into food webs remains unclear. We argue that coherent energetic patterns (biomass pyramids) occur in food webs that include parasites. These patterns typically evolve in stable associations when considered from the perspective of the parasite, and empirical evidence is provided of biomass patterns in a natural fresh water food web.

4.1 Introduction

It is clear that parasites are capable of exerting major effects on ecological interactions, and this is well documented in the growing mountain of theoretical and empirical evidence, much of it summarized within the pages of this very book. Yet, it is remarkable that several fundamental questions about the ecology of parasites still remain unanswered. For example, how do parasites fit into food web patterns?

This is a complex question at many levels, and it has been the subject of much debate since the 1920s. There have been several studies that have attempted to incorporate parasites into food web patterns (e.g. Elton 1927; Campbell *et al.* 1980; Goldwasser and Roughgarden 1993; Huxham *et al.* 1995, 1996; Leaper and Huxham 2002). However, few have provided verbal or theoretical models for thinking about how parasites might fit in food webs in the same way we visualize how predators or prey fit into food web patterns. Some of our best understanding of parasites in food web dynamics come from studies on the food web in the Ythan estuary, one of the largest and most completely documented food webs available, comprising 134 taxa of which 42 are metazoan parasites (Hall and Raffaelli 1991; Huxham *et al.* 1995; Leaper and

Department of Ecology, Evolution and Natural Resources, Rutgers University, Cook College, New Brunswick, NJ 08901.

Raffaelli 1999). Depending on whether parasites are included or excluded in the food web analysis, or whether they are included as biological species versus tropho species, there are significant changes in several food webs statistics including food chain length, omnivory, and connectance (Huxham *et al.* 1995; Leaper and Huxham 2002). However, no easy-to-recognize general pattern has emerged from these analyses. Indeed, the absence of a clear model of how parasites might fit into food webs is a hindrance for many ecologists working with parasites. Thus, although the dramatic effects of parasites on host behaviour and host population dynamics have been reported from numerous taxa almost invariably, the consequences of these effects on community processes are usually left to inference and speculation (Dobson and Hudson 1986; Minchella and Scott 1991; Marcogliese and Cone 1997; Poulin 1999; Moore 2002; Mouritsen and Poulin 2002, but see Chapter 8). The reasonable assumption by most investigators is that population effects will ultimately translate into system level effects.

In this chapter, we will briefly summarize some of the history of food webs and the debate about parasites. If one takes the view that the study of food webs is the study of energy flow through a community and the search for general patterns (Lindeman 1942; Odum 1953; Hairston and Hairston 1993; Winemiller and Polis 1996; Morin

1999), then it follows that before parasites can be considered legitimate actors in the food web drama, they must first be integrated into a general pattern of energy flow in the community. We will argue that coherent patterns may occur in evolutionary stable associations when viewed from the parasite's perspective, and we present some empirical evidence for a biomass pattern in a natural food chain.

4.2 A brief history of food webs and parasites

The food web literature is enormous and continues to grow exponentially, and it would be a formidable task to summarize all of the complex theoretical and experimental directions in modern food web studies. However, for the readers not familiar with this literature, an abbreviated history of the field might elucidate some of the principles and major perspectives that have guided progress in the field.

In 1927, Charles Elton was only 26 years old when he wrote *Animal Ecology*, which went on to become one of the most influential books in the field. Here he introduced the concept of food cycles (now food chains and food webs), and he argued that they provided the conceptual framework to understand species interactions in the context of complex ecosystems. Although other workers had used food cycle diagrams as an aid in the understanding of ecological systems, Elton's genius was in recognizing that food cycles play a central organizing role in ecology. One of his major insights on organisms was that 'food is the factor that plays the biggest parts in their lives, and it forms the connecting link between members of the community' (Elton 1927). Thus, food was the common currency of communities, and significantly, distinct patterns emerged from the feeding relationships in natural communities. He is most famous for '*Eltonian pyramids*', which argue that food cycles tended to be organized as a pyramid of numbers, where organisms at the bottom of the food chain tend to be very abundant, while predators on top are relatively few in numbers. Elton is now widely considered the father of community ecology, and his book helped formulate many of the basic questions that are still considered important by

ecologists today; the roles of competition, niche space, temporal and spatial heterogeneity, ecological succession, indirect effects, and body size on structuring communities, and the crucial importance of feedback between organisms and their environments (Elton 1927). His synthetic view of community interactions has continued to be a dominant paradigm in food web studies, and many of the basic lessons that Elton helped develop, like the importance of natural history and the use of multiple lines of inquiry, still resonate with ecologists.

Elton's most enduring gift to food web studies was the abstraction of complex community interactions in nature into simple patterns that were accessible to all. Eltonian pyramids provided a template for thinking about community structure both theoretically and empirically, and led to an explosion of food web studies. However, Elton was not always completely right. The pyramid of numbers, based on the relative size of food items, missed an important component of modern thinking, that is, the trophic inefficiency of energy transfer. Subsequently, Lindeman (1942) combined Elton's concepts with an energetic perspective from Lotka (1925), shifting the focus to pyramids of biomass rather than pyramids of numbers because biomass/productivity patterns better captured the dynamics of energy flow through the community. Lindeman's values of trophic efficiencies of ~5–15% are still widely accepted as an informal rule among ecologists (see Fig. 4.1). Odum (1953) further refined Lindeman's ideas (in a few marine communities biomass pyramids may overemphasize the importance of large organisms), and suggested that energy flow, that is, Productivity + Respiration, was the appropriate index for comparing any and all components of an ecosystem and across ecosystems. Thus, energy became the accepted currency of all community interactions, and this elementary perspective on the nature of food webs has remained unchanged for more than 50 years. The analysis of energy and material flow is still considered to be fundamental to understanding the patterns and dynamics in ecosystems and the way ecosystems are organized (DeAngelis 1992; DeRuiter *et al.* 1995). For most food chains, trophic biomass (standing crop) patterns are good

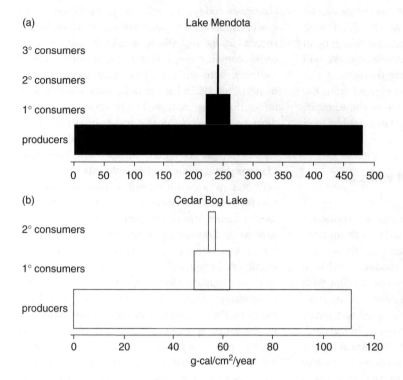

Figure 4.1 Biomass/productivity pyramids for the (a) Lake Mendota and (b) Cedar Bog Lake food webs.

Notes: Producers here are defined as macrophytic pondweed and microphytic phytoplankton, while primary consumers are the herbivores, or 'browsers', in aquatic ecosystems. Secondary and tertiary consumers are benthic and swimming predators, respectively.

Source: From data in Lindeman (1942).

surrogates of this energetic perspective (Lindeman 1942; Odum 1953).

Since these early studies, the food web field has grown enormously in size and complexity, and modern models often incorporate population dynamics as an important process in the structuring of communities and food webs (May 1972, 1973; Pimm and Lawton 1977, 1978; Pimm 1982; Paine 1988). Over the past few decades, several of the most gifted minds in ecology have devoted extraordinary intellectual effort towards capturing a model of natural webs. Investigators now have immediate access to more than 200 food webs in a shared database (ECOWeB, Cohen 1977), and it is probably fair to say that every interaction occurring between community members has been meticulously scrutinized and evaluated for pattern and meaning. Indeed, several patterns have emerged that seem to be repeatable across different food webs. The list includes, but is not limited to a constant ratio of predators to prey, food chains are short, three species loops are rare, omnivory is rare, and connectance and interaction strength vary from

food web to food web (Cohen 1977, 1978; Morin 1999). However, the ecological significance of these patterns remain controversial since many ecologists have serious reservations about the accuracy and completeness of food web descriptions in the database (Paine 1988; Polis 1991; Martinez 1991, 1992; Hall and Raffaelli 1993), and contrary to the strong assertions by many theorists, patterns from food webs of real communities generally do not support predictions arising from dynamic and graphic models of food web structure (Polis 1991; Leibold and Wooton 2001).

We now have a much better sense of the elegant mathematical complexity of natural food webs, but the synthesis that Elton hoped for has not yet materialized (Leibold and Wooton 2001). The field is now largely dominated by theorists, and dictated by statistical and methodological issues. The book 'Food Webs' edited by Winemiller and Polis (1996) is the published proceedings of a symposium to bring together viewpoints from theoretical ecology and empirical research in systems ranging from soil fauna to oceans. One need only peruse the titles of the 37 chapters (none dealing with parasites) to recognize the tremendous diversity of approaches, models, and analytical tools. The field is now so large it defies simple categorization, but it is abundantly clear that food web theory has far surpassed our experimental capabilities. This situation was already familiar to an older Elton, who in his fifties remarked that 'although the subject is partly illuminated, it is also greatly obscured for the ordinary ecologist by the brilliant cloud of mathematical theory that has evolved' (Elton and Miller 1954).

For the most part, parasitologists have not been very impressed with this remarkable effort, nor have they been overwhelmed by the conclusions. A clear point that emerged from this debate was that parasites had been ignored. Less than 15 webs in the literature contain parasites, and usually the data includes only small subsets of the parasite community (see table 1 in Marcogliese and Cone 1997). Parasitologists are appalled (or at least act appalled) that parasitism, a feeding strategy that is used by 50–70% of species on earth (Price 1980; Toft 1991), has been ignored in food webs. Ecologists respond that they do indeed recognize that parasites are ubiquitous in natural systems, that parasites can have significant effects on community structure (Dobson and Hudson 1986; Scott and Dobson 1989; Minchella and Scott 1991; Huxham *et al.* 1995), and even that parasites can act as 'keystone species' in some habitats (Curtis and Hurd 1983). In such cases, they acknowledge that omission of parasites from a community web is no more defensible than the omission of the principal vertebrate predator (Huxham *et al.* 1995). In addition, the inclusion of parasites obviates any size-based cascades (a popular analytical tool), and this by itself has led to several calls for analyses of webs that include parasites (Price *et al.* 1986; Lawton 1989; Polis 1991; Cohen *et al.* 1993; Goldwasser and Roughgarden 1993; Leaper and Huxham 2002). Nevertheless, it has been extremely difficult to retrofit parasites into eight decades of theoretical constructions and assumptions on the nature of food webs.

Parasitologists themselves have been very late entering into the food web debate. One reason is that the field of modern parasitology operates under an exacting biomedical rubric that favours extreme empiricism, and parasitologists were not trained to critique theoretical assumptions. Since parasitology began incorporating explicit ecological paradigms around mid-century (e.g. Park 1948; Haldane 1949; Holmes 1961; Schad 1963; Barbehenn 1969), parasite ecology has grown into an exciting discipline focusing on the rules of assembly of parasite communities within their hosts (see also Chapter 2), the roles of parasites in host population regulation (see Chapter 3), and the evolutionary and ecological implications of parasite mediation in trophic interactions (Esch 1977; Anderson and May 1979; May and Anderson 1979; Nickol 1979; Freeland 1983; Esch *et al.* 1990; Minchella and Scott 1991; Marcogliese and Cone 1997; Hudson *et al.* 1998; Lafferty 1999; Lafferty and Kuris 2002; Marcogliese 2002; Moore 2002; Torchin *et al.* 2003; Chapter 8). Most of the evidence for key roles of parasites in community structure is based on the differential susceptibility of host species to infection and its consequences, and although in principle these effects could greatly influence the structure of species assemblages, the roles of parasites in community dynamics are still being debated (Barbenhenn 1969; Holmes 1982; Freeland 1983; Price *et al.* 1986; Freeland and Boulton 1992; Holmes 1996; Poulin 1999; Thomas *et al.* 2000*a,b*; see Chapter 8). It is difficult to assess just how important parasitism is for food webs because few studies have been designed to specifically address the role of parasites at the community level (Mouritsen and Poulin 2002).

4.3 Visualizing parasites in food webs

Elton (1927) was the first to argue that food cycle pyramids could 'be fully understood only by bringing in parasites.' He had a good understanding of parasites, and even authored a classic '*parasitology*' paper on endoparasites in a mouse population (Elton *et al.* 1931). His famous book is peppered with references to host–parasite interactions and he devoted an entire chapter, simply entitled '*Parasites*', solely to discussing the role of parasites in food webs. He envisaged that parasites and hyper-parasites fit into food cycles in an inverted pyramid of numbers from the top predator. Elton's number-based analyses are now long fallen out of favour and we no longer think of communities in this context, but there have been few other ideas to supplant this inclusive perspective.

Modern ideas for including parasites into food webs fall into one of two general categories of argument, piggybacking or statistical. The most popular is the piggyback argument, which uses well-understood trophic patterns to show how parasites fit into the web. Typically, a complex (multi-host) parasite life cycle is depicted with arrows tracing the parasite's developmental path through successive hosts, and where invariably the intermediate hosts are eaten by the definitive hosts, for example, Fig. 4.2. Simple one host life cycles, or life cycles with hosts that do not interact trophically are not used as examples. Life cycle descriptions are the result of intensive field observations and experiments by

traditional parasitologists and they provide a natural history perspective that is persuasive to biologists. However, although they are powerful visual metaphors for the parasite's intimate connections with trophic interactions in the food web (Price 1980; Price *et al.* 1986; Marcogliese and Cone 1997; Marcogliese 2002), life cycle diagrams by themselves do not aid in our understanding of pattern and process.

There are several upgraded versions of the piggy-back argument, including some that verbally or diagrammatically map the parasite's movements up the trophic cascade. Data collected over a period of four years from streams in the New Jersey Pinelands were used to produce the examples of food webs in Fig. 4.3 (a), and the same food webs with parasites Fig. 4.3 (b). All of the interactions (arrows) in these webs came from studies reported in the literature, and represent probable interactions rather than

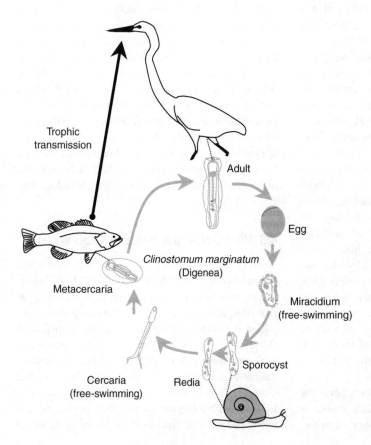

Figure 4.2 The life cycle of the digenean *Clinostomum marginatum*. Adult worms parasitize the intestine of egrets and other fish eating birds, where they produce eggs that are dropped into water with the bird's faeces. A miracidium hatches out of the egg and swims until it finds a snail and infects it. The miracidium sheds its cilia and develops into a sporocyst, which then produces multiple redia. The redia produce multiple cercaria, which leave the snail and swim until they find a fish to infect, and then develop into metacercaria. Predation of fish by birds facilitates the completion of the parasites life cycle.

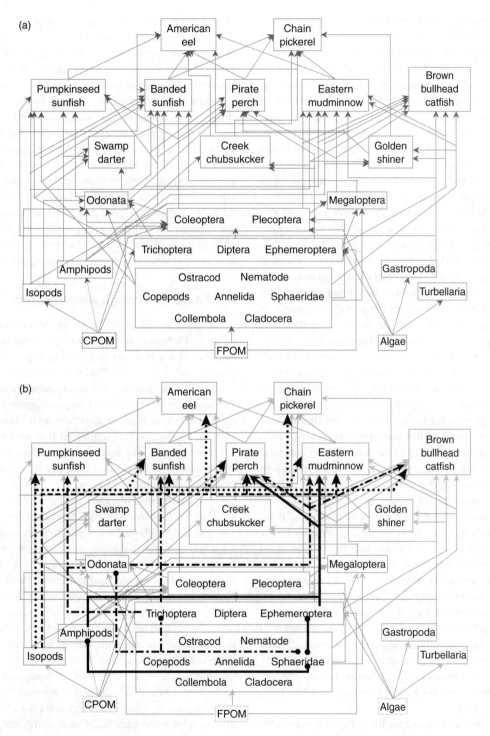

Figure 4.3 A topological food web from a New Jersey Pinelands stream during the fall season. (a) All possible trophic interactions between fish and macroinvertebrates, and between macroinvertebrates and their resource are shown. (b) Four helminth parasites whose adult stages parasitize fish are tracked through the food web.

Note: Dotted lines represent the path for *Fessisentis* sp., dash lines *Acanthocephalus* sp., dash-dot lines *Phyllodistomum* sp., and solid black lines *Crepidostomum* sp.

actual interactions. These topological diagrams help to visualize the complexity of natural systems, and the complexity is vastly increased when parasites are added. In the example in Fig. 4.3(b), only 4 of the 11 parasites in the system are mapped onto the web. There are even more sophisticated versions of analyses that piggyback onto food web relationships to examine parasite effects on life history traits, parasite mediation of interactions between competing species or to make predictions on the distribution and abundance of hosts and parasites (Curtis and Hurd 1983; Freeland 1983; Price *et al.* 1986; Marcogliese and Cone 1997; Hudson and Greenman 1998; Thomas *et al.* 2000*a*; Skorping and Hogstedt 2001; Marcogliese 2002; Mouritsen and Poulin 2002), but the community-level processes of many of these interactions remain elusive.

The statistical argument for including parasites in food webs is based on the popular approach of analysing real food webs to determine patterns in their properties. Food web theorists have developed sophisticated metrics to analyse food web structure, and in these analyses, parasites are generally treated as top predators, or trophospecies (omnivores feeding at multiple trophic level), although their hosts (and the parasites themselves) must fall prey to the predators. Typically, when parasites are included in the analyses, several food web statistics are altered, including increases in food chain length, increases in omnivorous links and increases in connectance (Huxham *et al.* 1995; Memmot *et al.* 2000; Leaper and Huxham 2002). Increases in food chain length and omnivory are relatively easy to explain, but the meaning of changes in connectance is less obvious. Food web theory centres around the idea of connectance, which in a mathematically profound way is a measure of system stability. It is the fraction of all potential trophic connections that are actually observed in the food web under study, and is calculated as the number of trophic links observed divided by the maximum number of possible binary connections in the species assemblage (Cohen 1978, 1989; Pimm 1982).

It is difficult to visualize the patterns that are illuminated by these web statistics (Raffaelli 2002). In addition, web statistics, including connectance,

may be too arbitrary to be useful in predicting patterns. These metrics are dependent on a variety of subjective decisions related to lumping different kinds of animals into single functional groups, and thus are subject to the idiosyncrasies of each investigator. Paine (1988) provides an example where five different analyses of the exact same food web yielded five different connectance values that varied by as much as 100%. Other investigators have raised similar concerns about the artefactual view of trophic interactions within communities that come from these analyses (Polis 1991, 1994; Tavares-Cromar and Williams 1996), and although the debate is beyond the scope of this chapter, it is clear that the search for a generalizable pattern that includes parasites has not yet been successful.

4.4 Patterns through simplification?

Most food webs in the literature are now thought of as incomplete caricatures of natural communities, and there have been several calls to improve them (Paine 1988; Pimm and Kitching 1988; Closs 1991; Polis 1991). In general, the recommendations call for two broad categories of improvements; more explicitness, and more exhaustiveness (Cohen *et al.* 1993). For example, it is suggested that the kinds of organisms in a food web should be reported by using units of observation that are as refined as possible (taxonomic resolution) to avoid the current problems with lumping animals (Polis 1991; Cohen *et al.* 1993; Goldwasser and Roughgarden 1993; Townsend *et al.* 1998). Previous studies are criticized for a failure to resolve webs at the highest possible level of taxonomic resolution, but increasing the resolution involves a considerable increase in the effort required (Lancaster and Robertson 1995). Recent studies have demonstrated that improving taxonomic resolution can have significant effects on several food web statistics, including connectance, but there is still some question as to whether rigour, or understanding, is improved (Thompson and Townsend 2000). Nevertheless, it appears that the proposed solution to the already complex situation of food webs is the addition of more complexity. The daunting diversity of parasitic strategies that exist is already a barrier to incorporating concepts about parasitism (Lafferty and Kuris 2002), and we would argue that, at least for parasites, the answer might better come from simplification.

Simplification of any aspect of complex food webs runs the risk of ignoring interactions that may be important in shaping web structure. Food web communities are envisaged as assemblages of species populations, many of which interact with each other to varying degrees. Thus, understanding the system requires understanding the dynamics of each species population, each pairwise interaction between species, and the direct and indirect influences of the whole web of community interactions on these relationships. If estimates of the number of parasites species are true (>50% of all species on Earth; Price 1980), then adding parasites to this mix could make these projects even more mind-boggling. On the other hand, if one tries to imagine the simplest food web possible, it would be a food chain. We would argue that it is at this basic operational unit that we should begin to examine the role of parasites.

In a sense, this approach is validated by the numerous empirical studies using simple food chains to extrapolate the basic rules of food webs, including those using artificially assembled protozoan food chains, or quasi-natural communities in pitcher plants (e.g. Gause 1934; Lawler 1993; Lawler and Morin 1993; Morin and Lawler 1996; Ellison *et al.* 2002). However, this approach also brings us back to the very beginnings of the discussion that was initiated by Elton and Lindeman, and this time, we can also look at it from the parasite's perspective.

4.5 Host specificity and food webs

One of the most important issues in this debate may be the question of specialists and generalists among parasites, because it causes the most confusion. All organisms specialize, and we share the view that the dichotomy created by the terms specialist and generalist is often artificial, and comes from consideration of only one or two resources axes along which an organism can specialize (Thompson 1982). Parasites are highly host-specific. The classic community-wide study of parasite specificity is

that of the feeding habits of parasitic beetle larvae on the seeds of dicotyledonous plants in the tropics (Janzen 1980), but there are numerous other examples (Price 1980). In Janzen's study, the vast majority (83 of 110) parasite species living in the forest fed exclusively on a single host species, 14 species fed on 2 host species, 9 on 3, 2 on 4, 1 on 6, and 1 on 8. A great deal of evidence in the literature suggests that phylogenetic backgrounds, body sizes, morphologies, and feeding behaviours of both players serve to promote parasite–host specificity (Freeland 1983). The problem is usually not why so many parasites are highly host specific (we recognize the intimacy of the host–parasite relationship), but why some species appear to not specialize as consistently on a single host species (Thompson 1982).

Part of the problem relates to the nature of semantic definitions. Parasites are designated as specialists or generalists with respect to the number of hosts they use. However, these terms are still used in a variety of different ways. For example, some studies may base their definition of specificity on the number of higher taxa on which a parasite feeds, rather than on the number of host species it feeds on because this allows inferences about the extent to which a parasite is tracking it host taxon phylogenetically. A parasite feeding on two closely related hosts might be more tightly bound to its host genome than a parasite feeding on two hosts from different families or orders. However, as Thompson (1982) argues, by this logic, a parasite feeding on 12 hosts (many) in the same host genus can be considered more host specific than a parasite feeding on two hosts (fewer) in different orders.

The second and more important concern is a methodological problem related to how data on parasite host distributions are recorded, and how these data might be used in the construction of food webs. Consider the situation in the North Atlantic food web, one of the most complex webs in the world, and one that features the cod, an important human food source (Yodzis 1996). Marine systems are thought to have the greatest numbers of generalist parasites because it is an adaptation for completing their life cycles in a dilute open system (Polyanski 1961; Bush 1990; Holmes 1990), and it is no surprise to find that cod are host to 105 parasites, most of which are generalists (Marcogliese 2002). This sort of data causes food webologists to tear their hair out because of all the lines they must now add to their interaction diagrams. However, in fact, very few of these parasites may be relevant to the energetic interactions occurring within specific food chains in the web. The majority of members on this list are paratenic generalists, and only seven of these species are actually specific to the cod (Marcogliese 2002). Paratenic stages enter fish species nonspecifically, and the parasites undergo no development in the paratenic hosts, tend to use few host resources, and produce little pathology. It is an opportunistic relationship, and the parasites simply use these hosts for transport up the food chain and for prolonged transmission in the large ocean. Essentially, this situation is no different from the strategy of parasites like the common liver fluke *Fasciola hepatica*, whose infective stage encysts on vegetation and awaits ingestion by the definitive host. This strategy does not make the plant an important part of the energy flow between the hosts and the parasite. Thus, for the most part, paratenic parasites tend to be unimportant in the food webs of their hosts, and the important hosts are the obligate hosts with whom they share evolutionary history.

Reducing the number down to seven obligate parasites in cod significantly eases the analytical effort, but the web can still be complicated. Again however, not all of these parasites might deserve their own lines on the food web. There is an astonishing amount of regional and local specialization by parasites. It has long been recognized in studies of diverse groups of parasites that in species with impressively long lists of recorded hosts, different populations of parasites often specialize on different host species (Gilbert and Singer 1975; Cates 1980; Fox and Morrow 1981; Thompson 1982). For example, resource partitioning in the ocean may contribute to the formation of distinct regional parasite assemblages, and within a single fish species, the parasites can vary significantly between inshore and offshore stocks (Polyanski 1961; Thoney 1993; Hemmingsen and McKenzie 2001; Marcogliese 2002). It is for this reason that parasites are often employed as indicators of fish stocks or populations

(Williams *et al.* 1992; Arthur 1997). Therefore, before making any inferences on the roles of parasites, it is important to identify the parasite species cycling in the specific food web of interest, rather than relying on the historical data of host records.

4.6 The parasite's perspective on host specificity

At the simplest level, individual organisms form the basic units of species interactions, and the organismal perspective of parasites has not been generally considered in the host-centric debate on food webs. We cannot truly 'see' the world the way any other animal sees it, but its perception can often be inferred from its responses to specific conditions in the environment (Von Uexkull 1934; Lorenz and Tinbergen 1957). In this regard, we are interested in the parasite's viewpoint on the question of host specificity.

The parasite's perspective on host specificity is perhaps best inferred from the numerous studies in immunology, taxonomy, biochemistry, behaviour, and physiology coming from the laboratories of traditional parasitologists. These data paint a picture of exquisite and intimate specificity in parasite interactions with their hosts. The finely tuned processes that parasites use to evade their hosts' immune responses, and blend unobtrusively into their hosts' physiological and metabolic activities, are the stuff of every textbook on the subject. Parasites are not just specific to their hosts, but they are often extremely specific to certain tissues and organs within their hosts. The classic example of parasite microhabitat specificity is a study that describes eight pinworm species that co-occur in the tiny rectum of the land turtle *Testudo graeca*, yet there is no overlapping in the parasites' microhabitats (Schad 1963). Parasites often make tortuous migrations through their hosts, using complex locomotor patterns and responses to get to these precise microhabitats. However, despite the apparent complexity of these responses, neurobiological, neuroanatomical, and behavioural studies on diverse parasite taxa have demonstrated that these behaviours are fixed, or genetically programmed activities that are hardwired in the organisms (Sukhdeo and

Sukhdeo 1994, 2002). There is no plasticity in these responses! Indeed, there is no need for behavioural plasticity because parasites are extraordinarily well adapted to their specific hosts (infections of the wrong host often results in the condition of 'larva migrans' where the parasites wander aimlessly until they die). Thus, parasitologists are quite comfortable with the notion that the relationship between a parasite and its host can be so *precise* that parasite behaviours will become fixed over evolutionarily long periods of interaction with their hosts.

For our purposes, the more interesting viewpoint is the perception of parasites moving about in the free-living world of the food web, as they leave one host and search for their next obligate host. The evidence suggests that the free-living stages of parasites searching for their hosts have narrow perceptual worlds, and that host-finding behaviours are genetically hardwired and fixed. The first illustrations of a parasite's perceptual world came from studies on ticks (Von Uexkull 1934), but modern examples come from the trematodes, a group that has been the focus of more behavioural and ecological study than any other group of endoparasites. As part of the trematode life cycle (Fig. 4.2), the cercarial stage emerges from the snail host and searches for the next host, usually in an aquatic environment. It is difficult to do justice to the huge diversity of extraordinary and unique search patterns that are found among the 40,000 species in this group (Schell 1970; Kearn 1998; Combes *et al.* 2002), but host-finding activities almost always include intricate patterns of behaviours which bring the cercariae to the host, where they attach and penetrate while discarding their tails (Sukhdeo and Mettrick 1987). Cercarial bodies are all morphologically similar, but cercarial tails reflect the diversity of shapes and sizes and other adaptations to get to their hosts (Schell 1970, Kearn 1998). In species where these host-finding behaviours have been studied in detail, a common finding is that the behaviours occur in repeatable fixed patterns (Graefe *et al.* 1967; Nollen 1968; Rees 1971; Chapman and Wilson 1973; Haas 1974, 1976, 1992; Whitfield *et al.* 1977; Combes 1980; Bundy 1981; Combes *et al.* 1994, 2002). Interestingly, these complex host-finding programmes are generated entirely by the tail, and the tail will execute the programmes even if the cercarial body is removed (Chapman and Wilson 1973; Prior and Uglem 1979; Uglem and Prior 1983). Neurophysiological recordings confirm that this activity is initiated in the tail and that sensory feedback from the cercarial brain in the body is not required (Prior and Uglem 1979). Tails have no organized ganglia or brains, and are innervated only by primitive fixed neural networks that must generate these complex patterns (Brownlee *et al.* 1995; Solis-Soto and De Jon-Brink 1994, 1995; McMichael-Phillips *et al.* 1996; Zuwaroski *et al.* 2001; Sukhdeo and Sukhdeo 2004), and this is hard evidence that these host-finding behaviours are hardwired! From the parasite's point of view, the host is a 'predictable' resource in the space and time of the food web (Combes *et al.* 2002), and therefore, there is no need for plasticity in the responses.

The significance of these observations on the hardwired nature of host-finding behaviour is that they argue that parasites are not just specialized within each of their hosts, but they are equally specialized to deal with the environmental space that connects their hosts. So, for example, in parasites with multi-host life cycles, it suggests that there must be long-term ecological stability in the interactions between the hosts, and between the hosts and their environment, before the parasite can establish a life cycle and specialize to the degree we now see. From the parasites' perspective, these interacting hosts would represent stable evolutionary units (Thompson 1982) in which they can profitably invest in obligate relationships that lead to specialization and fixation.

This view of the intimate relationship between parasites and their obligate hosts and their host's environment share the same underlying principles with arguments used by investigators who have proposed using parasites to track consistent trophic feeding relationships in food webs (Campbell *et al.* 1980; Campbell 1983; George-Nascimento 1987; Bush and Holmes 1986*a,b*; Gardiner and Campbell 1992; Bush *et al.* 1993; Marcogliese and Cone 1997; Marcogliese 2002). This view is also implicit in explanations of enemy release for the numerous examples where invaders do better than native competitors because the invaders are able to leave their parasites behind (Holt and Lawton 1994; Torchin *et al.* 2003, see also Chapter 3 and 7). It is also the view that derives from studies of coevolution in mutualistic and parasitic associations. Since the earliest ideas of gene-for-gene interactions between plant parasites and their host (Flor 1942, 1955), the elegant study on coevolution in butterflies and their host plants (Erlich and Raven 1964) and the development of the first mathematical model of coevolution (Mode 1958), the field of coevolutionary studies has grown considerably. Mutualists and parasites have been the focus of much of the debate, and the basic mechanisms of the coevolutionary process have been well described in numerous books and reviews (e.g. Price 1980; Thompson 1982; Price *et al.* 1986; Lively *et al.* 1990; Toft and Karter 1990; Freeland and Boulton 1992; Thompson 1994; Clayton *et al.* 2003). Very briefly, coevolution is reciprocal evolutionary change in interacting species, and an important part of the definition is that it involves the partial coordination of non-mixing gene pools. The evolution of an interaction between two host species can become a focal point around which parasites evolve and become part of the interaction (Price 1980; Thompson 1982, 1994). It is generally accepted that a basic evolutionary building block in many communities is a unit group of species within which selection acts on all participants (Gilbert 1975, 1977, 1979; Thompson 1982), and several investigators have reasoned that obligate parasites may effectively identify these unit groups (Price 1980; Thompson 1982; Marcogliese and Cone 1997; Combes 2001; Marcogliese 2002).

Consequently, when selecting a food chain for study on how parasites might fit into trophic patterns of energy flow through the system, it is important to include only those obligate parasites that are hardwired to the hosts and to the dynamic interactions occurring in the system, and which comprise stable unit groups of species. In these food chains, one might expect the energetic organization of trophic structure to stabilize into natural patterns that include these parasites.

We will describe an example from a model system in the New Jersey Pinelands, and as a first assessment of trophic relationships, we will use biomass patterns as the surrogate for energetic patterns.

4.7 New Jersey pine barrens

We have been studying helminth endoparasites (Acanthocephala, Cestoda, Nematoda, and Trematoda) that infect fish from streams in the Mullica River watershed, which is located within the New Jersey Pinelands (USA). The Pinelands is a region of over 1 million acres of pine–oak forests and sandy soil of the coastal plains. It is nestled in the centre of one of North America's most populous regions, the New York City and Philadelphia metropolitan area, but a significant fraction is under government regulation and is characterized by minimally disturbed habitats with relatively few farms and developed areas. High acidity and low concentrations of dissolved solids characterize undisturbed streams in this region (Morgan and Good 1988; Zampella 1994), and these streams typically support native fish communities (Zampella and Bunnell 1998; Zampella *et al.* 2001). The advantage to working with food webs in the Pinelands is that the naturally acidic system and low productivity contributes to lower overall biodiversity (Zampella and Laidig 1997; Zampella and Bunnell 1998), and this constrains the lengths and complexity of local food webs.

In fall 2002, we collected biomass data from four trophic levels in a second-order stream, Muskingum Brook, and this included leaf detritus falling into the stream, macroinvertebrates, fish, and parasites. Preliminary studies had previously identified two acanthocephalan species, *Fessisentis* sp. and *Acanthocephalus* sp. as the most abundant adult parasites infecting fish in this stream (66% of all adult parasites). The life cycles of both of these species occur between the same two obligate hosts: isopods, which are one of the most numerous detritivorous macroinvertebrates in the system (32% of detritivores), and pirate perch, which are one of the most abundant fish species in the system, with up to 50% of all fish caught at every sampling effort. Isopods and pirate perch are both native to this region (Peckarsky *et al.* 1990; Zampella and Bunnell 1998), and pirate perch feed primarily on isopods (Hernandez, unpublished data). The acanthocephalan parasites do not infect other macroinvertebrates in the stream, and both of these parasite species show high specificity for pirate perch over other fish species in the system (Fig. 4.4).

The amount of leaf detritus (dry weight, in grams) that fell into the stream was measured by

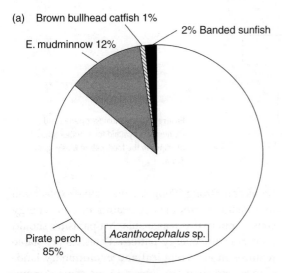

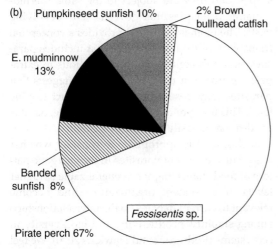

Figure 4.4 Distribution of adult worms of (a) *Acanthocephalus* sp. and (b) *Fessisentis* sp. within fish species in New Jersey Pinelands streams.

Note: The primary definitive host of these species is the Pirate Perch, a native species.

placing five plastic milk crates (surface area = 0.146 m²/sample) at random locations within a 25-m transect of the stream (surface area ~154 m²), one month prior to the collection of macro-invertebrates. The macroinvertebrates were collected by taking 10 samples, using a Hess-Sampler (surface area = 0.126 m²/sample) (modified from Hauer and Lamberti 1996; Jaarsma *et al.* 1998), and samples were fixed in 70% alcohol and brought back to the lab for sorting and counting under a dissecting microscope. Isopods were one of the commonest macroinvertebrates, and they were sexed, weighed (mg) and examined for the juvenile stages of acanthocephalan parasites. These juvenile parasites were then identified to genus and weighed (mg). Fish were collected from all habitat types for one hour within the same transect of the stream using a 4 mm seine (after methods in Zampella and Bunnell 1998), and then brought back to the lab where they were killed and frozen until examined for internal parasites using standard parasitological methods. Adult acanthocephalans were identified to genus and weighed.

The two acanthocephalan species used in this analysis constituted the vast majority of the obligate parasites of the fish in this food chain. Adults of only two other parasite species were recovered from these fish hosts, *Crepidostomum* sp., a trematode parasite of the intestine (55% prevalence), and *Phyllodistomum* sp. a trematode parasite of the urinary tract (45% prevalence). The intermediate hosts of these trematode parasites are amphipods, caddisflies, and dragonflies. Thus, these parasites do not cycle through the isopod food chain, and their biomass data were not included in the analysis of this food chain. Biomass values for parasites were the total of all stages of the parasites, and the biomass values of fish included all species that can act as host. The biomass pyramids from these data are shown in Fig. 4.5(a). These results show that in this simple food chain, obligate parasites fall into a natural pattern at the top of the biomass pyramid. The estimate of energy that makes its way into total parasite biomass is ~4.5% of the predators' trophic level, and this is consistent with thermodynamic principles of trophic inefficiencies first elaborated by Lindeman (1942). The pyramids of numbers for this data set is shown in Fig. 4.5(b), and it validates

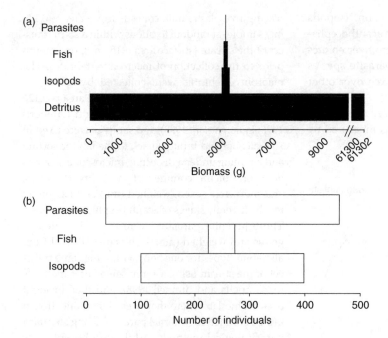

Figure 4.5 (a) Biomass pyramid and (b) numbers pyramid for acanthocephalan parasites in the food web of Muskingum Brook.

Elton's ideas of an inverted pyramid of numbers above the level of predator. Clearly more work needs to be done; this study was a first try, and there are probably several ways to improve it. Nevertheless, it is still interesting that this simple biomass pattern has emerged, especially given the notorious quirkiness of field studies on natural systems.

4.8 Concluding remarks

In the search for fundamental patterns that include parasites in food webs, we have argued the need for simplification, and suggested that simple natural food chains and their obligate parasites may form stable evolutionary units as conceived by Thompson (1982), and that these should be the units of study when looking at parasites in food webs. Our results on biomass pyramids from a simple aquatic food chain in the Pinelands suggests that parasites might fit into a meaningful pattern of energy flow through successive trophic levels in the food chain. Parasites are not normally considered to be important in the web's energy flow, and there has been some conjecture on how parasites can exert major regulatory effects on host population and yet do this without consuming much energy

(Polis and Strong 1996). An alternative explanation may relate to trophic constraints in their energy consumption. It is reasonable that parasites should be part of the energy pattern. Parasites are stable features in all ecological and evolutionary landscapes, and they are subject to the same thermodynamic principles that govern their coevolving hosts. The biomass pyramid provides a conceptual framework of trophic structure that includes parasites, and although we must be cautious in extrapolating from our data, the pyramid suggests that parasites may deserve to have a distinct trophic level. This interpretation may be debatable, but it is an idea that is easily testable in the field.

In closing, it is appropriate to consider whether this inclusive view of parasites in the energetic pattern of food chains might be generalizable to higher levels of ecosystem organization, and become relevant to current debates on the interrelationships among stability, productivity, and biodiversity in ecosystems (Rodriguez and Hawkins 2000). We feel that the answer to this question is yes, and that the energetic pattern of parasites in food chains will be reflected at the ecosystem level. As a shortcut to synthesizing a view on the complex nature of modern food web studies, we will defer to the opinions of

Dave Raffaelli of Ythan estuary fame, an accredited world leader in theoretical and experimental studies on food webs, and possessing the accumulated wisdom of decades of excellent work in the field. In a recent *Science* article, Raffaelli (2002) bemoaned the complexity of mathematical and modelling efforts that has blurred and obscured fundamental patterns in food webs, and he argued for the return towards more 'Eltonian' principles in the search for basic patterns in food webs. As a way of conceptualizing simple patterns he created 'Raffaelli's pyramid'; species in a food web are grouped into functionally similar trophic levels, and the biomass of each level is used to construct a pyramid whose slope may reflect system stability. We have replicated the original diagram of 'Raffaelli's pyramid' almost exactly in Fig. 4.6, with the exception that we have added the parasite's trophic level. The implications of this pyramid are enormous! At the very least, adding parasites will add to the rigor of estimating the pyramid's slope. If this pattern holds true at higher levels of ecosystem organization, it may explain why we are not being overrun by parasites even though parasitism is such a popular feeding strategy. Again, it may not be difficult to test these ideas in the field, especially since the taxonomic identity of the parasites need not be precise

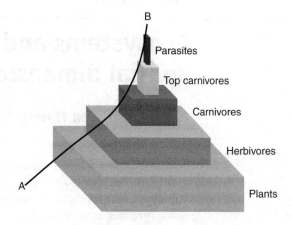

Figure 4.6 A hypothetical pyramid of biomass first described by Rafaelli (2002), but modified to include parasites.

Note: Biomass decreases, as trophic level increases, and the slope of the line AB may be a reflection of system stability.

at this level, and we would need only good estimates of their biomass.

There are many exciting directions in which we can take this discussion, but at this time it would consist mostly of speculation in the absence of data. Thus, we will conclude this chapter by saying that much work needs to be done, and that traditional parasitologists and ecologists might benefit from increased interactions in food web studies.

CHAPTER 5

Ecosystems and parasitism: the spatial dimension

Robert Holt[1] and Thierry Boulinier[2]

Within ecosystems, parasites and hosts often occur in multiple, spatially distributed sites, dispersed over large distances. The interplay of local interactions and dispersal at various spatial scales leads to a rich array of spatial dynamical processes, with important implications for the ecology and evolution of host–parasite interactions and the ecosystem impacts of parasitism.

5.1 Introduction

Most species of pathogens and their hosts occur in multiple, spatially distributed sites, dispersed over much larger distances than the spatial arena that circumscribes the lives of individuals and interactions between pathogens and hosts. Yet, some pathogens have tiny spores and can move great distances via aerial dispersal, which potentially leads to coupling of host–parasite systems at continental and even global spatial scales (Brown and Hovmoller 2002). The combination of local interactions and dispersal at various scales leads to a range of important implications of spatial dynamics for the ecology and evolution of host–pathogen interactions, with consequences for the ecosystem impacts of parasitism (Hochberg and Holt 2002). There is a vast literature on the dynamics of infectious disease in well-mixed populations (Anderson and May 1991). For many years, many authors have recognized the importance of space for the ecological and evolutionary dynamics of host–pathogen systems (e.g. Cliff *et al.* 1981; Bolker and Grenfell 1995) (see also Chapter 2). Mollison and Levin (1995) provide a useful review of earlier work. We

[1] Department of Zoology, 223 Bartram Hall, P.O. Box 118525, University of Florida, Gainesville, Florida 32611-8525.
[2] Laboratoire d'Ecologie, CNRS-UMR 7625, Université Pierre & Marie Curie, 7 Quai St Bernard F-75005 Paris, France.

will not attempt to synthesize this far-flung literature (which would take a volume all by itself), but instead draw out some highlights that seem most pertinent to ecosystem issues.

5.2 Ecosystem implications of parasitism: some potential major effects

From an ecosystem perspective, parasites have several distinct consequences; spatial dynamics can matter for each of these.

5.2.1 Population limitation and regulation

Understanding how populations are limited in abundance, and bounded in their fluctuations, is essential to understanding how ecosystems as a whole are governed. Although not sufficient for understanding ecosystem processes, the standard concerns of population and community ecology—the focus on processes of positive and negative feedbacks arising from density dependence, resource dependencies, and interspecific interactions—are necessary for understanding patterns of energy and nutrient flows in ecosystems, and the responses of ecosystems to disturbance and secular environmental change. Demographic effects of parasites leading to population limitation and regulation, when quantitatively strong, can affect many aspects of ecosystem dynamics. For instance, one hypothesis for the rich diversity of trees in tropical rain forest is that species experience strong, intraspecific density dependence due to host-specific pathogens acting on seedling plants (the Janzen–Connell hypothesis, see also Chapter 8). Given that diversity is maintained (for whatever reason), there are likely to be numerous other differences present among species. Such differences could be important at buffering the system from environmental change (the 'insurance hypothesis', Loreau *et al.* 2003)

Spatial dynamics can influence the ability of parasites to limit and regulate their hosts, over both ecological and evolutionary time-scales; theoretical reasons why this is to be expected are given below. It is thus likely that the ecosystem roles of parasites have important spatial dimensions.

5.2.2 Energy and nutrient flows

A core concern of ecosystem ecology is to understand the fluxes of energy and material through a given population that is potentially available to the rest of the ecosystem. Flux rates are closely related to the death rate of the population, which governs the provisioning of biomass either for consumption by higher trophic levels, or for decomposition. Pathogens that directly increase the death rate of their hosts will thereby facilitate entry of nutrients into the decomposer food web. Pathogens that make their hosts more vulnerable to predation will alter the strength of trophic interactions, and thus the channelling of energy and nutrients through food webs. Parasites which lead to morbidity in their hosts may make those hosts less capable as consumers, making these species less significant factors in the dynamics of their resources (both biotic and abiotic). Spatial heterogeneity and dynamics which influence the average parasite 'load' of a host population can thus have profound ecosystem consequences.

As an example of how to place host–pathogen interactions into a canonical ecosystem context, consider a simple host–pathogen interaction with classic SI dynamics (e.g. Anderson and May 1981), in which the host is regulated entirely by the pathogen.

$$\frac{dS}{dt} = (b - d)S - \beta SI + \delta I,$$

$$\frac{dI}{dt} = \beta SI - (d' + \delta)I.$$

Here, S is the density of healthy hosts, and I the density of infected hosts. Alternatively, if hosts have a given biomass, these equations could describe changes in biomass. Healthy hosts give birth at a per capita rate b, and die at a rate d. Infected hosts die at a rate d', do not give birth, and recover at a rate δ. There is no permanent immunity, or lingering demographic consequence of having once been infected, following recovery. The disease transmission process is the usual mass action term, with transmission scaled by β. As long as the recovery rate is non-zero, the population will reach an equilibrial abundance. The equilibrial densities of healthy and infected hosts are respectively

$$S^* = (d' + \delta)/\beta, \text{ and } I^* = (b - d)(d' + \delta)/\beta d'.$$

A principal concern of ecosystem ecology is characterizing and interpreting the causes of flux rates among compartments. The total rate of production of biomass by the host population at equilibrium must equal the rate at which biomass enters other compartments in the ecosystem (e.g. the food web). Adding dS/dt and dI/dt, at equilibrium we have total deaths equal to total production, or

$$dS^* + d'I^* = bS^* = b(d' + \delta)/\beta.$$

Note that this measure of production does not depend upon the basic death rate of the host, when healthy, but instead depends upon the death rate of infected hosts. Given that the pathogen regulates host numbers, one ecosystem 'function' performed by that host (namely, its production) appears to be governed by the death rate of infected hosts. However, note that an alternative parameterization of the model is to write the death rate of infected hosts as the basic death rate of healthy hosts, plus a difference term, that is, $d' = d + q$, where q measures the mortality effect of the pathogen. So the basic death rate of the host is not necessarily irrelevant, but environmental factors which may affect the death rates of healthy hosts but not the death rate of infected hosts will not alter the productivity of the population, and thus not change the flux of materials and energy it provides via deaths to other ecosystem compartments. This is not an implausible scenario. For instance, if deaths arise due to aggressive contest competition, and infected individuals avoid such aggressive encounters, spatial variation in the intensity of competition will not influence the death rate of infected individuals and so would not be expressed in ecosystem fluxes through the population. This model of course does not directly consider space, but it does suggest some hypotheses regarding spatial effects that could be assessed in more complex models. For instance, if spatial dynamics tends to produce systems with overall lower transmission rates because of the spatial localization of interactions (see below), this should increase host population size, and thus enhance the particular ecosystem process of production.

If spatial dynamics lead to shifts in virulence (as measured by d'), then this will likewise alter the contribution of this host species to ecosystem productivity. Some models discussed below suggest that spatial dynamics can characteristically produce systems with lower levels of virulence (viz., lower d'). If so, then the total production of the host population in the ecosystem context will be *reduced*. This may seem counterintuitive. The reason for this is that with lower virulence, the host population will equilibrate with fewer healthy hosts, and more infected hosts. In other words, with lower virulence the host carries a heavier load of parasites. We have assumed that infected individuals do not reproduce, and that the parasite is the sole factor regulating host numbers; hence, this decrease in virulence can shift individuals from productive to nonproductive states, and so depress host population productivity. Also, somewhat counterintuitively, an increase in the recovery rate for individuals at the level of the population translates into an increase in total death rate (for all individuals). Any ecosystem factors that might influence recovery rates (e.g. the presence of bioaccumulated toxins) could thus indirectly alter population productivity and flux rates to other ecosystem compartments.

Finally (and to return to the spatial theme of this chapter), assume that the above model applies in each of a number of distinct habitats, which each reach their own respective demographic equilibria.

The habitats differ in one or more parameters in a fixed manner (e.g. due to topographic, edaphic, or climatic factors, or because they are at different stages of plant succession). An expression for the average productivity per habitat is given by taking the expectation of the above expression, or $E[b(d' + \delta)/\beta]$. If the parameters vary across space independently, then by using Jensen's inequality it is immediately apparent that the only parameter for which spatial variation affects the mean is the transmission rate. A host with a spatially varying transmission rate has a higher production, averaged over a landscape, than does a similar host with the same average transmission rate, but one which is spatially invariant.

For most of the remainder of this chapter, we will not directly consider ecosystem processes, but rather focus on the population and evolutionary dynamics of the host–pathogen system, and how these are influenced by space. Throughout, however, there is assumed to be an implicit link to ecosystem function, via impacts of pathogens on host abundance and stability. Moreover, even if this link is not of direct interest, ecosystem context (e.g. habitat productivity, patterns of spatial connectivity) can be of great importance in determining the population and evolutionary dynamics of host–parasite systems.

5.3 Spatial variability in empirical patterns of parasite distribution within ecosystems

Before reviewing models of spatial dynamics in host–parasite systems and their implications for the understanding of epidemiology and evolution, we present some empirical patterns that stress the role of space at various levels within ecosystems and highlight a series of factors that have been considered or need to be considered in theoretical studies.

5.3.1 Geographical distribution of parasites species: availability of hosts and opportunity for transmission

Parasites need their host(s) to complete their life cycle, either as an important source of nutrients (for example, for many ectoparasites like fleas, mosquitoes, or ticks parasitizing vertebrates) or as a habitat to live and reproduce (for example, for some helminthes and microparasites such as bacteria and viruses). The distribution of hosts in the environment will thus condition the distribution of their parasites. This constraint is especially strong as most parasites are specialized to a limited number of hosts, and also because some parasite life cycles are complex and involve series of hosts, with some playing the role of vectors or of intermediate hosts (Combes 2001). Before considering the factors affecting the spatial variability in the distribution of parasites within a given host population, a first step is thus to see how heterogeneity in the spatial distribution of parasite species within ecosystems relates to the spatial distribution of their hosts.

As most species are parasitized by several parasites, most of which are specialized to a given host species, the diversity of parasite fauna is spatially constrained, within and among ecosystems, by the diversity and ecology of their component species. The spatial distribution of parasites is also affected by the opportunities for completing their cycle which can be prevented by abiotic conditions outside their hosts. An example of this involves arctic ecosystems (see also Chapter 6) where an important component of the parasite fauna of seabirds are the flukes (Digenea), and where a detailed study of such parasites compared their distribution between two intermediate host species and among spatial locations (Galatkionov and Bustnes 1999). Different species of digeneans have life cycles which may consist of one intermediate host and no free-living larval stages, two intermediate hosts and one free-living stage, or two intermediate hosts and two free-living larval stages. The study examined the distribution of such parasites in the intertidal zones of the southern coast of the Barents Sea (northwestern Russia and northern Norway) by investigating two species of periwinkles (*Littorina saxatilis* and *L. obtusata*) which are intermediate hosts of many species of digeneans. A total of 26,020 snails from 134 sampling stations were collected. The study area was divided into five regions, and the number of species, frequency of occurrence, and prevalence of different digenean species and groups of species

(depending on life cycle complexity) were compared among these regions, statistically controlling for environmental exposure. The authors found 14 species of digeneans, of which 13 have marine birds as final hosts. The number of species per sampling station increased westwards, and was higher on the Norwegian coast than on the Russian coast. The frequency of occurrence of digeneans with more than one intermediate host increased westwards, making up a larger proportion of the digeneans among infected snails. The prevalence of different species showed the same pattern, and significantly more snails of both species were infected with digeneans with complicated life cycles in the western regions. The authors concluded that the causes of changing species composition between regions are probably (1) the harsh climate in the eastern part of the study area reducing the probability of successful transmission of digeneans with complicated life cycles; and (2) the distribution of different final hosts (Galatkionov and Bustnes 1999).

The combined effect of spatial variability in host availability and abiotic conditions on levels of parasite infestation has also been addressed for other ecosystems. Ecosystems at tropical latitudes are well known for harbouring much higher number of animal and plant species than at higher latitudes (e.g. Rosenzweig 1995), and these areas are thus expected to harbour more parasite species. The picture is not that simple though. For instance, despite some evidence of higher parasite richness in marine fish ectoparasites (Rohde and Heap 1998), field studies conducted on communities of endoparasites of freshwater fish as a function of latitude have reported lower richness in host species living in tropical areas than at higher latitudes (Choudhury and Dick 2000). This result holds even after controlling for potential confounding effects such as sampling effort, host body size, and phylogenic relationships among host species (Poulin 2001). Geographical differences in the diet of related host species are likely to affect the richness of the endoparasite fauna, whereas latitudinal effects in ectoparasite richness may be more related to the abiotic characteristics of the environment in which the host species live (Rohde and Heap 1998). Other factors such as spatial variability in the seasonality of host reproduction and the biogeographic history and diversity of actual or potential host species also have to be considered.

The geographic range of a host and its specialist parasites may thus differ, with the geographic range of the host being usually larger than that of the parasite (though interesting exceptions to this generalization may occur if parasites have widely ranging transmission stages in their life histories). Some parasite species with complex life histories and intermediate host species may actually show a much broader apparent geographic distribution than any one of their host species. In such cases the parasite geographic distribution may nevertheless be strongly constrained by one of the host species, as for instance when the transmission among final hosts needs to be done through a species playing the role of a vector. Vertebrate species affected by blood parasites such as *Plasmodium* spp. may carry the parasite all over the world in their bodies (e.g. during seasonal migration or business travel) but the transmission of the parasite to another final host is nevertheless constrained by the need for a local host species that will play the role of a competent vector (Kiple 1993). The type of life cycle will thus affect greatly the geographic distribution of a parasite and its ecological meaning: some microparasites which are directly transmitted among hosts will be found everywhere hosts are found in sufficient density, but some parasites with extensive free living stages or complex life cycles involving different host species may on the contrary have geographic distributions that do not match tightly those of their hosts.

A low richness of parasite and host communities in some areas, for example, at very high latitudes (see Chapter 6), does not mean that parasites are of negligible importance in these systems. For instance, an extremely high prevalence of infestation of the sibling vole *Microtus rossiaemeridionalis* by the taeniid tapeworm *Echinococcus multilocularis* has recently been reported in a geographically isolated and very small population of that small mammal in the Svalbard archipaelago (Henttonen *et al.* 2001). The life cycle of the parasite involves the arctic fox *Alopex lagopus* as final host; long distance movements by foxes between Siberia and Svalbard,

together with the human-mediated introduction of the vole to Svalbard, are likely responsible for the presence of the parasite in such a remote population of the intermediate host (Henttonen *et al.* 2001). *E. multilocularis* is the agent of a life-threatening zoonosis. Thus, this example highlights the different roles that humans can play in a spatial context, sometimes being inadvertently efficient at changing the spatial availability of hosts and facilitating the completion of life cycles (see also Chapter 10). A better understanding of the role of space in the dynamics of host–parasite interactions can be gained by considering the processes responsible for the distribution of parasites among hosts at different scales.

5.3.2 Aggregation of parasites among hosts and spatial distribution of hosts and parasites

A striking and taxonomically widespread pattern is that the distribution of parasites among hosts within populations is typically aggregated, that is, most host individuals have no parasites but a few hosts are infested by many parasites (Shaw and Dobson 1995). Another often reported pattern [when the spatial locations where the host individuals were sampled are known] is that the proportion (prevalence) of parasitized hosts varies among areas (Wilson *et al.* 2002). These observations are key to understanding the importance of spatial variability in host–parasite interactions within ecosystems. Analyses of spatial aggregation have mostly been done for macroparasites (Hudson and Dobson 1995). Little explicit attention has been given to this in microparasites, the abundances of which are usually not quantified within individual hosts, and for which the reporting of prevalence is often linked with information on their rate of spread in the host population (see below). The existence of latent periods in infection, and asymptomatic infected host individuals, is however consistent with heterogeneity among host individuals in the abundance of pathogens within them. A concern with overdispersion and aggregation as defining attributes of the distribution of parasites among hosts is important, as this form of spatial heterogeneity has been identified as a key factor for the stability of the dynamics of host and parasite populations (Anderson and May 1978; Jaenike 1996).

Many factors can contribute to generating an aggregated distribution of parasites among hosts, including hosts with different histories of exposure of hosts to parasites, and differential susceptibility of host individuals to parasites. These factors may be structured in space at different scales, and this structuring may contribute to the aggregated distributions of parasites among hosts. Indeed, pooling of individuals from locations with different levels of infestation into single combined analyses can generate an overall aggregated distribution. Classically, explicit information on the relative spatial location of the host individuals has seldom been considered in the analysis of aggregative distribution patterns. It is nevertheless interesting to measure aggregation at different spatial scales to attempt to identify the spatial scale at which the aggregative process is occurring (Boulinier et al. 1996). The tick *Ixodes uriae* has for instance been found aggregated among nestlings of its seabird host, the black-legged kittiwake (*Rissa tridactyla*), when all samples were pooled, but when the level of aggregation was quantified both within-nest among nestlings and among nests within an area, the ticks were found aggregated among nests but not among nestlings within nests (Boulinier et al. 1996). This pattern is not surprising, given the specific features of the system considered; the ectoparasite, which has limited mobility, infests the nesting substrate of the breeding colonies of its host, and nestlings within each nest share traits in common likely to affect their levels of tick infestation.

In this species, aggregation is evident among nests. This pattern may be due to a combined effect of the correlated age of the nestlings within each nest, some genetic basis for susceptibility to the ticks (Boulinier et al. 1997) and spatial heterogeneity in local exposure to the ticks which overwinter in the nesting substratum. Within a breeding cliff, spatial autocorrelation in the level of infestation of nestlings has for instance been reported (McCoy et al. 1999). Another potentially important factor is an induced maternal response by females against local parasites (Gasparini et al. 2001). A comparable approach to partitioning aggregation at different scales has been conducted with other systems (e.g.

Haukisalmi and Henttonen 1999; Elston et al. 2001; Poulin and Rate 2001; Latham and Poulin 2003) and often a minimal spatial scale is identified at which little aggregation is found (Jaenike 1994). In such cases, the identification of the spatial scale at which aggregation occurs helps to identify which processes are potentially important in the transmission of the parasites and the maintenance of infestation. It can also facilitate the identification of environmental variables possibly responsible for the variable levels of local infestation, as is done in the field of landscape epidemiology. In landscape epidemiology, geographic information systems (GIS) combining field data with remotely collected data on climate and landscape attributes (Hess et al. 2002), and geostatistic modelling (e.g. Kitron et al. 1996; Diggle et al. 2002; Srividya et al. 2002), are being increasingly applied. When the spatial distributions studied are those to vectors for parasites, host aggregation is especially important to consider; the reason for this is that aggregative responses of vectors to hosts can define foci of transmission (e.g. Perkins et al. 2003).

It should further be noted that spatial structure in infestation levels is not static, because it arises from the interplay of the local population dynamics of hosts and parasites, and dispersal by each species, potentially at different scales. This can be especially evident when studies are conducted at different time intervals, permitting the temporal dynamics of the spatial distribution of infestation to be apprehended. The role of space in this context is highlighted in the study of the spatial dynamics of epidemics in ecological landscapes (see below).

5.3.3 Determinants of dispersal and host–parasite interactions

Dispersal is now recognized to be a major factor affecting the spatial dynamics of populations. This is especially likely for parasites, as hosts and groups of hosts can be considered as islands among which parasites must disperse, to persist (Boulinier et al. 2001). Dispersal can enable individual parasites (either within or outside of host bodies) to reach groups of hosts that are susceptible and/or uninfected. It will also lead to gene flow.

Dispersal can be linked with transmission, but dispersal and transmission can also occur independently. Dispersal is usually modelled as a rate that can vary with habitat, but factors directly related to the host–parasite interaction can be involved as well. For instance, dispersal of hosts could be affected by the local level of parasite infestation. In colonial birds, increased natal dispersal of cliff swallows (*Hirundo pyrrhonota*) (Brown and Brown 1992) and breeding dispersal in black-legged kittiwakes have been associated with higher level of ectoparasite infestation of nesting areas (Boulinier *et al.* 2001). Nevertheless, few studies have addressed the question of how dispersal rates and transmission rates and patterns are interrelated.

If parasite infestation can lead to host dispersal, movement of hosts can also be responsible for the dispersal of parasites. And indeed, the few studies that looked at the population genetic structure of parasites as a function of their host ecology have shown patterns suggesting that host movement is responsible for parasite dispersal (in large herbivores, Blouin *et al.* 1995; seabirds, McCoy *et al.* 2003; and salmonids, Criscione and Blouin 2004). Movement of hosts leading to parasite dispersal may not imply host dispersal. Making such a distinction is important as it will affect differently the spread of parasites, and thus epidemiology, but also the relative gene flow of host and parasites, and thus the dynamics of host–parasite coevolution (see Holt and Hochberg 2002 and below).

Keeping in mind the potential need to incorporate these complex spatial processes when considering some specific host–parasite systems, simple modelling approaches can nevertheless capture the main properties of the dynamics of host–parasite interactions in a spatial context.

5.4 Host–parasite interactions in coupled, heterogeneous patches

One straightforward way to incorporate space into host–pathogen systems is to imagine that the environment is comprised a number of distinct habitats. In each of these, there is a well-mixed population of hosts and pathogens, described by standard epidemiological models. The habitats are then coupled

by dispersal of hosts, pathogens, or both. Examples of authors who have explored such models include Post *et al.* (1983), Rodriguez and Torres-Sorando (2001), Hethcote and Ark (1987), Diekmann *et al.* (1990), Sattenspiel and Dietz (1995) and Sattenspiel and Simon (1988). In general, the conditions for establishment of the parasite, and its equilibrial incidence, can be strongly influenced by heterogeneity among habitats. Sattenspiel and Dietz (1995) provide an expression for establishment of an infectious disease in a spatially heterogeneous population, where total numbers are fixed by factors other than the disease (as appropriate for many infections of humans, for example). Rodriguez and Torres-Sorando (2001) develop comparable models for malarial infections of humans, and Diekmann *et al.* (1990) discuss the general issue of calculating the basic reproductive ratio R_0 in heterogeneous populations.

In principle, there are no conceptual complexities in this, but in practice it can be difficult to wade through the algebraic tangles which arise when analysing models with multiple patches and non-uniform mixing. As a relatively simple example, we consider the problem of initial establishment of an infectious disease in a landscape consisting of just two distinct habitats (one with area A_1, the other with area A_2) coupled by host movement. In the absence of the disease, we assume that the host in each habitat equilibrates at a carrying capacity, that the disease has direct transmission, and that both healthy and infected hosts can move between habitats at constant per capita rates (though possibly at different rates in the two habitats). A model for this scenario which describes the initial stage of infection is given by duplicating the above SI model, with two pools of infected individuals (recall, for the moment we are assuming that the host population is initially all healthy hosts, with fixed densities). The model describing the initial stages of infection is

$$\frac{dI_1}{dt} = \beta_1 S_1 I_1 - (d_1' + \delta_1)I_1 - m_{12}I_1 + m_{21}I_2(A_2/A_1),$$

$$\frac{dI_2}{dt} = \beta_2 S_2 I_2 - (d_2' + \delta_2)I_2 - m_{21}I_2 + m_{12}I_1(A_1/A_2).$$

All the parameters in the above earlier *SI* model have now been made habitat-specific. In addition, we have assumed that infected individuals can move between habitats, at rates that are also potentially habitat-specific. (Healthy hosts may also be moving, but if so, they are assumed to do so in a manner that does not alter the pattern of abundances between habitats). Because the variables are cast in terms of density, the fluxes between habitats have to be adjusted to account for the fact that the number of individuals per unit time moving from habitat i to habitat j equals the product of the density in habitat i, the area of habitat i, and the migration rate; whereas, the impact this influx of individuals has on density in habitat j has to be scaled against the area of habitat j.

The above model is a pair of coupled linear differential equations, so it can be fully analysed. In particular, the dominant eigenvalue of the characteristic matrix defines the growth rate of the infection over both habitats, after an initial transient phase. For simplicity, we combine infection, death, and recovery into a habitat-specific intrinsic growth rate for the infection, when rare:

$$r_i = \beta_i S_i - (d_i' + \delta_i).$$

A habitat may foster a high initial growth rate for the parasite simply because there is a high density of hosts there, or instead because of individual impacts of the infection upon hosts (e.g. locally low death rates of infected individuals). We assume that habitat 1 has the higher growth rate. With this notation, the growth rate of the infection can be shown to be

$$\tfrac{1}{2}[-m_{12}-m_{21}+r_1+r_2$$
$$+\sqrt{(m_{12}+m_{21}-r_1-r_2)^2-4(-m_{21}r_1-m_{12}r_2+r_1r_2)}].$$

(see Holt 1985 for an analogous treatment of population increase in a two-habitat environment, albeit with symmetrical movement).

This expression can be manipulated to make some general statements about how habitat heterogeneity influences the establishment of a disease.

1 Note that the relative habitat areas drop out. The slightly counterintuitive result is that a combination of intrinsic habitat qualities influences invasion rates, but relative habitat areas do not. The reason for this is basically that with reciprocal movement, the descendents of any given individual cycle through both habitats. Overall, the asymptotic growth rate reflects a nonlinear averaging of the growth rates of the two habitats.

2 If we let both movement rates be equal, and then take the limit as they get very large in the above expression, the growth rate simply becomes the arithmetic average over the two habitats, $(r_1 + r_2)/2$. In this limit, the landscape is actually just one habitat with internal heterogeneity. Given rapid movement of infected individuals, landscapes with the same average growth rate (averaged among habitats) but different degrees of internal heterogeneity, nonetheless should have the same growth rate for the infection.

3 If both intrinsic growth rates are positive, so is the overall growth rate; conversely, if both growth rates are negative, the overall growth is also negative. Parasite invasion requires that there be at least one habitat in the landscape which is intrinsically favourable, and could potentially sustain an invasion on its own (were infected individuals not to move).

4 If we let m_{21} approach zero, while keeping the other movement rate fixed, all movements will be from habitat 1 into habitat 2. In this case, the growth rate overall converges on the growth rate of the better habitat (which we have assumed to be habitat 1) minus losses to emigration. Thus, an infectious disease may grow in habitats where it has an intrinsic growth rate less than zero, provided it is maintained in habitats where it has a positive growth rate. All else being equal, such 'spillover' modes of invasion by an infection should be most noticeable in habitats with lower than average host densities (this result emerges from inspecting the eigenvector describing the distribution of individuals between the two habitats, when the invasion has settled into its equilibrial rate of change).

The above bit of theory provides insight into how spatial heterogeneity can influence parasite establishment. We would be the first to admit that this provides just a first pass through this problem. A full analysis of this issue would require one to analyse more complex landscapes, alternative transmission dynamics, additional classes (e.g. hosts with acquired immunity) and so on. Moreover, we have not paid attention to feedbacks via depression of healthy host numbers, or to transient dynamics.

Some insights into the consequences of habitat heterogeneity for disease dynamics can be also gleaned from parallel studies of the classical Lotka–Volterra predator–prey model, in which prey grow exponentially in the absence of predation, predators die at a constant rate, and the two are coupled by a mass action term describing predator attacks. This familiar model emerges as a limiting case of the standard SI epidemiological model (Holt and Pickering 1985), when infected individuals have very low recovery and birth rates. Holt (1984) analysed two habitat patches with Lotka–Volterra interactions occurring in each, and predator movement. With some simple reinterpretation of parameter definitions, this is also the SI host–parasite model for two habitats, with movement of infected individuals. (A number of authors have found general Lotka–Volterra models to provide useful limiting cases for explorations of host–parasite ecology and evolutionary dynamics, for example, Frank 1997). Spatial heterogeneity is broadly stabilizing, because it permits source–sink populations to develop, in which some host species are more heavily exploited than expected from just local conditions alone.

When healthy hosts are also allowed to move (Nisbet *et al.* 1993), more complex scenarios are feasible in heterogeneous landscapes. The basic point is that in some circumstances spatial heterogeneity coupled with dispersal can lead to stability, and in others, it leads to instability (Holt 2002). Despite this potential for a diversity of effects of spatial heterogeneity, our sense is that more often than not, spatial heterogeneity is broadly stabilizing (Hoopes *et al.* in press).

The above remarks pertain to heterogeneity that arises from variance in local conditions (e.g. productivity, or rates of infection). In other systems, heterogeneity in rates of movement or mixing can itself also promote stability (Holt 1984 notes this case for the Lotka–Volterra model). Gubbins and Gilligan (1997) carried out an experiment and demonstrated that heterogeneity in parasite establishment (due to incomplete mixing) promoted the persistence of the mycoparasite *Sporidesmium sclerotivorum*, a biological control agent on *Sclerotina minor* (a fungus on lettuce).

In a heterogeneous landscape, flows among habitats can permit greater parasite loads to be maintained in some habitats than would be expected just from local dynamics. Hochberg and Ives (1999) show that if there is substantial spatial variation in host productivity, flows of natural enemies (e.g. pathogen transmission stages among habitats) can lead to restriction of species from particular habitats, and even define the edges of geographical ranges for hosts. The model of Holt (1984; see also Holt 2002) mentioned above shows that 'spillover' limitation of a host in a low productivity habitat can readily arise, if this habitat is coupled with a high productivity habitat. This can be viewed as a kind of apparent competition, linking the dynamics of host populations that live in different habitats. Hosts that occupy low productivity habitats are vulnerable to the impact of parasites maintained in more productive habitats. This is true regardless of whether the populations are the same biological species of host, or are different host species. In the latter case, one may observe indirect exclusion of one host by another via shared parasitism (Holt and Pickering 1985), even though the two hosts never cooccur in the same habitat patch (a specific metapopulation model of this effect is in Holt 1997). All of these landscape effects depend on the movement of parasites across space, either because infected hosts themselves move, or because the parasites have free-living, mobile life history stages (e.g. aerial spores), or because movement is provided by the behaviour of vectors.

5.5 Epidemics in a spatial context

One broad class of examples of spatial host–parasite interactions, to which we will not attempt to do real justice, is the study of epidemic waves across space (e.g. of dengue fever emanating from foci in Thailand, Cummings *et al.* 2004). The theoretical models which have most often been used in this context are partial differential equations describing invasive waves of epidemic disease. Just to mention one interesting example, Murray *et al.* (2003) explored a spatially distributed mass mortality event in an Australian pilchard (*Sardinops sagax*) population. Their model tracked susceptible, infected and latent, infected and infectious, and finally removed (dead or recovered) individuals. As in most such models, the wave velocity is sensitive to diffusion coefficients, viral transmission rates (which enter into local intrinsic growth rates, see above), and latency period. Large-scale spatial heterogeneity in these parameters can help explain differences among regions in the time course, spatial development, and intensity of the epidemic. The broad ecosystem consequence of these epidemic waves is that they act as major disturbances in the ecosystem, with potential ripple effects on many other species. An example of such a perturbation is provided by the chestnut blight fungus (*Cryphonetria parasitica*), which decimated the American chestnut (*Castanea dentata*) throughout the eastern deciduous biome of the United States in the early decades of the twentieth century. This species was once the dominant tree in this biome, but the fungus destroyed approximately 3.5 billion trees (Taylor 2002). Other species of tree have filled in the gaps left by the demise of this species. Its wood is resistant to decay, and there are still many places in the southern Appalachians where chestnut logs are prominent features of the understory (personal observation). Although the ecosystem consequences of this epidemic (say on soil properties) are not well documented, they are doubtless profound and long-lasting in these systems.

5.6 Effects of space in homogeneous environments

Spatial dynamics can influence stability and persistence of host–parasite systems, even in the absence of heterogeneity. For instance, Jansen and de Roos (2000) analysed the Lotka–Volterra model for two coupled habitats, with uniform predator movement, and no parameter differences between the two habitats. Ultimately, these systems settle into spatially uniform, neutrally stable oscillations. In a single patch, these oscillations can be arbitrarily large. But the transient dynamics (which may be very long) in the two patch model can be very different. In particular, Jansen and de Roos show that in the long run, fluctuations of large amplitude will not be observed, for nearly all inhomogeneous

starting conditions. Such stability effects become greatly amplified when one pays due attention to the discreteness of individuals. Because interactions are spatially localized, and occur between individuals who experience the chance vicissitudes of birth, death, and movements, stochastic variation alone can lead to a shifting pattern of heterogeneity in host–pathogen interactions even in a homogeneous environment. This basic insight underlies a vast array of recent studies of space in ecological systems (Tilman and Kareiva 1997). Space becomes particularly important when infection occurs only over short distances among individuals who themselves do not move (e.g. plant populations). Keeling *et al.* (2000) provide a general argument on how limited movement in natural-enemy victim systems generically leads to spatial structure, which in effect provides refuges for the host/prey, and generates exploitative competition among parasites/predators.

Introducing demographic stochasticity creates many challenging mathematical problems, but also a consideration of demographic stochasticity points to some important potential implications of spatially localized infection processes. Rand *et al.* (1995) and Keeling (1999, 2000) considered a system in which a virulent disease is spreading through a slowly growing, sessile population (e.g. a fungal pathogen on a plant population). The model is a probabilistic cellular automata model for a host–parasite system, which attempts to capture in a simple way the consequences of localized infection and host renewal processes. The system consists of a lattice of sites, each of which can be empty, occupied by a healthy host, or, occupied by a parasitized host. What happens at each site depends upon its current state and that of its immediate neighbours. Healthy hosts send offspring into empty adjacent sites (e.g. a plant sending out seeds over a short distance); if healthy hosts are next to an infected host, they can be infected, with a fixed probability. For simplicity, Keeling assumes that infection is lethal.

This model suggests a number of important messages that appear to characterize a much broader range of models. First, there is a range of transmissibilities, within which the pathogen persists, and outside of which it cannot. If transmission rates are too low, then the 'birth' rate of new infections will not exceed the rate at which infections are lost to mortality. This of course describes nonspatial infection dynamics, too. More interestingly, if transmission rates are too high, then persistence may also be unlikely. The reason for this is that the pathogen in effect overexploits its hosts in localized arenas, and then itself is vulnerable to extinction. The interaction between the host and pathogen leads to a fracturing of both populations into small isolated patches; the pathogen can then easily disappear locally due to demographic stochasticity. Second, given that the interaction persists, it can do so at very low levels of overall prevalence, compared to expectations drawn from homogeneous, mean-field models, in effect because the localization of interactions and dispersal permits the emergence of ephemeral transient refuges. In a sense, the localization of interactions can be viewed as a reduction in the overall rate of infection, per host, so that the impact of the parasite upon the population dynamics of its host is reduced. Third, because the system is probabilistic and tracks integer numbers of individuals, the local environment is often found in a state which is very far from the global average, and there are dramatic fluctuations in infection at a local level. Fourth, there is an emergent spatial structure, in which parasites spread as wave fronts through the susceptible hosts, with patches empty of hosts left behind in the wake of the wave.

Many of these results appear in a wide range of models. Haraguchi and Sasaki (2000) examined how spatial structure influenced the evolution of virulence and transmission rate in a parasite interacting with a host in a lattice. The host was assumed not to be evolving. Constraining transmission so that it only occurs through local contact leads to evolutionarily stable traits of parasites that are completely different than expected with complete mixing. Viscosity tends to select for an intermediate ESS rate of transmission, even without the classical tradeoff between transmission and virulence. They found an interaction between the host growth rate and parasite evolution; at low host growth rates, the parasite had difficulty persisting near its ESS, whereas at high host growth rates, the

parasite could overexploit its host, with both risking extinction. This is an evolutionarily driven analogue of the classical 'paradox of enrichment'. Analysis of a similar system by Rauch *et al.* (2003) using techniques which tracked genetic phylogenies revealed some interesting features. Mutant strains continually arise with higher transmission and virulence. However, these strains, after a period of growth, deplete hosts within regions, and then themselves go extinct. This leaves behind pathogens with intermediate virulence. Thus, the evolution of the whole system reflects a self-organized spatial structure (which amounts to a kind of group selection).

If these results prove to be general, they obviously have important consequences for ecosystems. For instance, the spatial localization of interactions may mean that some pathogens may be less important factors regulating population size (and thus exert a relatively weak influence on biomass production, nutrient pool fluxes, and so on) than expected judging from the direct impact of the pathogen upon individual hosts. Such ecosystem effects, moreover, may be highly heterogeneous across space and through time.

Properties of the ecosystem may in turn feed back on the host–pathogen interactions observed. One general finding in lattice models (which is usually treated as an inconvenience by theoreticians) is that the size of the lattice can influence the probability of persistence of the infection. There are two distinct reasons for this. First, there is often a characteristic length scale describing the correlation among nearby cells, reflecting the emergence of spatial asynchrony in dynamics. A lattice that is smaller than this characteristic scale will not contain sufficient spatial heterogeneity among patches in the phase of the host–parasite interaction to persist. Second, in models which are stochastic (e.g. individual-based models), the probability of randomly fluctuating to extinction over a given time frame goes up rapidly as the maximal number of individuals declines (a result which in general ecology goes back at least to MacArthur and Wilson 1967). This is closely related to the concept of a 'critical community size' in epidemiology, which is defined as the minimum size of a population required for a disease to persist (Bartlett 1957). Wilson *et al.* (1998) explored a tritrophic host–parasitoid–hyperparasitoid interaction. Within each cell in a lattice, interactions tended to be unstable, and dispersal occurred among adjacent cells. There was a very strong lattice size effect on the persistence of the system, and the full tritrophic interaction required a much larger lattice to persist than did the host–parasitoid interaction along. It is often difficult to gauge the critical size theoretically (Dye 1995), but it is clear that ecosystem size is an important ecosystem factor which can influence the character of the host–parasite interactions one might observe in natural systems. We suspect that characterizing the effects of ecosystem size on host–parasite systems is a topic that will receive much more attention in the future (for related thoughts on how ecosystem size governs food web attributes, see Holt *et al.* 1999, Post 2002, and Holt and Hoopes, in press).

The generalization that the spatial localization of interactions may quite broadly facilitate the persistence of parasites in ecosystems emerges in many situations. For instance, Grenfell *et al.* (1995, see also Keeling 1997) modelled the dynamics of measles. Here, the patches are cities or large towns, and spatial coupling reflects traffic among towns. With 10 such identical cities, a very weak amount of coupling (0.1% individuals moving) was shown to increase the persistence of the disease overall. The reduction in extinction rate provided by spatial localization of interactions in large measure reflects the rescue effect (Brown and Kodric-Brown 1977). The decorrelation of dynamics among different sites permits some populations to be large, even when others are quite rare; the former can then provide immigrants which boost numbers in the latter, preventing local extinctions. This effect tends to increase when spatial localization is assumed (rather than weak global mixing among patches), as well as when the model explicitly considers birth, death, and movement at the level of individuals (demographic stochasticity). Both factors tend to decrease the correlation among sites in population dynamics.

In the real world, spatial localization of interactions may make it more difficult for many pathogens to persist. Phocine distemper in the harbour seal (*Phoca vitulina*) in the North Sea provides a potential example. This species is distributed in well-defined local populations, separated by unsuitable habitat. Colonies can go extinct and then become re-established. From the point of view of the virus, each group of seals is a patch in a metapopulation. Swinton *et al.* (1998) parameterized a model, so as to analyse conditions for persistence. They concluded that a very large population, indeed one larger than the entire population of seals in the North Sea, was required to maintain the disease. The reason is that within each local population, there is a rapid fadeout of the disease, followed by a slow entry of new susceptibles via birth. This suggests that even larger spatial scales must be considered if one is to understand the origin and maintenance of this disease.

A consideration of extinctions and patchy populations leads naturally to the theme of metapopulation perspectives on host–parasite dynamics. Often, local disease dynamics seem to imply that extinction is expected (e.g. due to 'fade-out'), but persistence actually occurs. Analyses of plant–pathogen interactions at landscape scales can reveal considerable stochasticity, suggesting the importance of recurrent colonization and extinction events (Burdon and Thrall 2001). Often, host–pathogen systems exist in ecosystems where there are other drivers that determine local extinctions (e.g. episodic disturbances). In this case, metapopulation perspectives should be particularly useful.

This observation suggests that the pattern of connectivity may be critical in governing the importance of parasites in ecosytems. In conservation biology, for many years there has been a concern with how fragmentation reduces connectivity and thus may foster the erosion of biodiversity (ranging from the loss of genetic diversity within species, to extinctions of entire clades of extinction-prone species). This in turn has led to considerable attention being given to the potential value of corridors linking habitat patches. Hess (1994, 1996*a,b*) pointed out that there was a dark dimension to corridors, namely that they might promote the spread of infectious diseases, which could in turn reduce the conservation value of the habitat patches themselves. He developed metapopulation models to explore this idea. These models suggest that if a disease is highly contagious, and moderately severe (in terms of enhanced local extinction risks), it could become widespread in strongly connected landscapes, thus increasing the probability that the host would go extinct.

One metapopulation model considered by Hess (1996a) has the following simple form:

$$dS/dt = mS(1-I-S)-eS - mpIS,$$
$$dI/dt = mI(1-I-S)-e'I + mpIS.$$

Here, S is the fraction of patches occupied by disease-free populations, and I is the fraction containing the disease. The fraction of patches that are empty is $1-I-S$. This model assumes that empty patches are equally likely to be colonized from either healthy or infected patches, and that cross-infection (scaled by p) can occur, leading to the conversion of healthy into infected patches. In the absence of the disease, the model reduces to the familiar Levins formulation, with an equilibrial occupancy of $S^* = 1-e/m$. It should be noted that formally, this model is an example of 'intraguild predation' (Holt and Polis 1997). Healthy and infected patches both compete for empty patches, and in addition healthy patches can be exploited by infected patches. There is a general tendency for the top predator to exclude the intermediate predator in intraguild predation; in this model, the infection can dominate the population if the extinction rate of infected patches is not elevated too much.

A simpler version of the model is to assume that infected patches can only infect healthy patches. In this case, the model becomes identical to a special case of a predator–prey metapopulation model considered in Holt (1997), where a model is sketched for two host species occupying distinct habitats, but sharing a parasite that both increases local extinction rates and can colonize across habitats. Holt (1997) demonstrates that indirect competitive exclusion can occur in this system. The species which occupies the rarer habitat is particularly vulnerable to exclusion.

5.7 Spatial dimensions of host–parasite evolution

There is a rich and rapidly growing literature on the implications of space for genetic and evolutionary aspects of host–pathogen interactions (e.g. Hochberg and Holt 2002, and references cited therein). Although of considerable intrinsic interest, it should be cautioned that many evolutionary studies may not actually directly bear on ecosystem processes. If all pathogens do is alter relative fitnesses of individuals within host species, without any overall impact upon population size, turnover, or stability, it is not clear that these studies have direct implications for ecosystem processes. For instance, in theories of sexual selection in which mate choice is based upon parasite load, an assumption is often made of 'soft selection', in which parasites and hosts reciprocally determine fitnesses in each other, but parasites do not directly affect host population size. The most useful class of models for the purposes of ecosystem ecology are those which simultaneously examine population dynamics and evolutionary genetic processes, and in particular those which elucidate the interplay of ecological and evolutionary phenomena. In the next few paragraphs, we discuss some major themes in host–parasite coevolutionary dynamics. Essentially none of the literature we consider is directly concerned with the ecosystem implication of these dynamics.

There is considerable evidence for spatial variability in adaptation in host–parasite systems (a very useful review is by Dybdahl and Storfer 2003). Local adaptation may be defined as occurring when the mean fitness of a population when measured in sympatry is greater than in allopatry (Gandon and Michalakis 2002); operationally, one can carry out a series of cross-population infectivity studies, and assess relative performance of parasites and hosts, when paired with the population they normally encounter, compared with 'foreign' populations. There is a remarkable variety of patterns that have been reported in empirical studies of host–parasite coadaptation.

One popular theory of host–parasite coevolution leads to the expectation that parasites should be locally adapted to their hosts (with greater replication rates and prevalence in sympatric hosts, to which they have evolved) than to allopatric hosts (to which they have not evolved). This could arise for instance because parasites often have large effective population sizes and short generation lengths, relative to their hosts, and so should be able to track slow shifts in the genetic composition of local host populations (Dybdahl and Lively 1998; Lively 1999). There are some excellent examples of local adaptation by parasites to their hosts (e.g. Morand *et al.* 1996). Thrall and Burdon (2003), for instance, show that virulent pathogens dominate in host populations with resistant hosts, whereas avirulent pathogens characterize host populations with more susceptible hosts.

But in other systems, exactly the opposite patterns are found. For instance, Altizer (2001) examined variation among geographical races of the Monarch butterfly and its protozoan parasite *Ophryocystis elektroscirrha*. The prevalence of this parasite varies dramatically among populations, as does host resistance and parasite virulence. The migratory populations tend to have higher resistance and experience lower virulence. The parasite is not in this case more infectious to their native hosts, and indeed may be more maladapted. Altizer proposes that this pattern is due to selection being strong in the migratory population (where small parasite loads could translate to large fitness disadvantages, due to the energetic requirements of migration), and to correlated shifts in the relative importance of horizontal and vertical transmission routes. Another potential cause is that migratory populations may experience more effective gene flow, and so tend to have the genetic variation needed to mount strong adaptive responses to parasitism.

Theoretical studies of host–parasite coevolution suggest that which patterns are observed depends on a number of factors, and in particular on the relative rates of dispersal of the interacting species, and the presence of spatial differences in patch quality (Gandon *et al.* 1996; Nuismer *et al.* 1999, 2000; Gomulkiewicz *et al.* 2000; Gandon 2002). Lively (1999) also notes that the pattern one observes is likely to shift with time; because the systems are expected to be dynamic at both a local and global level, local adaptation at any given site, for either species, is likely to wax and wane with time.

As with analyses of spatial effects on host–parasite ecology, it is useful to distinguish scenarios in which space solely matters because interactions and dispersal are localized, and those in which there exists spatial heterogeneity in extrinsic environmental factors. Biotic and abiotic factors are rarely uniform across a species' range, and such variation has implications for the strength and even direction of selection (Thompson 1994, 1999). Thompson (1994) refers to sites where each of a pair of species has strong, reciprocal effects upon the other's fitness as coevolutionary 'hotspots'. Typically, for a variety of reasons such hotspots will be embedded in landscapes with many coevolutionary 'coldspots', where just one species responds to the other, or evolution is locally decoupled (Thompson 1994, 1999). Environmental gradients in climate or resource availability may for instance account for spatial variance in the virulence of parasitoids of *Drosophila melanogaster* (Kraiijeveld and van Alphen 1995). Patterns of local adaptation can be strongly influenced by the spatial mixture of hot and cold spots (Gomulkiewicz *et al.* 2000).

The primary ecosystem driver of environmental productivity in particular has been identified as a factor that indirectly governs the strength of coevolution between hosts and pathogens (Hochberg and van Baalen 1998; Hochberg and Holt 2002), leading to the prediction that virulence should decline to lower levels when productivity is lower. The mechanistic reason for this is that in host–pathogen systems (as for resource–consumer interactions in general), an increase in local productivity indirectly increases the abundance of the pathogen. This automatically increases the strength of selection on the host via selection to reduce attack rates. In the simple *SI* model discussed above, note that as host intrinsic growth rate increases ($b–d$), which should be facilitated by increased productivity, the relative number of hosts that are infected, versus healthy, will increase. This in turn implies that the strength of selection on withstanding the infection (e.g. by recovering) will increase, as gauged against potential costs of such defence for reproduction by healthy hosts. In systems with top-down regulation of hosts by parasites, increases in productivity also tend to reduce the strength of density dependence. This makes the relative advantage of defence against parasitism increase, and so indirectly also increases the likelihood of an evolutionary response.

Depending upon the details, these effects on the host can in turn alter evolution in the parasite. In some (though not all) situations, selection on the parasite to overcome host resistance will increase. Moreover, if the increase in host productivity translates into greater parasite numbers, more genetic variation should become available via mutation, upon which selection can then act. Thus, high productivity sites should be coevolutionary 'hotspots'.

Patterns of dispersal should also include coevolution between hosts and parasites. Complex patterns of local maladaptation and adaptation may arise, because of the mixing of traits among populations that are pulled in different evolutionary directions (Thompson et al. 2002). The idea that gene flow can hamper local selection is a familiar, old idea in evolutionary biology. A countervailing factor is that gene flow permits an infusion of genetic variation, providing the raw material for evolution by natural selection (Gomulkiewicz et al. 1999; Holt et al. 2003). The degree of local adaptation should strongly depend on the magnitude of dispersal relative to the strength of selection. It is an open question whether mismatching, or matching, of coevolved traits is the typical condition of natural systems.

Dispersal should interact with productivity. If there is weak dispersal, then in effect sites with different conditions provide distinct, largely closed arenas for adaptive evolution. Moreover the only sites where the species will be present will be those where they can sustain viable populations. Small amounts of dispersal in this case may mainly matter as providing avenues for the infusion of useful genetic variation; in any case, dispersal is not likely to create local maladaptation, if it is sufficiently weak. Indeed, in this case high immigration tends to corrode local adaptation. However, if the population is unable to persist without immigration (i.e. the habitat is a sink), dispersal will tend to enhance the initial stages of adaptation to the environment (Gomulkiewicz et al. 1999; Holt et al. 2003). Whether it does so may depend upon the quantitative magnitude of mutational effects upon fitness (Kawecki and Holt 2002), and the impact of dispersal upon fitness (given intraspecific density dependence, Gomulkiewicz et al. 1999). In severe sinks, it may be difficult for evolutionary responses to occur at all (Holt and Hochberg 2002). In heterogeneous environments, dispersal is likely to be asymmetric between environments varying in productivity. Environments with low productivity and population sizes are likely to be net recipients, rather than sources, of immigrants. All else being equal, these are also potentially sites where a host, or pathogen, can be maladapted (Holt and Hochberg 2002; Nuismer et al. 2003).

5.8 Ecosystem drivers of host–parasite interactions: from parasitism to mutualism

Because parasites live intimately in the bodies of their hosts, there is the potential for selection to favour avirulence, or even the transformation of a parasitic association into a mutualism. Different environmental conditions can shift the balance between mutualisms and antagonistic interactions (e.g. Herre et al. 1999). Hochberg et al. (2000) explored how this transition from negative to positive interactions might be modulated by demographic differences among locations. They found that virulence tended to emerge most often in habitats where host population growth was highest. The reason for this was twofold. First, total host numbers could be higher there, so if virulence was associated with transmission, it could be selected. Moreover, there was greater opportunity for cross-infection among strains, so that virulent strains could replace avirulent strains within hosts. Conversely, when host populations were unproductive, as in marginal habitats, avirulence and mutualisms could be favoured.

This relationship between levels of productivity and the shift between parasitism and mutualism has potentially significant implications for ecosystem processes. Many mutualisms are associated with resource acquisition (e.g. nitrogen fixation in plants). If parasitic associations are labile, and tend to evolve towards mutualism in unproductive environments, this provides a kind of buffering in terms of ecosystem productivity.

5.9 On the topic of maladaptation

One important implication of spatial flows is that moderate (and at times severe) degrees of maladaptation are to be expected. Thompson et al.

(2002) discuss in particular how geographical structuring in coevolutionary systems can lead to a substantial incidence of maladaptation.

There are several basic reasons to expect maladaptation in geographically structured coevolving systems.

1 Change leads to evolutionary lags. When the selective environment changes, there will usually be an evolutionary lag before a focal species settles into a new adaptive equilibrium. The length of time of the lag depends upon a variety of factors, such as the degree to which adaptation depends upon standing variation available at the time of the environmental change, or instead upon novel variation generated by mutation. Such lags are expected even to changes in the physical environment. Host–parasite interactions can exhibit dynamical instability in the selective environments faced by one or both species, either because of fluctuations in population size, or because evolution in one species in effect amounts to a deterioration in the environment for the other species.

2 Gene flow perturbs local adaptation. If there are fixed differences among populations, such differences could lead to differences in fixed local outcomes (Hochberg and Holt 2002). In this case, movement among populations can displace species from their local adaptive optima, because of the interplay of gene flow and selection.

3 Gene flow provides adaptive genetic variation. Adaptation by natural selection depends upon genetic variation; the migrational input of variation can permit one species to evolve more rapidly or effectively than another. Mathematical models of coevolution with migration at different rates for the interacting species reveal an interesting pattern. Gandon *et al.* (1996) predicted that if parasites migrate much more than do hosts, the parasites should be locally adapted. Conversely, if hosts migrate more than do parasites, hosts may be better adapted. The latter can arise for instance if the dispersing stage of the life history is different than the one harbouring the parasite. Oppliger *et al.* (1999) tested this prediction with a lacertid lizard (*Gallotia gallota*) from the Canaries, and a haemogregarine blood parasite. Juveniles appear to be parasite free,

and are the life stage when dispersal occurs. Cross-infection experiments revealed that parasites performed better on hosts from allopatric populations, revealing that the parasites are maladapted to the hosts with which they live.

4 Hosts and parasites have different spatial ranges. In this case, different geographical arenas determine evolution for hosts and parasites. Nuismer *et al.* (2003) have recently explored implications of the general pattern that hosts and parasites do not typically have completely congruent geographical ranges. Their model suggests that spatial zones of maladaptation in one or both species are to be expected in spatially distributed host–parasite systems.

As we noted earlier, at the ecosystem level, these emergent patterns of adaptation and maladaptation will be reflected in overall rates of death for hosts, which in turn should alter the ecosystem fluxes associated with particular host species. Spatial dynamics could have an important and underappreciated impact on ecosystem processes, mediated through the realized death rates of hosts across spatially heterogeneous landscapes.

5.10 Concluding remarks

We have argued that the ecosystem implications of host–pathogen interactions are likely to have important spatial dimensions. The empirical examples we reviewed above suggest that there is often substantial variation in host–pathogen interactions, for instance because of nonconcordant geographical ranges, and differences among sites in the degree of aggregation. If pathogens can regulate host abundance and productivity, spatial variation in the impact of such regulation can lead to corresponding spatial variation in biomass and productivity. The theoretical studies of host–pathogen interactions in spatially distributed systems we have reviewed reveal that in many situations, patchiness and spatial heterogeneity can help stabilize otherwise unstable dynamics, and lead to some sites with much heavier loads of parasites than can be sustained by local dynamics alone. The spatial localization of interactions between parasites and hosts can lead to systematic deviations from the expectations of nonspatial theory (e.g. as in the evolution of virulence, see above), and make it likely that ecosystem area or volume is a critical aspect of the persistence of host–pathogen systems. Together with the potential for rapid evolution, spatial localization can also lead to complex spatial patterns of local adaptation and maladaptation in these systems. All of these points bear on current concerns about the implications of global change for the viability of ecological systems.

Most components of global change involve a spatial dimension and this could result in parasitism playing an especially important role via these spatial dimensions (see also Chapter 7). Increased fragmentation due to human activities can affect dispersal of host and parasites, and thus influence the persistence of their interactions at different scales, as well as the emergence of new diseases. Climate change should lead to geographic change in species' ranges and habitat productivity which, as seen above, are critical for the dynamics of host–parasite interactions. Finally, increasing rates of introduction of species to foreign ecosystems can also result in dramatic shifts in the role of parasitism, either directly via the introduction of parasitic species, but also more indirectly by the introduction of potential hosts for parasites. The spatial dimension of the role of parasitism in ecosystems should thus become even more relevant in future attempts to understand and predict the ecological effects of global change (see also Chapter 7).

CHAPTER 6

Parasitism and hostile environments

Richard C. Tinsley

*Parasite transmission involves location of the host in an external
environment typically regarded as hostile (where parasite survival is
precluded) and the host is considered the safe 'patch'. However, lethal factors
operate within the host, provoked by the parasite's presence, and the
interaction represents a distinguishing feature of parasitism. The parasite's
response to hostile conditions determines whether infection causes pathogenic
disease, coexists asymptomatically, or becomes eliminated. What is the
empirical evidence of parasite mortality within hosts? What circumstances
lead to the ultimate expression of a hostile environment—extinction?*

6.1 Introduction

Parasites occur in the widest diversity of
environments: they are recorded not only in all the
inhabited regions of earth, but also within the environments created by the free-living organisms that
inhabit those regions. It is axiomatic that hosts represent favourable 'patches' on which parasite
growth and reproduction occur; the spaces between
these 'patches' may typically represent adverse
conditions where parasites, with an obligatory
need for the host, cannot survive. The difficulties
inherent in patch location are traditionally associated with the notorious fecundity of parasites.
However, the degree of hostility created by these
gaps between hosts is variable. In many life cycles,
the medium for transmission is generally not intrinsically hostile to 'off-host' stages. This is illustrated
by life cycles involving water-borne infection, especially in marine environments where osmotic and
other conditions create relatively little physiological stress. Often, the major losses of offspring are
determined by the limited lifespan of the transmission stage, governed by its fixed energy reserves, in
relation to availability of hosts. Life cycles where
parasites are 'free-living' in the external environment

School of Biological Sciences, University of Bristol, Bristol BS8
1UG, UK.

typically incorporate a dispersal stage that is not
equipped to exploit exogenous nutrients. So, the
strategy of mass production of energetically cheap
infection stages has low survival probability not
because of hazardous environmental conditions
but because of time. Where transmission stages
are quiescent eggs or protected larvae, then other
determinants such as frequency of encounters
with prospective hosts—always a key factor in
life cycle strategies—assume greater importance.
This, too, is the major influence in life cycles based
on the feeding activities of other organisms (intermediate hosts involved in food chains, arthropod
vectors that effect transfer through their blood-
feeding on successive hosts). In many cases,
the transferred stages occur exclusively within
environments created by hosts. For these, hostile
conditions are experienced through the body of
the host (such as temperature for parasites in
ectotherms) or are determined by the hazards to
survival faced by that host.

Parasite life inevitably involves an interaction
with adverse conditions, as is the case for free-living
organisms. These are influences that have a fundamental role in regulating animal populations. This
chapter begins with this broad remit but then
focuses specifically on elements of the environment
that are particular to parasitism, emphasizing the

85

distinctive characteristic of this way of life—the nature of the conditions encountered within the host.

6.2 Adaptation to hostile environments: abiotic factors

There are two types of adaptation to adverse conditions. Capacity adaptation enables an organism to grow and reproduce whilst experiencing conditions that would be regarded as extreme for a majority of other species. Resistance adaptation enables an organism to suspend normal activity and survive environmental extremes until the return of favourable conditions when growth and development can resume (see reviews by Perry 1999; Wharton 1999, 2002).

There is extensive documentation of the adaptations of parasites to survive in environments that may be judged—by a range of criteria—to be extreme: representing conditions at the limits for the survival of life when protein structures and cellular functions should be damaged beyond repair. One of the key survival mechanisms involves exclusion of water that is, of course, a basic requirement of life but whose physical properties create great vulnerability. Anhydrobiosis enables survival of almost total loss of body water (with water content only 1–5%) (Perry 1999): metabolism is brought virtually to a standstill with metabolic rate of desiccated *Ditylenchus dipsaci* below 1/10,000 of that of hydrated larvae (Wharton 2002). Many of the adaptations are linked. Thus, nematode stages that are tolerant of desiccation, in anhydrobiosis, will also survive low temperatures since there is no water to freeze. There are also interactions with survival mechanisms to osmotic and oxygen stress that may operate simultaneously under natural conditions. Nematodes in an anhydrobiotic state can survive exposure to radiation and chemicals that would kill a hydrated nematode (Wharton 2002). The records are remarkable for demonstrating the ability of living organisms to tolerate the most extreme conditions (appropriate for survival during transport through space!). Trichostrongyle nematode larvae have been reported to remain infective after years of storage in liquid nitrogen (–196 °C) (Halvorsen and

Bye 1999). Records for long-term survival during desiccation include 32 years for the plant parasite *Anguina tritici* (by Wharton 2002) and 39 years for the free-living nematode *Filenchus polyhypnus* (by McSorley 2003).

Wharton (1999) has reviewed the adaptive significance of cold tolerance in parasites, for which most data concern nematodes and ticks. There is little information for other parasite groups. Parasites of endotherms are protected by the behavioural responses and thermoregulation of their hosts. However, parasites of ectotherms may be exposed to lethal low temperatures within their hosts. For amphibians and reptiles that tolerate subzero temperatures, their nematodes are also subjected to freezing. Experimental studies on *Wetanema* sp., a parasite of a New Zealand insect, the alpine weta, show that this nematode can survive freezing to – 61 °C within its host (Wharton 2002). So, the survival mechanisms of the host may intensify selection of corresponding adaptations by parasites. Nematodes may also survive freezing in the carcasses of their hosts; this is a valuable trait for parasites such as *Trichinella nativa* transmitted to scavengers in Arctic environments (see below).

In general, the survival mechanisms are not special to parasites, they are found also across a series of free-living invertebrate groups (including rotifers, tardigrades, and nematodes), and represent adaptations to a range of unstable, fluctuating environments. Nevertheless, the survival strategies are important for parasites in providing the means of 'patch location' where the spaces between patches are exceptionally hostile (as in the Arctic environments considered below) and where timing to exploit availability of patches is crucial to life cycle success.

In nature, adaptations to environmental constraints may involve complex interactions in which timing is a key factor, enabling synchrony of development of infective stages with favourable environmental conditions, including host availability (see Wharton 2002). In *Nematodirus battus*, autumn conditions induce eggs to enter diapause; chilling overwinter prior to a temperature increase in spring is required to stimulate mass hatching of larvae, and this coincides with grazing by the new cohort of lambs. Trichostrongyle larvae may survive winter conditions as infective 3rd stage larvae on pasture but, in some species, larvae acquired in autumn may become arrested within the host (hypobiosis) and delay development to adults until spring. Deposition of eggs onto pasture and development of infective larvae then coincides with the availability of new-born susceptible hosts. This pattern may be a response by the preceding free-living stages exposed to low temperatures in autumn that trigger arrest following host invasion, but the host's immune response has also been implicated (Wharton 2002). In these cases, survival of specific hostile factors by parasites becomes part of a wider strategy for exploitation of hosts.

6.3 Adaptation to hostile environments: ecosystems

6.3.1 The deep sea

The deep sea may sometimes be overlooked in assessing the extent of parasite infection; however, over half the earth's surface is covered by water with a depth of over 3200 m, and there is enormous faunal diversity (Bray *et al.* 1999). Bray *et al.* emphasized that, on the basis of its total volume, the abyssal ocean and its inhabitants represent the most typical environment of this planet. From the perspective of life near to sea level, the constraints are exceptional, including physical conditions such as high pressure, low temperature, and lack of light. Bray *et al.* (1999) listed the physiological effects of high pressure on living organisms, noting that animal and bacterial life occurs even at pressures of 1100 atmospheres. For parasites, there are additional factors concerning spatial distribution of hosts, the nature of food chains and the characteristic of pelagic animals that there is no 'platform' on which transmission can occur.

In a review of digenean parasites of teleost fishes in the deep sea (defined as below 200 m depth), Bray *et al.* (1999) recorded over 200 species from 18 families. Certain of these appear to be adapted to a very wide depth (and hence pressure) range: for instance, *Gonocerca phycidis* occurs from 200 to 4850 m. Prevalence of infection at depths below

4000 m seems remarkably high, ranging from over 30% to nearly 60%. Bray *et al.* (1999) noted the relatively great diversity and density of potential intermediate hosts for digenean life cycles in continental slope samples at depths of 1500–2500 m: one study reported 798 species in a 21 m² surface area including 106 mollusc species, 385 annelid species, and 185 arthropod species. Certainly there is no suggestion of a restricted fauna 'struggling' to survive in limiting conditions.

Hydrothermal vents on the ocean floor represent a further amplification of the constraints of the deep sea making this one of the planet's most extreme environments (Lee 2003): there may be extreme temperature variation (from 400 °C within vents to 1–2 °C in surrounding areas) and high concentrations of specific chemical elements (Rothschild and Mancinelli 2001). Buron and Morand (2004) documented 126 species of parasites found at depths of 1000 m or below, not associated with hydrothermal vents. Digeneans and copepods represented 65% of this total; cestodes, digeneans, and acanthocephalans were recorded down to about 5000 m and copepods to 7000 m. In contrast, few parasites are known to occur in hosts associated with deep-sea vents: Buron and Morand (2004) reported records of a leech (at 2500 m), a nematode (at 3600 m) and several species of digeneans (2600–3600 m), acanthocephalans (2300–2600 m), and parasitic copepods (2250–2600 m), but these scattered data probably represent only the beginnings of parasitological documentation. There is a relatively low diversity of potential host species at vents, but their high population densities might favour exploitation by co-adapted parasites (Buron and Morand 2004). The records give little idea of specialist adaptation to these conditions.

6.3.2 Deserts

Desert ecosystems exemplify environmental conditions that are hostile for parasite life cycles. Indeed, deserts provide some of the harshest tests of life in general, defined by extremes of water deficit (a combination of low precipitation and high evaporation) and temperature (above and below survival thresholds for most living organisms). For prolonged periods of each year, the harshest environmental conditions may preclude activity by most desert organisms, and their survival typically involves corresponding periods of dormancy. As a consequence, deserts are characterized by very low food availability. However, in most of the world's deserts, especially the hot deserts, periodic weather patterns produce rainstorms that create favourable conditions for life. It is well documented that rainfall in deserts may be followed by a sudden pulse of activity, both of plants and animals. The organisms that exploit these conditions generally have exceptional abilities to tolerate or avoid drought, temperature fluctuations, and prolonged starvation. In addition, their suite of adaptations typically includes a quick response to unpredictable opportunities and a lifestyle geared to rapid growth, reproduction, and accumulation of reserves during favourable periods that then enables survival through the next period of hostile conditions (Tinsley 1999a,b).

Given the constraints on life in general, it might be predicted that desert ecosystems would impose a severe challenge for host-to-host transmission. Free-living stages, released into the external environment, may experience hostile conditions directly (especially extremes of temperature and desiccation). For helminths with indirect life cycles, there may be limited opportunities during the brief activity season for infective stages to enter and complete development within an intermediate host and then to be transferred to another definitive host. Life cycles employing arthropod intermediate hosts may benefit from their abundance during favourable conditions; however, transmission may have the disadvantage of low probability that intermediate hosts, carrying parasites at an infective developmental stage, are ingested by appropriate final hosts. In deserts where there is, typically, a relatively short activity season, indirect life cycles must rely on rapid development of successive stages (for within-season transmission) or extended survival in intermediate hosts (permitting infection in one year and onward transmission in the next). A strategy of between-season transfer is vulnerable to intermediate host mortality, both by parasite-induced pathology and by environmental effects

during the long period of hostile conditions. Some parasite life cycles are virtually precluded by failure of intermediate host groups to tolerate desert conditions: thus, digeneans are typically rare or absent in desert ecosystems because of the absence of snails. These factors contribute to a low species diversity of parasites in desert ecosystems, a feature characteristic of extreme environments in general (see Combes and Morand 1999).

Paradoxically, there is comprehensive information for life cycle adaptations in a host–parasite system that should not exist in a hot desert ecosystem (Fig. 6.1). *Pseudodiplorchis americanus* belongs to a platyhelminth group, the Monogenea, that principally comprises parasites of fishes. Typically, these possess a ciliated, swimming infective larva, lack resistant stages, and have no tolerance of desiccation. Host-to-host transfer in the direct life cycle occurs exclusively in water. The host is an amphibian, an equally improbable inhabitant of hot deserts, with the typical life history requirement that breeding occurs in water. In the southwestern deserts of North America, the toad, *Scaphiopus couchii*, spends the major part of each year dormant underground to escape the harshest desert conditions. Emergence is triggered by torrential rainfall, normally beginning in early July. The toads breed in newly created ephemeral pools and are then active on the desert surface when they feed on desert invertebrates and accumulate energy reserves. By early September, increasing drought and declining temperatures force a return to dormancy. This annual schedule, of about 2 months activity and 10 months inactivity, is actually more restricted. *S. couchii* is nocturnal, so activity is confined to the period between 21.00 h (dusk) and 04.00 h (dawn). Foraging occurs only when the desert surface is relatively damp, after rainfall, and is probably limited to fewer than 20 nights each year (Tinsley 1999b). Spawning typically occurs on only a single night each year (sometimes two or three nights for a given population in a 'wet' summer), and toads are otherwise entirely terrestrial. These strict constraints on host ecology are reflected in a very limited parasite fauna: *S. couchii* is infected by only five species of helminths and, for four of these, prevalence is less than 5% with very low worm burdens (Tinsley 1990a).

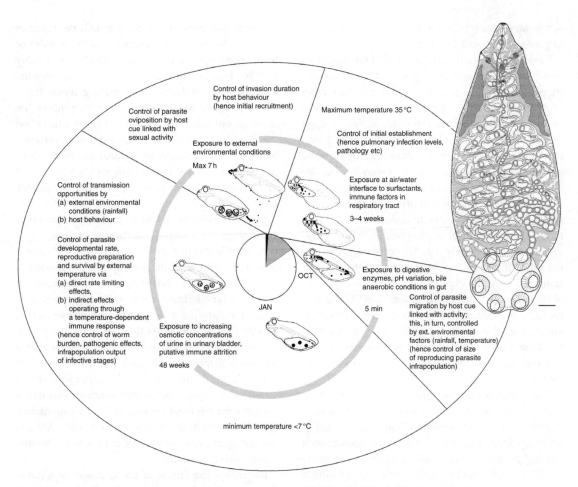

Figure 6.1 Summary of the constraints operating during successive phases of the life cycle of *Pseudodiplorchis americanus*.

Notes: Inner ring of annotations lists the variations in environmental conditions experienced by the parasite; outermost annotations identify the accompanying environmental controls on parasite biology. Central diagram correlates parasite life cycle events with the annual activity cycle of the host, *S. couchii*: transmission occurs during host spawning (●); juvenile development and internal migration during host feeding (◉); maturation and accumulation of embryos *in utero* during host hibernation (○). The cycle follows one cohort of parasites, but adult worms (⊕) producing infective stages (∴) in one season's transmission may also survive to reproduce in the next year(s). Additionally, juveniles failing to migrate may remain in the respiratory tract throughout hibernation and then migrate to the bladder when the host becomes active. (Reproduced from Tinsley 1999*b*).

Field data show that, during exposure to external environmental conditions (the period of host-to-host transmission), about 30% of infective larvae succeed in invading a host. However, between initial establishment and the opportunity for transmission, only 3% of worms survive, reflecting the hostility of the internal host environment (see text).

Paradoxically, the helminth that is most common in this desert host is the monogenean that relies on water for transmission. Patent infections of *P. americanus* occur in around 50% of *S. couchii* with a mean intensity of around 6 worms/host. Alongside these, post-infection juvenile parasites have a prevalence of 100% and a mean intensity typically about 100 worms/host (Tinsley 1999*b*).

The knowledge of host ecology allows these infection levels to be linked precisely to the period of transmission: invasion by ciliated oncomiracidia is possible only when *S. couchii* enters water to spawn, usually a maximum of 7 h but even in 'wet' summers always less than 24 h in the entire year. The effectiveness of this transmission period, probably briefer than for any other platyhelminth, is attributable to a series of exceptional specializations of parasite reproductive biology.

Clearly, deposition into the external environment of recently-assembled egg capsules (as in the majority of monogeneans and digeneans) could not meet the demands of this desert system. However, there are many platyhelminth examples in which eggs are retained and complete development *in utero*, hatching usually just after release into the external environment. This strategy, reviewed by Tinsley (1983), avoids the hazards of development away from the host and eliminates the time delay between oviposition and invasion. In a diverse range of circumstances in platyhelminth life cycles, release of immediately infective offspring permits the transmission process to be concentrated in both time and space to exploit periods of host vulnerability (Tinsley 1990*b*). In *P. americanus*, the key reproductive adaptation is a long uterus in which encapsulated embryos develop to the point of hatching and are then stored for mass release during the once-per-year transmission opportunity (Fig. 6.1).

The adaptations for production of infective stages by *P. americanus* have no close parallels elsewhere in the Platyhelminthes. The egg capsule is built up from around 12 layers of membranes derived from the uterus wall, forming a thin elastic 'shell' which is able to expand as the larva inside it develops. Nutrients are supplied directly to the embryo via a system of cytoplasmic processes connecting the lining of the egg capsule to the tegument of the developing larva. This placenta-like system ensures that oncomiracidia are maintained in a continuous state of readiness for discharge and that, despite the unpredictable timing of transmission, they have maximum energy reserves for host invasion (Cable and Tinsley 1991). A suite of other adaptations complements the demands of the life cycle (reviewed by Tinsley 1999*b*, 2004).

Alongside these parasite adaptations for instantaneous transmission in a desert environment, the key determinant of the numerical success of invasion is host behaviour. The 'explosive' breeding of *S. couchii* ensures that, when the once-per-year opportunity arises, maximum numbers of potential hosts are concentrated together in confined bodies of water. Samples of mating populations taken at intervals during the 7h exposure show a more or less exponential increase in infection levels (Tinsley 1999*b*). However, host behaviour also regulates the overall input of larvae: the mating toads leave water at dawn and this abruptly terminates the invasion episode. A series of field studies based on the *Pseudodiplorchis/Scaphiopus* system in Arizona have shown a consistent 'saturation' of the host population, with all individuals infected at the end of each transmission season (reviewed by Tinsley 1999*b*).

Viewed critically, while this life cycle involves adaptations in a host–parasite system found in a very hostile environment, the process of host-to-host transfer is actually accomplished remarkably easily. The precise targeting of mass release of infective larvae into water with high densities of host results in a very efficient transmission. Tinsley (1999*b*) calculated that larvae have a probability of about 0.3 of successfully invading a host, a success rate probably higher than for any other platyhelminth. This process depends absolutely on an infallible trigger factor that synchronizes larval release with the brief periods of host vulnerability, but the parasite's excursion into the external environment does not involve exposure to 'hostile' conditions (Fig. 6.1).

Regarding the effects of harsh desert conditions during the other 364 days each year, the most important constraint imposed by the external environment involves low temperatures rather than high. Winters in the Sonoran Desert are cold: soil surface temperatures (weekly minima) are below freezing from November to March. In the buffered environment 15 cm below the soil surface, temperatures are below 10 °C for 4 months and below 15 °C for 6–7 months each year. Laboratory experiments by Tocque and Tinsley (1991*a*) showed that parasite development and reproduction are strongly temperature-dependent and are virtually halted at 16 °C, that is, from October to April. So, although *P. americanus* has one year for preparation from one transmission opportunity to the next, optimum temperatures for reproductive preparation (25 °C) occur for only about 12 weeks each year and this allows 1st year worms to make only a minor contribution to transmission. Older worms, having completed initial body development, can produce progressively greater numbers of infective stages during this restricted period, but numbers of parasites decline significantly after 2 years and survival to 4 years post-infection is exceptional (Tinsley 1999*b*). These interactions, including time, temperature, and parasite survivorship, are responsible for limiting mean lifetime reproductive output to only around 150 offspring per worm (Tocque and Tinsley 1991*b*).

6.3.3 Arctic ecosystems

Svalbard reindeer in the high arctic experience a long cold winter, with the upper metre of ground frozen and snow-covered from October to the end of May or early June. Halvorsen and Bye (1999) showed that the nematode species of Svalbard reindeer are not specialists of Arctic environments but instead are generalists, widely distributed outside the Arctic, that are capable of sustaining populations under relatively more extreme environmental conditions. It could be predicted that transmission might be restricted to the period of more favourable conditions, in summer. However, Halvorsen *et al.* (1999) demonstrated that infection occurs throughout the year, even in the most severe conditions of mid-winter. These authors cited previous studies where larvae remained infective on herbage from autumn until the following summer. Indeed, larvae of the nematodes *Marshallagia marshalli* and *Teladorsagia circumcincta* survive storage in liquid nitrogen (see above and Halvorsen and Bye 1999).

An illustration of parasite establishment *de novo* in a hostile environment is provided by the records of Henttonen *et al.* (2001) of introduction of *Echinococcus multilocularis* into Svalbard. Arctic foxes, final host of *E. multilocularis*, range widely across the pack ice between Svalbard and the Siberian mainland where infection is present. However, the parasite could not have become established naturally on Svalbard because of the absence of native rodents. The sibling vole, *Microtus rossiaemeridionalis*, was introduced accidentally by humans in the early to mid-twentieth century and is assumed to have acquired infection

from migrating arctic foxes (Henttonen *et al.* 2001). In this severe environment, in which major fluctuations in vole populations are driven by winter mortality, *E. multilocularis* is maintained in a very simple cycle with only a single species of both final and intermediate hosts. Prevalence in the intermediate host approaches 100% in older animals and field data suggest that transmission occurs throughout the year, even in winter (Henttonen *et al.* 2001).

Trichinella nativa is superbly adapted to Arctic ecosystems. The carnivore/scavenger transmission route means that there are no free-living stages passed into the external environment to encounter hostile abiotic conditions directly. However, this feature—a key adaptation in the Arctic—is also found in all other ecosystems exploited by *Trichinella* species, including mesic and tropical environments. Kapel *et al.* (1999) found that larvae of *T. nativa* from Arctic foxes (from Greenland and Svalbard) remained infective for 4 years at −18 °C and suggested that freeze resistance is positively correlated with geographical latitude. Survival ability also varies in different host species: parasite isolates from arctic hosts (polar bear and wolf) survived many months at −15 to −20 °C but failed to survive relatively short exposure (5–7 days) at −10 to −15 °C in mice. Kapel *et al.* (1999) attributed differences in freeze tolerance to dissimilar development of the nurse cell–larva complex in different host tissues.

6.3.4 Ecosystems showing extreme variation

In considering the adaptations of parasites to hostile conditions, it is instructive to examine the success of parasites that thrive in a wide diversity of environments. *Fasciola hepatica* is considered to be native to Europe and it represents a remarkable test of adaptation that fascioliasis currently creates a public health problem in 51 countries on five continents (Mas-Coma *et al.* 2001). Indeed, it has the widest latitudinal, longitudinal and altitudinal distribution known amongst vector-borne diseases. Thus, hyperendemic human infection extends from below sea level (in Iran) to altitudes exceeding 4000 m (in Venezuela, Peru, and Bolivia) (Mas-Coma *et al.* 2003).

The epidemiology of *F. hepatica* is apparently strictly constrained by temperature-dependency of the life cycle stages external to the mammalian host. Development of both eggs and intramolluscan stages ceases below 9 °C, so transmission is virtually arrested for a major part of each year in temperate climates. In the United Kingdom, this period extends from October to May. Even with this 6-month limitation, life cycle efficiency permits epidemics with high mortality to develop within a single season in cool wet conditions in the British Isles (Ollerenshaw 1974). Nevertheless, year-round low temperatures would be expected to preclude life cycle completion. Mas-Coma *et al.* (2003) provided summary data for the temperature regimes at sites in South America where fascioliasis causes human disease at very high altitude. At a series of localities in the Andes, monthly and annual temperatures exhibit means consistently below the 9 °C minimum for free-living and intra-molluscan stages. Thus, at Mérida, Venezuela, mean annual temperature is only 6.1 °C with monthly range from 5.0 °C (in January) to 6.6 °C (in September); at Cotopaxi, Ecuador, mean annual temperature is 7.8 °C, monthly range 7.4 °C (June and July) to 8.1 °C (March); at Puno, Peru, mean annual temperature is 7.0 °C, monthly range 3.6 °C (July) to 8.8 °C (December). These data suggest that *F. hepatica* should not survive. Mas-Coma *et al.* measured life cycle characteristics of Northern Bolivian high altitude isolates maintained under controlled environmental conditions and compared these with equivalent data for European isolates. At the selected temperature, 20 °C, there were no rate differences in developmental periods and infectivity. This could indicate that there has been no significant evolution, within the past few hundred years, of the major parameters that might permit parasite development at temperatures below the limits for European isolates. However, critically, there are no data for the developmental biology of parasites exposed to field conditions in these extreme environments. Thus, while temperatures remain consistently below 9 °C throughout the year, as represented by monthly means, there is no information on effects of regular short periods of higher temperatures such as would occur daily around mid-day. Diurnal

fluctuations may allow development to proceed in pulses: the gradual accumulation of these increments could permit survival at average temperatures far below recognized minima. Diurnal temperature variation may be more pronounced in the very shallow water typical of wet pasture where *F. hepatica* transmission is so effective; indeed, parasite stages within the intermediate host may benefit additionally from heat absorption by snails exposed to sunshine above the water surface.

It is possible that the temperature constraints on parasites infecting fish may be more predictable: diurnal temperature fluctuations should have limited influence in relatively deep water. So, for the salmonid parasite, *Discocotyle sagittata*, the inhibitory effect of environmental temperatures below 10 °C, predicted by laboratory studies (Gannicott and Tinsley 1998), is likely to be more directly relevant to transmission patterns in nature. A series of other systems have well-documented temperature thresholds that determine inhibition of parasite transmission for a significant part of each year, as in *P. americanus* where, paradoxically for a hot desert, winter temperatures represent the major constraint on life cycle preparation (Tinsley 1999*b*). For *Ascaris suum,* egg development ceases below 15 °C (Larsen and Roepstorff 1999). These conditions cannot be considered 'hostile', yet they actually have the most serious effect on parasite population dynamics by precluding transmission.

6.4 Hostile environments created by parasites

6.4.1 Competition

Hostile conditions for parasites within hosts may also be created by other parasites. One of the most violent interactions is demonstrated by echinostome rediae that feed upon larval stages of other digeneans within the tissues of mollusc intermediate hosts (for this and other competitive interactions, see Esch *et al.* 2001).

Intraspecific competition has been shown to affect a range of parasite life history parameters including pre-patent period, egg production rate, and life cycle span, all indicative of suboptimal conditions. For measures such as reproductive output, it is rare to be able to establish how these constraints affect individuals within parasite infrapopulations: most data relate to mean rates of output averaged across the total worm burden within a given host. However, the reproductive specializations of *P. americanus*, involving retention of an entire year's offspring production *in utero*, provide this detailed insight. Tocque and Tinsley (1991*b*) recorded a significant density-dependent reduction in the per capita rate of reproduction and, in particular, demonstrated unequal effects on individual worms: the majority of parasites produced offspring at reduced rate as density increased, but a few worms retained high rates of production even in large burdens. These intraspecific effects may have a significant impact on parasite population regulation, but hostile interactions are illustrated more dramatically by interspecific competition.

Studies on the reproductive biology of *Protopolystoma* species, parasites of the urinary bladder of *Xenopus* species in Africa (reviewed by Tinsley 2004), illustrate a series of negative interactions in mixed species infections. Reproductive interference has been demonstrated in concurrent infections of *P. fissilis* and *P. ramulosus* in *X. fraseri* in central Africa. Viability of eggs produced from two-species infections was significantly less than that from single species infections, suggesting genetic incompatibility following interspecific cross-fertilisation (Jackson and Tinsley 1998*a*). In another two-species combination, antagonistic interactions between *P. fissilis* and *P. simplicis* in *X. wittei* result in the absence of concurrent patent infections in the same host individual: expulsion of established infections of one species coincided with arrival of the second species in the bladder habitat (Jackson *et al.* 1998). This mutual exclusion would avoid the reproductive wastage (and potential resource competition) that accompanies the two-species infections found in *X. fraseri* (above). Strict host specificity is normally observed in polystomatid monogeneans, including *Protopolystoma* where single parasite species are generally restricted to single species of *Xenopus* (Tinsley and Jackson 1998, 2002). However, different *Xenopus* species frequently occur

in sympatry (see Tinsley *et al.* 1996) and might be expected to experience accidental invasion by foreign *Protopolystoma species*. Jackson and Tinsley (1998*b*, 2003*a*) showed that infective larvae do not discriminate between normal (compatible) and non-normal (incompatible) *Xenopus* species but, following invasion, failed to survive in an inappropriate host species. Nevertheless, in experimental infections of *P. xenopodis* (specific to *X. laevis*) and *P. orientalis* (specific to *X. muelleri*), abortive invasion of an incompatible *Xenopus* species induced resistance to subsequent infection by the natural parasite species. Thus, in the absence of effective host species discrimination, two negative effects of the host–parasite interactions could be envisaged. First, accidental cross-infection by heterospecific larvae may induce cross-resistance, reducing future reproducing populations of the host-specific parasite species. Second, incompatible hosts may represent a 'trap' for infective larvae of parasites specific to other host species, reducing transmission efficiency (Tinsley and Jackson 2002: Jackson and Tinsley 2003*a*) (Fig. 6.2). These negative effects are avoided in cases where parasite specificity is determined by precise host species recognition by infective stages.

Comprehensive evidence of competitive interactions comes from very elegant field and experimental studies of schistosomes by Southgate, Tchuem Tchuenté, Jourdane, and colleagues. Male and female schistosomes locate one another in the hepatic portal system of the definitive host and form pairs as a prelude to maturation and egg laying. There are no barriers to interspecific pairing and this can lead to hybridization between closely related species. Experiments involving sequential infections of two different schistosome species have revealed a world in which usurping males can pull females away from their established mate and pair with them. Hostile interactions may also be mediated via the host's immune response: an initial infection may induce resistance that reduces survival of invading worms from subsequent exposure. This immunity, important in single species infections, also confers protection in experimental two-species interactions. Among several potential advantages (discussed by Cosgrove and Southgate 2002), one effect of this heterologous immunity is

Table 6.1 Survival of *Protopolystoma*

(A) Hosts in which parasites survive and reproduce

In single species infections of hosts, parasites establish, develop to maturity, and contribute to onward transmission.   *But*

◆ Survival to reproduce is influenced by compatibility (host × parasite genotype interactions, 'strain' differences). For *P. xenopodis* in *X. laevis* (under standard conditions of experimental exposure, in full-sib, naïve hosts) prevalence at patency varies from 15% to 70% (Jackson and Tinsley 2003b). Adult burdens are generally 1–2 (max 6) worms/host (Tinsley, 1995). Following successful establishment, parasite mortality before first reproduction is >95% (Jackson and Tinsley 2003c).

◆ In successful adult infrapopulations, in primary infections, reproductive performance shows wide variation, attributable principally to host x parasite interactions:

(a) Pre-patent period: median 9 weeks, max 19 weeks;

(b) Rates of egg production/host: range of max output 6–66 e/h/d;

(c) Reproductive lifespan, 1–27 months (Jackson and Tinsley 2001).

◆ In secondary infections, survival is reduced (see C) and in surviving adult infrapopulations reproductive performance is significantly depressed in comparison with primary infections in the same hosts:

(a) Pre-patent period: median 21 weeks, max 28 weeks;

(b) Rates of egg production/host, max output 6–13 e/h/d;

(c) Reproductive lifespan, max 10 months (Jackson and Tinsley 2001).

(B) Hosts in which negative parasite–parasite interactions affect reproduction

◆ Reproductive output is affected by intraspecific density-dependent reduction in egg production (Tinsley 2004).

◆ In some combinations, two species infections may co-occur in the same host, but reproductive interference leads to reduced egg viability (failure of interspecies cross-fertilization?) (Jackson and Tinsley 1998a).

◆ Prior abortive infection by heterospecific larvae results in reduced egg production by the normal species subsequently establishing in this host environment (Jackson and Tinsley 2003a).

(C) Hosts in which parasite survival is blocked

◆ Converse of A above, in primary infections of normal host species, 30–85% of worm burdens are eliminated before reproduction (zero survival to patency) (Jackson and Tinsley 2003b).

◆ In secondary infections of normal host species, acquired immunity eliminates 85% of worm burdens (compared with failure of 6% of primary infections (Jackson and Tinsley 2001).

◆ All parasites invading a non-normal host species are killed in early development (Jackson and Tinsley 1998b, 2003b).

◆ Parasites invading a normal host species that has previously experienced an abortive heterospecific infection show reduced survival (heterologous immunity) (Jackson and Tinsley 2003a).

◆ Parasites successfully established may be displaced/eliminated by the arrival of a second species in the same infection site (Jackson, Tinsley, and Hinkel 1998).

(D) External environmental factors affecting parasite reproduction and survival

◆ Increased temperature accelerates developmental and reproductive rates but survival is significantly reduced leading to lower prevalence and intensity (see Fig. 6.5) (Jackson and Tinsley 2002).

◆ Low temperatures permit increased parasite survival in hosts with suppressed immunity, but reproductive contribution is delayed: prepatent period is extended (see Fig. 6.5) (Jackson and Tinsley 2002), and reproductive output reduced (Tinsley and Jackson 2002).

(E) Host population factors affecting parasite population biology

◆ Resistance

non-normal host species, and

normal host species with acquired immunity from normal species infection, and

normal host species with experience of non-normal parasite species

represent a trap for infective stages: these show no discrimination at the point of invasion but are killed during juvenile development.

Continued overleaf

Table 6.1 *Continued*

♦ Susceptibility

Host reproduction generates cohorts of naïve individuals that 'flood' the population with susceptible targets for invasion (but, parasite mortality between invasion and first reproduction is still >95% in primary infection of juvenile hosts, and even the youngest host age groups can develop strong acquired immunity (Jackson and Tinsley 2003a)). However, primary infections in these young cohorts support a reproducing parasite population that will contribute to transmission for up to 2 years during which further host cohorts are normally generated.

♦ Host survivorship

Factors that prejudice host survival also affect parasite survival and reproduction, including parasite-induced pathology. *Protopolystoma* infections are strongly regulated but repeated infection of juveniles in the kidneys may be pathogenic (Tinsley and Jackson 2002). Other parasite species, including the nematode *Pseudocapillaroides xenopodis*, may cause host mortality and prejudice concurrent *Protopolystoma* infection (Tinsley 1995).

Notes: Experimental data, above, are based on exposures of *Xenopus laevis* (naïve, full-siblings) to *Protopolystoma xenopodis* under standard conditions; interspecfic data based on *P. fissilis, P. ramulosus, P. simplicis, P. orientalis* (see references cited).

A. Hosts in which parasites survive and reproduce

B. Hosts in which negative parasite–parasite interactions affect reproduction

C. Hosts in which parasite survival is blocked

D. External environmental factors affecting parasite reproduction and survival

E. Host population factors affecting parasite population biology

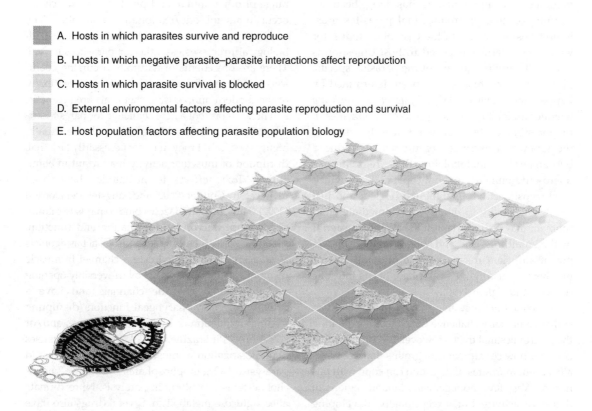

Figure 6.2 Confrontation between species of *Protopolystoma* and hosts *Xenopus* spp. leads to successful invasion: consistently in laboratory exposures ~30% of each infection dose establishes up to 1 week p.i. (There is no host species discrimination by infective larvae.) However, subsequent survival is influenced by a complex of host effects, past experience of infection, parasite interactions, and biotic factors. The reactive environment is made more or less hostile by increased or decreased temperature respectively.

Note: The outcome of infection of different categories of hosts and interacting factors is listed in Table 6.1 (opposite).

to create an interspecies reproductive isolating mechanism (as in the *Protopolystoma* species, above).

These case studies demonstrate that the environment inhabited by parasites may acquire hostile characteristics because of co-infections and this may be moderated, in part, via the immune response.

6.5 Hostile environments created by hosts

6.5.1 The host gut

The vertebrate alimentary tract, especially stomach and intestine, constitutes a very hostile environment for parasites. Function is specifically designed to reduce complex biological molecules into small monomeric components. It has long been an assumption that gastrointestinal parasites must inhibit host enzymes. In the case of cestodes, for which the interface exposed to host digestion is also the absorptive surface of the parasite, specific protective mechanisms have been determined by Pappas and Uglem (1990). *Hymenolepis diminuta* liberates an inhibitor of the proteolytic enzymes in the intestinal contents of its rat host. In addition, the tapeworm excretes organic acids produced by anaerobic metabolism that regulate its micro-environment to about pH 5.0, at which level trypsin activity is inhibited (Uglem and Just 1983). This ability to modify the immediate environment may be important in the alkaline posterior of the small intestine and where tapeworms are not of sufficient mass (because of small size or numbers) to acidify the bulk-phase of the intestine (Pappas and Uglem 1990).

Protection from digestive attack will have been an early evolutionary challenge for parasites exploiting the gastrointestinal tract. However, it is intriguing to consider how certain parasite groups that are typically absent from sites of digestion can cope with this hazard. Very few monogeneans inhabit regions of digestive activity but, exceptionally, developing juveniles of *P. americanus* migrate from the host's respiratory tract to the urinary bladder along the length of the alimentary canal. Protection is mediated by membrane-bound vesicles that accumulate in the tegument prior to migration and are discharged at the surface during migration (Fig. 6.1). If migrants

are restrained in gut contents for some hours, the store of vesicles is eventually exhausted, the tegument disintegrates and the worms die. However, movement (by looping locomotion) is rapid, and migration along 80–100 mm of gut typically takes only 5 min (Cable and Tinsley 1992). The nature of the secretions and the cue for their release have not been determined (Tinsley 1999*b*).

6.5.2 Man-made hostile environments: chemotherapeutic drugs

Parasite evolution has recently been confronted by an entirely novel factor: one host species has devised hostile environmental conditions specifically to eliminate parasites, both from itself and from a wide range of other animal and plant species which co-occur in a symbiotic/exploitative association. The hostile agents are chemicals that interfere with biological processes essential for parasite survival. There are precedents of naturally occurring chemicals that have a parasiticidal effect; however, many of the modern antiparasitic chemicals are synthetic, and hence represent a new challenge for parasite survival. Target effects of these chemicals are usually highly specific but they may not necessarily be lethal: disruption of muscular activity may result in elimination. Toxic effects in nematodes have been reviewed by Conder (2002) and Sangster and Dobson (2002). Thus, benzimidazoles bind to parasite tubulin and disrupt microtubule formation and function. Imidothiazoles, including levamisole, act as agonists at an acetylcholine-gated cation channel in muscle membranes. Avermectins act by irreversibly opening glutamate-gated chloride channels and have a primary effect on pharyngeal function (disrupting feeding, regulation of hydrostatic pressure, and/or secretion). Piperazine, a GABA-agonist, causes hyperpolarization of muscle membrane and flaccid paralysis. Organophosphates inhibit acetylcholinesterase, resulting in paralysis. Nitroscannate affects glucose metabolism. Several drugs also have immunostimulant effects (imidothiazoles) or increase susceptibility to immunological attack (diethylcarbamazine) (Conder 2002; Sangster and Dobson 2002).

The creation of host environments pervaded by these novel hostile factors has generated an 'experimental system' in which parasite adaptation to environmental conditions can be followed. In nematodes, selection and evolution of drug resistance has taken place over less than 40 years and now affects every modern anthelmintic (Sangster and Dobson 2002). Even those drugs considered highly effective because they eliminate 'almost all' parasites provide the potential for resistance to develop—among the very few surviving. Indeed, the most effective drugs create the most intense selection pressure because only worms carrying resistance alleles survive and reproduce. Sangster and Dobson (2002) emphasized this impact of selection pressure in the evolution of resistance. Resistance is more likely to develop in large parasite populations (with higher genetic diversity); it is favoured in parasites with direct life cycles and short generation time; selection pressure is increased by frequent treatment and by underdosing (when parasites carrying recessive resistance alleles as heterozygotes may survive treatment).

This hostile environment can be made even more relentless (in an effort to limit drug resistance) by treatment with combinations of drugs and with annual rotation of drug compounds. Then, only worms simultaneously resistant to both drugs survive. However, Sangster and Dobson (2002) noted that multiple resistance has developed in sheep nematode infections in South Africa and Australia, and triple drug resistance has been reported in the United Kingdom (Bartley *et al.* 2001; Yue *et al.* 2003). In worms resistant to levamisole, higher concentrations of cholinergic compounds are necessary to cause the contraction response shown by susceptible worms, up to 13-fold higher for the most resistant isolates. The change in the ACh receptor which occurs in levamisole resistance also confers resistance to organophosphate compounds that cause ACh accumulation. The review by Sangster and Dobson (2002) emphasizes lucidly that each new strategy, with improved drug and treatment regimens, simply selects ultimately for *greater* resistance. This is a clear demonstration of the ability, and remarkable speed, of parasites to adapt to hostile conditions.

6.5.3 Other hostile factors: predation

Hostile factors affecting ectoparasites include deliberate removal through grooming, by the host individual itself and conspecifics, and by heterospecific cleaner organisms (birds on terrestrial mammals, cleaner symbionts on client fish). Among monogenean skin parasites, there is evidence that this selection pressure may have led to the development of camouflage (pigmentation) or body transparency as a means of evasion (Whittington 1996; Kearn 1999).

6.5.4 The host immune response

There is extensive documentation of the functioning of the vertebrate immune system against protozoan and metazoan parasites, principally concerning mammals, but also in other tetrapods and fish. There is also much information on the internal defence systems of invertebrates, including molluscs and arthropods, protecting against parasitic infection.

The hostility of this natural factor in the parasite's environment represents a distinctive feature of parasitism: the reciprocal interaction with another living organism. A major approach of current research aims to elucidate the complex machinery of the host-immune response, and this area is well reviewed in the immunological literature. This chapter takes a parallel approach in assessing some of the evidence that quantifies the outcome of interactions within this hostile environment—the scale of host-induced mortality in parasite populations. In most host-parasite systems, there is relatively little quantitative information on these lethal effects, measured as the proportion of parasites successfully invading but not surviving to reproduce within the host. This chapter explores, in particular, a series of naturally occurring host–parasite associations, rather than familiar laboratory models, in which field and lab evidence gives unambiguous evidence of the extent of parasite mortality post-invasion.

6.5.4.1 Regulation of polystomatid populations in amphibians

Empirical studies on the polystomatid monogeneans described above have the advantages first, that they concern entirely natural host–parasite systems for which exact and comprehensive field data can be obtained by dissection and, second, that laboratory experiments can be carried out under controlled environmental conditions closely simulating natural events.

Complementary field and laboratory studies indicate that the reproducing populations of most polystomatid monogeneans are strongly regulated. Despite a wide diversity of life cycle patterns that might, intuitively, be expected to produce variations in transmission dynamics, infection levels of adult worms are universally low (Tinsley 2003). Tinsley (2004) summarized studies that confirm the influence on parasite population biology of external environmental factors (including temperature), parasite factors (including age and density), and host ecology (including behaviour and life history schedules). However, these studies also provide exact data quantifying the scale of host-induced parasite mortality: it emerges that immune attack represents the dominant influence. Field data for *P. americanus* demonstrate that the typical single annual wave of invasion (described above) produces consistently high infection levels. The clear-cut schedule of events in this life cycle allows the fate of these infrapopulations to be followed in detail. In the period until the first opportunity for the newly established cohort to contribute to transmission, parasite mortality is around 97% (Tinsley 1999*b*).

In the case of *P. xenopodis*, with a permanently aquatic host (*X. laevis*), transmission is more or less continuous and host individuals would be expected to carry a succession of generations of parasites. Field data confirm the predicted relatively high prevalence and intensity of post-invasion juveniles. However, very few survive to maturity: the prevalence of adult worms is consistently around 40%; about 80% of infected hosts carry only one or two mature worms and maximum intensity rarely exceeds 6 worms/host (Tinsley 1995). Experimental evidence, reviewed by Jackson and Tinsley (2001, 2002, 2003*a,b,c*) and Tinsley and Jackson (2002), indicates that the major part of this attrition is host-mediated. In naïve *X. laevis*, initial establishment is highly successful: laboratory exposures typically produce 100% prevalence measured immediately after invasion and around 30% of each exposure dose establishes successfully. However, in hosts exposed to the same larval doses and examined at the time of parasite maturity (90 days p.i. at 25 °C), less than 3% of the worms that initially

established survived to maturity (Jackson and Tinsley 2003*c*). Consistently in these experimental infections, the numbers of worms surviving to begin reproduction were only 1–3, most commonly a single worm per host. The loss of established worms continues in adult burdens. Although the maximum life span of *P. xenopodis* is 2.5 years, studies recording lifetime reproduction in laboratory infections showed that median duration of survival was only 6 months p.i., and only 16% of infections survived 12 months or more (Jackson and Tinsley 2001; Tinsley and Jackson 2002).

These experimental data relate to single exposures of previously naïve hosts, but continuous invasion in nature might be expected to result in progressive addition to worm burdens. In the study of Jackson and Tinsley (2001), the survival of a secondary infection was followed in the same host individuals from which primary infection characteristics had been monitored. In comparison with 94% prevalence in the primary infection, only 15% of these hosts retained a secondary infection until patency. The experimental evidence of strong acquired immunity indicates that most of the individuals recorded as uninfected in natural populations are likely to be resistant to re-infection and this immunity may be relatively long lasting (Jackson and Tinsley 2001) (Fig. 6.2).

These findings lead to the prediction that highest infection levels in nature should occur in juvenile *X. laevis*, experiencing their first invasions. This is supported by field studies on feral populations of *X. laevis* in the United States and the United Kingdom where there is a strong age-dependent effect on prevalence of patent infection (unpublished). However, even the youngest stages of post-metamorphic *X. laevis* are capable of developing acquired immunity to secondary infection (Jackson and Tinsley 2003*a*).

Further indication of the hostile nature of the environment in hosts with prior experience of infection is provided by measures of reproductive performance in secondary infections. In the small proportion of hosts in which secondary infections survive, the period to patency is more than doubled (median 9 weeks in a primary infection, 21 weeks in a secondary infection of the same host individuals), and rates of egg production are much less than half (Jackson and Tinsley 2001).

Further studies based on field isolates have demonstrated a complex pattern of parasite infectivity and host susceptibility in natural populations in Africa: different parasite isolates showed consistent differences in infection success in a single group of full-sibling laboratory-raised hosts. Infectivity was found to change during laboratory passage, especially when bottle-necked through a small number of host individuals. These and other studies indicate that the key factors influencing survival in this host–parasite relationship are highly heterogeneous and emphasize the importance of compatibility between host and parasite 'strains' in determining infection success. In other words, the nature of the host environment—reflecting its hostility to parasite infection—varies in different host and parasite genotype combinations (Jackson and Tinsley 2003*b*) (Fig. 6.2).

6.5.4.2 Dynamic environmental change affecting Gyrodactylus *species*

In most host–parasite interactions demonstrating acquired immunity, the host environment may be considered 'benign' at the time of initial invasion but there is a change, over a variable time course, to a profoundly hostile environment. This is illustrated very clearly by the infrapopulation dynamics of *Gyrodactylus* species. For these fish ectoparasites, worm numbers can be counted on individual living hosts as the interaction proceeds. Gyrodactylids have a unique form of viviparity in which an offspring developing within the uterus of a parent worm has, in its uterus, another developing embryo and this, in turn, may carry another. The reproductive processes (reviewed by Cable and Harris 2002) result in a sequence of parasites that are already adult at birth, released directly onto the infection site. This generates the potential for exponential population growth that may have serious pathogenic effects. However, the host has the capacity to limit and eventually eliminate infection by an immune response. A typical pattern of infection is shown in Fig. 6.3 (curve a) for *G. turnbulli* parasitic on guppies (*Poecilia reticulata*). In this tropical system, at 25 °C, mean parasite burden rises to a

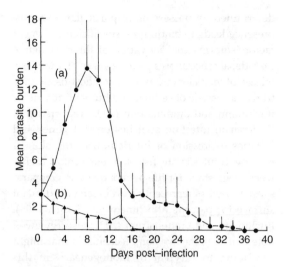

Figure 6.3 Population dynamics of *Gyrodactylus turnbulli* on guppies, *P. reticulata*, following experimental infection of naïve fish (a) Parasite burdens rise rapidly to a peak and then decline; (b) re-infection of the same fish after the primary infection leads to rapid elimination of the challenge infection confirming acquired immunity.

Note: Bars represent one standard error.
Source: Reproduced, with permission, from Scott (1985).

peak at about 10 days post-experimental infection and then falls, initially very rapidly, leading to extinction after 38 days (Scott 1985). Fish that have eliminated a primary infection are refractory to challenge infection (Fig. 6.3 curve b), and their acquired immunity is maintained for at least 6 weeks. Equivalent characteristics have been demonstrated in many *Gyrodactylus* species (with variations in time course attributable in part to species-specific natural temperature ranges). However, there is also significant variation between hosts of the same species in susceptibility to infection, and this is heritable (Madhavi and Anderson 1985).

Buchmann and Lindenstrøm (2002) reviewed the range and action of a series of host molecules interacting with infecting skin monogeneans, including lectins, complement factors, antibodies, acute phase proteins, lysozyme, and anti-microbial peptides. They interpreted the variation in the response of the host environment to gyrodactylids particularly in terms of the binding affinities of different isoforms of complement factors. Parasite survival may be determined by possession of particular epitopes: presence leads to binding by specific complement factor isoforms and activation of the destructive cascade; absence allows parasite survival. Likewise, release of mediators from damaged host tissue may trigger a cascade of cellular and humoral reactions: Buchmann and Lindenstrøm (2002) reported that *G. derjavini* infection switches on IL-I genes and initiates expression of interleukin in the skin of rainbow trout. Clearly, fish skin represents a highly reactive substrate for life by monogenean ectoparasites, as well as representing an inherently difficult surface for parasite attachment (see Kearn 1999). Buchmann and Lindenstrøm (2002) also reviewed potential immune evasion mechanisms that might contribute to survival of monogeneans in this hostile environment, but it is significant that the response is typically local and temporary respite can be gained by migration of worms to other body areas.

The unstable environment that is encountered by a growing population of fish ectoparasites over a relatively short time period is illustrated by many studies of *Gyrodactylus* infection dynamics, demonstrating variable outcomes through subtle differences in factors. Cable *et al.* (2000) compared reproductive characteristics of experimental infections of *G. salaris* on different stocks of Atlantic salmon (*Salmo salar*), originating from two rivers in Norway and one in the Baltic. Detailed measurements of life history schedules revealed small differences in parasite fecundity, developmental time and mortality between infections on these host genetic strains. On hosts of Baltic origin, fewer parasites (45%) survived to give birth once in comparison with hosts of Norwegian origin (60–63% survival); maximum number of births per worm was only two on Baltic hosts (mean 0.6) compared with four on Norwegian stocks (means 1 and 1.3 respectively); mean developmental period of the first offspring was 2.3 days on Baltic and 1.8 days on Norwegian hosts. These and other variations were sufficient to cause fundamental differences in infection dynamics, with predicted exponential growth on Norwegian fish (R_0 = 1.07 and 1.08) but negative growth and eventual extinction on Baltic fish (R_0 = 0.94). This work demonstrates very

precisely the small scale of the host effects that can determine whether parasite populations may increase rapidly to cause pathogenic disease, coexist asymptomatically, or become eliminated.

The studies of Bakke, Cable, Harris, and co-workers on *G. salaris* provide further rich insight into variations in the effects of host environment on parasites because this gyrodactylid species has an exceptionally wide host species range, adding the extra dimension of host specificity. Salmonid species range from highly resistant (brown trout, whitefish), through moderately resistant (char, grayling, rainbow trout), to highly susceptible (Norwegian strains of salmon). Bakke *et al.* (2002) suggested that these states may be part of a single spectrum of response to gyrodactylid infection, and that the same mechanism, probably polygenic, controls resistance throughout the range of different host-parasite interactions.

For a parasite individual, faced with a natural assemblage of fish species in a given habitat, the alternative host environments that it encounters will determine the potential for survival and reproduction (Fig. 6.4). At one end of the range, non-salmonids may act as transport hosts maintaining worms (probably without feeding) for several days. Resistant hosts may permit a brief 'round' of reproduction and onward transmission. The distinctive feature of gyrodactylids, that transmission is effected by established adult worms, means that parasites eliminated from one host may survive to infect others. Highly susceptible hosts that present the most favourable conditions for parasite population growth may be overwhelmed by infection and die, producing a massive short-term boost to the transmission potential of the suprapopulation.

One implication of the experimental studies of *Gyrodactylus* infection dynamics is that the host environment is entirely permissive during the initial exponential phase of parasite population growth. Then, after a time delay, the host response develops and infection levels begin to fall. This is illustrated by the 'classic' pattern shown in Fig. 6.3. However, the very detailed analyses of *G. salaris* population biology by Bakke *et al.* (2002) revealed, remarkably, that parasite reproductive rate declines throughout an infection and not, as would be

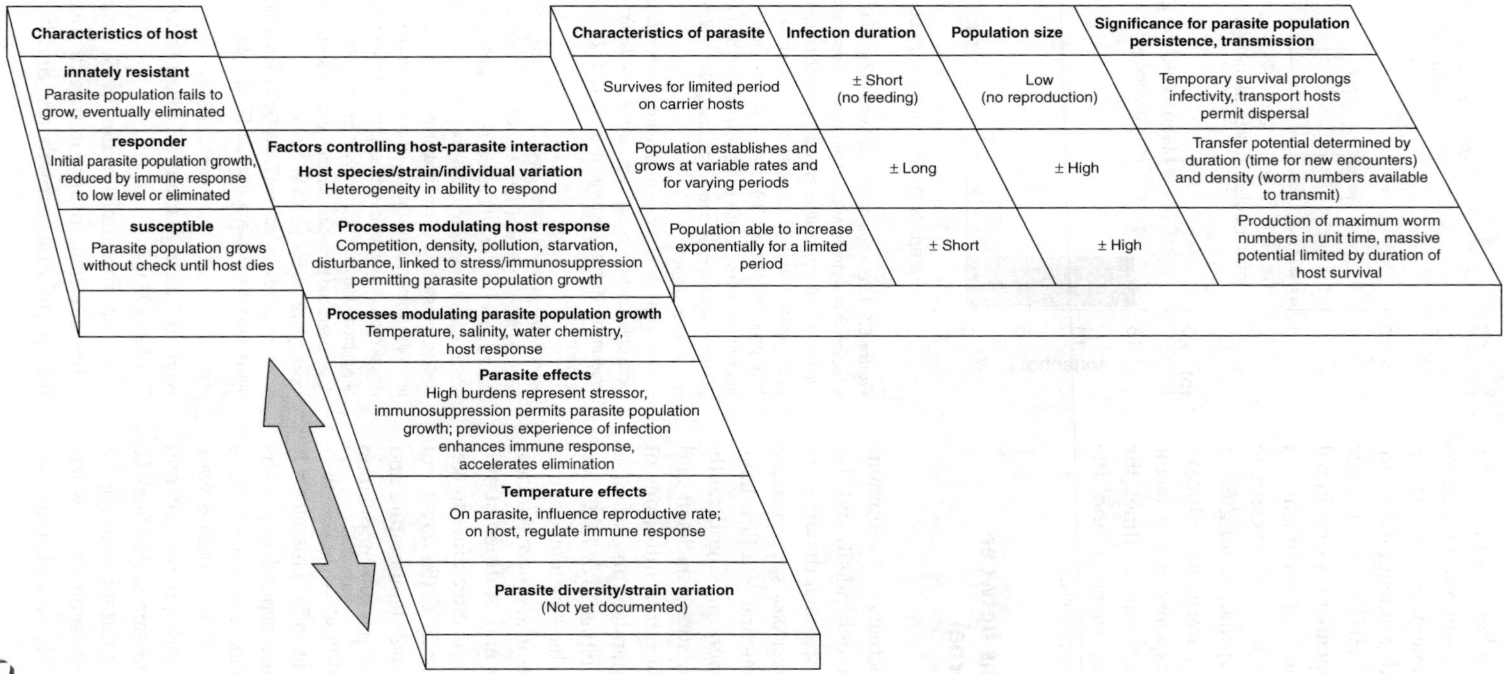

Characteristics of host

innately resistant
Parasite population fails to grow, eventually eliminated

responder
Initial parasite population growth, reduced by immune response to low level or eliminated

susceptible
Parasite population grows without check until host dies

Factors controlling host-parasite interaction

Host species/strain/individual variation
Heterogeneity in ability to respond

Processes modulating host response
Competition, density, pollution, starvation, disturbance, linked to stress/immunosuppression permitting parasite population growth

Processes modulating parasite population growth
Temperature, salinity, water chemistry, host response

Parasite effects
High burdens represent stressor, immunosuppression permits parasite population growth; previous experience of infection enhances immune response, accelerates elimination

Temperature effects
On parasite, influence reproductive rate; on host, regulate immune response

Parasite diversity/strain variation
(Not yet documented)

Characteristics of parasite	Infection duration	Population size	Significance for parasite population persistence, transmission
Survives for limited period on carrier hosts	± Short (no feeding)	Low (no reproduction)	Temporary survival prolongs infectivity, transport hosts permit dispersal
Population establishes and grows at variable rates and for varying periods	± Long	± High	Transfer potential determined by duration (time for new encounters) and density (worm numbers available to transmit)
Population able to increase exponentially for a limited period	± Short	± High	Production of maximum worm numbers in unit time, massive potential limited by duration of host survival

Figure 6.4 Variability of the host environment for *Gyrodactylus salaris* infecting the spectrum of salmonid and non-salmonid fish available in natural habitats. Categories based on the overview by Bakke et al. (2002). Conditions on resistant hosts preclude feeding and reproduction (whilst permitting survival); responder hosts have inherent capacity to create hostile conditions but the course of the interaction is highly variable, subject to effects of abiotic, biotic, host and parasite factors, and these determine the extent of parasite population growth and transmission potential; conditions on completely susceptible fish provide no restraint, permitting runaway parasite population growth that kills the host. The factors controlling the interaction (second column) have the effect of shifting the outcome along the gradient of host and parasite characteristics, changing the potential for parasite population growth and transmission.

expected, from the time of the host response. This suggests that the host environment reacts (becomes hostile) from the initial encounter with the parasite. The response is not density-dependent (and hence not influenced by factors including pathogenic effects) but instead is time-dependent. So, the initial process of parasite infection and establishment immediately ignites a fuse leading, in most cases, to extinction on that host individual. Among other interacting factors, especially parasite reproductive rate and intensity of host response, a key factor determining the overall outcome is time: the duration of infrapopulation survival and the opportunities for onward transmission.

6.6 Complex interactions between parasite, host, and external environmental factors

For ectothermic vertebrates, activity of the immune system is highly temperature-dependent, and this has a fundamental effect on seasonal dynamics of parasite infection. For *Protopolystoma* spp. infecting species of *Xenopus*, experimental studies have shown that increase in temperature significantly reduces adult parasite populations. Jackson and Tinsley (2002) compared laboratory infections of *P. xenopodis* resulting from a standard dose of infective larvae and maintained at either 15 or 25 °C during development to maturity. These are temperatures experienced in natural habitats in southern Africa at different seasons of the year and at different altitudes. The period to patency was three times longer at 15 °C (150 days p.i.) than at 25 °C (50 days), but prevalence was 96% at the lower temperature and only 38% at 25 °C. The overall parasite population surviving to begin reproduction at 25 °C was only one-third of that at 15 °C (Fig. 6.5). This effect is consistent with the temperature dependency of the immune response in amphibians. Equivalent effects were evident when infections of *P. americanus* were maintained in *S. couchii* hibernating under different temperature regimes in the laboratory. At 15–20 °C, parasites survived without significant decrease for over one year. However, at constant 25 °C, worm burdens declined progressively and all parasites

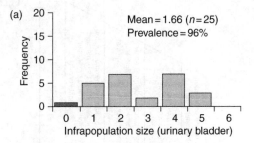

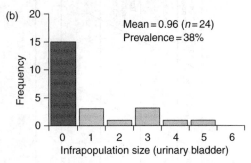

Figure 6.5 Interaction of multiple factors determining the outcome of infection in an ectothermic host, illustrating direct effects and indirect significance: (i) temperature affecting parasite processes (feeding, developmental rate), (ii) temperature affecting host immunity (ability to respond to infection), (iii) parasite affecting host (worm burden and nutrient drain determining pathology), and (iv) host affecting parasite (immune regulation of burden size, and probably also growth rate, reproductive output).

Histograms show frequency distributions of *Protopolystoma xenopodis* surviving to start reproduction in *X. laevis* following exposure to 20 infective stages at (a) 15 or (b) 25 °C (*n* = 25 and 24, respectively).

Notes: Prepatent period was 50 days at 25 °C, 150 days at 15 °C; 62% of hosts lost their infections at 25 °C, only 4% at 15 °C; total parasite population was 2.8 times higher at 15 °C than at 25 °C. Regarding potential pathogenic effects, at lower temperatures, while most hosts are infected and have higher burdens, parasite feeding is reduced; at higher temperatures, when blood removal is greatest, worm burdens are reduced. Regarding parasite reproduction, at lower temperatures, slow development delays start of egg production and this proceeds at low rates; at higher temperatures, development and per capita reproductive output are greater but the surviving adult parasite population is small.

Source: Figure from Jackson and Tinsley (2002), with additional interpretation from Tinsley and Jackson (2002).

were lost after 10 months (Tocque and Tinsley 1994; Tinsley 1995).

In experimental studies of *G. salaris* infecting salmonid fish, different trajectories of parasite infrapopulation growth are highly sensitive to differences in abiotic environmental factors (temperature, salinity, water chemistry, and pollution), biotic and host genetic influences. Bakke *et al.* (2002) reviewed the key effects of temperature controlling the strict seasonality of population dynamics. For *G. salaris*, maximum life span is negatively correlated with temperature (8 days at 19 °C but 53 days at 2.5 °C) while reproductive rate is positively correlated ($r = 0.02$ at 2.6 °C, 0.22 at 19 °C). So, parasites survive at low infection levels in winter, but populations increase with rising spring temperatures. However, while elevated temperature increases parasite reproductive rate, it has a relatively greater effect on population dynamics by enhancing the host immune response. Therefore, when water temperatures are highest in late summer, and parasite populations should be largest, low prevalence and intensity is characteristic (Bakke *et al.* 2002). There are other significant interactions between environmental factors, especially those conditions that cause stress (pollution, fish density, starvation) and reduce host immunocompetence (Fig. 6.4). Most importantly, a major determinant of parasite population dynamics is past experience of infection. Extrapolating from the pattern shown in Fig. 6.3 to natural host populations, individuals that have recently eliminated an infection would be unavailable to the parasite suprapopulation for some weeks or months. It remains to be established how repeated stimulation from attempted invasion might affect this refractory period.

Another impact of the parasite in affecting this spectrum of environmental constraints is recorded by Bakke *et al.* (2002). A rapidly growing gyrodactylid infrapopulation can potentially elevate host blood cortisol, causing immunosuppression. This could act as a positive feedback mechanism, leading to runaway parasite population growth and host death.

Studies on the nematode, *Strongyloides ratti* by Viney (2002) provide further insight into the complex interactions of environmental factors that affect parasite life cycles. Adult female worms in the rat intestine produce eggs that develop in the external environment along two alternative pathways. One produces larvae that directly re-infect the host population and develop into adult females that are genetically identical to each other and to their mother. The other produces larvae that develop into free-living adult males and females that reproduce sexually: eggs from these matings develop into infective larvae that invade the rat host. The proportions of offspring developing by these two routes, involving asexual or sexual reproduction, changes during the course of infection in individual rats. Viney (2002 and references) established that the developmental switch is determined by the host immune response. This leads to an increase in the frequency of sexual reproduction over the direct developmental route that involves mitotic parthenogenesis. External environmental temperature also affects parasite development: low temperatures favour direct development to infective larvae that reinvade the rat; high temperatures favour indirect development to free-living adult females that reproduce sexually. These two environmental effects, host immunity and temperature, interact during the course of an infection so that sensitivity of developing larvae to temperature increases as a result of their experience of the immune response. This implies that larvae developing in the external environment must have a 'memory' of the immune status of the host from which they were passed (Viney 2002).

Viney noted that the developmental switch in *S. ratti* conforms to the general trend that sexual reproduction occurs at times of environmental stress. In the case of *S. ratti*, the host immune response constitutes this environmental stressor for parasites within the host, but the stages that undergo sexual reproduction are in the external environment. So, Viney suggested that the immune response is used as a predictor of the host environments that may be encountered by future infective larvae.

Viney (2002) argued that the developmental switch in the *S. ratti* life cycle is unlikely to be caused directly by increasing hostility of the parasite's environment as the immune response develops. Certainly the change involves a specific anti-*S. ratti* immune response, but the parasite may assess a change in its internal physiology affected by host immunity as a measure of the intensity of the immune response. Ultimately, the changes in the parasite's environment completely eliminate the infection, after which rats are strongly immune to re-infection.

6.7 Host control of hostility

Creation of the hostile environment within the host's body carries a cost in terms of resources: diversion of energy and materials to the immune response may reduce the host's investment in growth and reproduction (i.e. its fitness). The costs of defence influence the evolution of resistance, especially the relationship between the fitness reduction imposed by immunity and the negative effects of infection in a susceptible host (Kraaijeveld et al. 2002; Medley 2002).

Medley (2002) considered this interaction from the perspective of an individual host, modelling how the host should optimally distribute resources between reproduction, survival and immunity to maximize fitness. Results showed that, under conditions of continuous infection, optimum resource allocation by the host allows tolerance of some parasite infections. In other words, the hostile environment created by the immune response may be regulated by the host, permitting a proportion of invading parasites to survive. However, Medley (2002) also demonstrated the sensitivity of this state to variations in resource acquisition: with poorer nutrition, hosts should invest less in immunity and parasite burdens increase. Each host has a different optimal response, determined especially by different resource bases, and this translates into each host having its own optimum immune response and, therefore, its own parasite burden. This perspective attributes variations in the hostility of the parasite's environment to a 'decision' (optimization) by the host. However, the costs of immunity (which determine the parasite burden tolerated) are subject to manipulation by the parasite. Medley (2002) raised the complicating factor that resources devoted to immunity may contribute to both protective and non-protective responses. A parasite that induces production of non-protective antibodies increases the resources required for a host defence of reduced effectiveness: this would tend to increase the optimum burden tolerated. Clearly, the hostile

environment encountered by parasites within their hosts is in a dynamic state, influenced by the interaction of a complex of factors.

6.8 Parasite control of the hostile environment

The comprehensive studies on Gyrodactylus and Strongyloides species concern systems where the parasites appear to be 'on the run' from an environment that becomes increasingly hostile. The life cycle strategy of both can be interpreted in terms of achieving maximum reproduction while the environment permits survival. However, significantly, the mode of reproduction involves initial asexual production of offspring, suited to the current environment, followed by sexual reproduction that generates diversity better equipped to meet variation in future environmental conditions. Nevertheless, the course of infection in the present host automatically provokes conditions eliminating infection in that microenvironment. In other host–parasite systems, it is well documented that parasites faced with a host immune response can evade or modulate the developing hostile conditions. For several intensively studied parasites, these interactions have been examined in very considerable detail: there are comprehensive reviews in the literature interpreting the mechanisms of immune evasion by schistosomes, malaria, trypanosomes, and others, and of modulation of immune attack. It is unlikely that these sophisticated processes have evolved in only a limited range of parasites (selected for intensive research because of their relevance to humans); it might be predicted that equivalent adaptations exist widely among parasites to counter what is, arguably, the dominant environmental factor influencing survival within the host (Tinsley 1999a).

Alongside the documentation of host immune mechanisms and parasite survival strategies based on parasites of medical and veterinary importance, it is interesting to consider interactions in other wildlife systems. Studies by Riley and co-workers provide this wider insight for members of the phylum Pentastomida, parasites whose adults infect the respiratory tract of tetrapod vertebrates, principally reptiles. Fossil representatives of this group occur in the early Palaeozoic, and molecular studies confirm its origin from crustacean ancestors (Lavrov *et al.* 2004). Riley and Henderson (1999) emphasized the ancient association of pentastomids with reptiles and documented complex survival adaptations that probably evolved very early in the relationship.

The vertebrate lung might seem to provide a favourable habitat for macroparasites, with its delicate pulmonary epithelium, rich blood supply, and direct entry and exit route for infective stages. However, relatively few helminths have colonized vertebrate lungs to reside in or on the respiratory epithelium (reviewed by Riley and Henderson 1999), and this is likely to reflect fundamental difficulties for survival in this highly specialized microenvironment. A key component regulating lung function is pulmonary surfactant, a complex mixture of phospholipids, neutral lipids, and proteins. This has an essential biophysical role and is also vitally important in protecting the lungs from infection. Lung-dwelling pentastomids are typically large (some are 5–10 cm in length), long-lived (in some cases many years), and feed on blood from the pulmonary capillaries; but, they cause little observable pathology. Riley and Henderson (1999) attributed the success of these parasites to their adaptations for immune evasion that involve secretion of their own surfactant. Pentastomids have a prominent system of glands that discharge membrane-dominated excretory/secretory (E/S) products onto the surface of their cuticle. This coating has a lipid composition very similar to that of lung surfactant and is therefore immunologically compatible with host secretions. Alongside its role in evading immune surveillance, pentastomid surfactant also reduces inflammatory reactions and this explains the characteristic lack of pathology in long-term infections (Riley and Henderson 1999).

Adults of *T. spiralis* occur in an increasingly hostile environment, in the intestine, that leads to parasite elimination and acquired immunity (see, for instance, Bell 1998). The life cycle requires that larvae, in the muscles of the same host, remain infective for months or even years. As in the case of limited exploitation of the vertebrate lung luminal environment (above), striated muscle tissue has a very restricted list of colonizing protozoans and helminths (Despommier 1998): this may reflect significant problems for parasite survival. The larva of *T. spiralis* occupies an individual muscle cell without killing it and remains metabolically active (unlike the encysted, dormant stages of other muscle parasites including cestode larvae). Exceptionally among helminths, *Trichinella* reprogrammes host genomic expression via secreted peptides to produce a nurse cell devoid of muscle proteins that supports growth and maintenance of the parasite (Despommier 1998). This achieves relatively long-term survival, but the combined effect of infection is to reduce muscle activity and induce fatigue, increasing the possibility of predation by the next host in the life cycle.

Among intracellular habitats exploited by protozoans, macrophages represent an exceedingly hostile microenvironment. Following uptake by phagocytosis, *Leishmania* species survive within phagolysosomes (although some parasites may enter cells such as fibroblasts which are less hostile than macrophages, and this strategy may contribute to long-term survival within the host (Rittig and Bogdan 2000)). Macrophages should, upon activation, be able to kill intracellular organisms, but *Leishmania* has mechanisms to subvert these lethal effects including inhibition of production of superoxide and hydrogen peroxide in the parasitophorous vacuole (Handman and Bullen 2002). While *Leishmania* can avoid direct attack by the host's immune system, the infected cell can still counteract resident parasites by initiating its own death via apoptosis. This defence mechanism is considered to have put selective pressure on intracellular parasites (viruses, bacteria, and intracellular protozoans) resulting in strategies to inhibit the apoptotic programme of the host cell (reviewed by Heussler *et al.* 2001). Such a dangerous manipulation may have lethal consequences. In immunocompromised hosts or in infections with virulent strains of *Leishmania*, the balance may be disturbed leading to rapid apoptosis of T cells, so that an effective immune response does not develop and the host is overwhelmed. Interference with apoptosis by *Theileria* results in proliferation of infected host cells that cannot be controlled by the immune system, setting off unspecific responses that result in host death (Heussler *et al.* 2001). In these cases, the escalation of the evolutionary arms race leads to mutual destruction.

6.9 The outcome of hostile environmental conditions: extinction

The preceding discussion of hostile environmental conditions has explored factors that limit parasite distribution—within geographical areas, within species, within populations, and within particular host individuals. Typically, it is possible to identify a gradient of unfavourable conditions and, ultimately, a threshold beyond which adaptations fail. The most hostile conditions experienced by organisms may be defined as those completely prohibiting survival, resulting in extinction. Hostile factors may cause mortality of organisms directly and their deaths accumulate to eliminate populations, or hostile conditions may prevent recruitment so that surviving organisms are not replaced when mortality occurs naturally.

6.9.1 Extinction risks from external environmental factors

A case study of local population extinction attributed to the influence of external environmental conditions was reviewed by Tinsley (1999*b*). Relatively long-term population studies on the *Pseudodiplorchis/Scaphiopus* system in the Sonoran Desert, Arizona, demonstrated that infection levels at specific field sites were stable from the early 1980s to early 1990s (Tinsley 1995). However, a series of severe summer droughts began in 1992 and continued until 1995. This provided a test of the ability of the host–parasite system, which is entirely dependent on water for reproduction, to tolerate exceptional perturbation. Age determination of all host and parasite individuals enabled reconstruction of population effects on age specific cohorts (details in Tinsley 1999*b*).

For the host populations, the weather conditions precluded almost all *Scaphiopus* recruitment in 1993–95 and again in 1997, either because poor rainfall prevented spawning (lack of suitable breeding sites) (1995, 1997), or because very brief rainfall triggered spawning but all tadpoles subsequently desiccated (1993, 1994). The outcome, assessed from counts of annual growth rings in toad bone, was that younger individuals corresponding to these recruitment years were entirely absent from populations surveyed in 1997. Instead, these populations comprised progressively ageing individuals: half of the toads in some populations were aged 10 years or more, up to a maximum age of 17 years, and none was aged less than 7 years. Heavy rainfall in 1996 and 1998 permitted successful host recruitment and the beginnings of recovery were documented at the conclusion of the study (Tinsley 1999*b*).

The effects on the parasite population were more complex. In the initial drought years, when brief isolated storms triggered host spawning, parasite transmission—geared to 'instantaneous' episodes (typically a few hours)—was successful. Although the tadpole populations subsequently died, the mating toads carried away the products of very efficient invasion. Thus, a mass input of larvae in 1994 was recognizable by high prevalence and intensity of 2-year-old adult parasites in 1996 and 3-year-olds in 1997. In 1995, the rainfall was insufficient to produce breeding sites so there was a failure both of host spawning and, therefore, of parasite transmission. In 1996, very heavy rainfall produced extensive flooding and, although host spawning occurred, breeding toads were dispersed in moving floodwaters. Infective stages that are typically released into small pools of standing water were washed away. This 1996 cohort of *P. americanus* was missing at most sites in the following years. The 1997 summer rainfall and environmental conditions resembled those in 1995 and, at most localities, parasite recruitment was reduced or absent for a third successive year (Tinsley 1999*b*). Maximum lifespan of *P. americanus* is generally 3 years (about 0.5% of worms may survive 4 years) (Tocque and Tinsley 1991*b*), and interruption of transmission for the three successive years (1995–97) could not be tolerated. At sites where *P. americanus* had been stable in previous long-term studies, the local suprapopulations were found to be extinct in 1998 (Tinsley 1999*b*).

6.9.2 Extinction involving parasite–parasite interactions

A dramatic illustration of the effects of interspecies hybridization, leading to replacement and local extinction, is provided by studies on schistosomiasis in Loum, Cameroon (updated by Webster *et al.* 2003). Until the late 1960s, human schistosomiasis in this area was caused by a single species, *S. intercalatum* (infecting intestinal blood vessels). However, records in 1972 revealed that *S. haematobium* (in urinary bladder blood vessels) had become established and that hybridization between the species was occurring. Subsequent studies showed that hybrids resulting from *S. haematobium* male X *S. intercalatum* female crosses were more viable than the reverse cross. Pairings with *S. haematobium* males had an advantage in transmission since the pair located in the vesical site and egg output more easily evaded basic sanitation controls. Several environmental changes were implicated in this introduction, including local forest clearance that favoured the snail host of *S. haematobium*. Parental parasite species were found to be strictly specific to single snail species, but their hybrids were able to utilize both snail species as intermediate hosts. Over the next 25 years, a series of field studies demonstrated a progressive change from intestinal to urinary schistosomiasis and, finally, the complete replacement of *S. intercalatum* through introgressive hybridization. In 1990, one-third of schistosomes were of hybrid origin, but molecular studies showed that the proportion of hybrids in the overall population fell to 5% in 1999 and 2000 with the overwhelming majority being *S. haematobium* (Webster *et al.* 2003).

This impressive documentation has revealed a rapid evolutionary process in which external environmental change (including forest clearance) created a zone of sympatry (both outside and inside the definitive host). The resulting interspecies interactions produced conditions detrimental to one of the species and, although genes of *S. intercalatum* may persist in recombinants, the outcome has been extinction. A similar risk of partial or total exclusion of one species by another has now been recognized in other two-species schistosome interactions, including the progressive replacement of *S. haematobium* by *S. mansoni* in the Fayoum, Egypt (Webster *et al.* 1999; Cosgrove and Southgate 2002).

6.9.3 Evidence for extinctions in ecosystems

Environmental conditions that are responsible for complete extinction may be considered to define the most hostile factors facing parasites (and all other organisms). Extinction of the host, by definition, removes the environment of a host-specific parasite and is equivalent to loss of an essential component of the ecosystem for free-living organisms. However, the occurrence of such extinction events in the past, involving total loss of genetic information from evolutionary lineages, is difficult to document for parasites. It would be entirely plausible to imagine that the dinosaurs, dominant land vertebrates for over 150 million years during the Triassic, Jurassic, and Cretaceous, possessed a parasite fauna that largely disappeared with their extinction. Indeed, the well-documented mass extinctions of major animal groups that have punctuated the history of life must inevitably have been accompanied by undocumented mass extinctions of their stenospecific parasites. With the general absence of fossil evidence, these extinctions remain speculation, but there are some exciting glimpses of this fauna in recently discovered fossils. Upeniece (1999) reported a series of putative parasite fossils, represented by groups of radially arranged hooks, on the bodies of Late Devonian fish belonging to the extinct Placodermi and Acanthodii. Poinar (2002, 2003) recorded well-preserved nematode parasites in ants and dipterans fossilized in Baltic amber (dated at 40 mya).

An alternative exploration of evidence for parasite extinction could be provided where host groups persist but have significant gaps in their parasite assemblages that might reflect past loss. The pipid amphibians have a long evolutionary history, with representatives amongst the earliest fossil anurans in the early Cretaceous. These are animals that overlapped the dinosaurs, but survived. The lineage including extant *Silurana* and *Xenopus* originated about 64 mya (Evans *et al.* 2004) and has a very rich parasite fauna with over 25 genera from 7 invertebrate groups (including protozoans, digeneans, cestodes, monogeneans, nematodes, a leech, and an acarine mite) almost all strictly specific to this anuran group. Tinsley (1996) interpreted two distinctive components contributing to this assemblage: one group comprises parasites related to those in other anurans, reflecting inheritance from a common stock; the other has nearest relatives among the parasite fauna typical of fishes, representing species transfers (through common ecology, shared habitats, and diet). Significantly, the very distinctive taxonomic position of almost all these parasites (in their own genera, families and higher taxonomic groupings specific to *Xenopus*) suggests an ancient origin (for both presumed sources) and a subsequent isolated evolution. Among this richness and diversity, it is conspicuous that the phylum Acanthocephala, with a strong representation in other anuran amphibians and in fish (i.e. both potential sources of origin), is entirely absent from *Xenopus*. This is despite ecological links (especially diet) that should provide an infection route. Similarly, the lungs of anuran amphibians are well known as sites for abundant metazoan parasite infections, including digeneans and nematodes, yet the well-developed lungs of *Xenopus* are unexploited (Tinsley 1996). It is entirely conjectural that these empty niches reflect past extinction events. The alternative explanation would be that prospective parasites 'missed the boat' (Paterson and Gray 1997), but it is surprising that they subsequently failed to 'get on board' throughout the long voyage, unless they were prevented.

Tinsley (2003) assembled evidence to suggest that the present distribution of polystomatid monogeneans infecting anuran amphibians may be indicative of widespread extinction. The evolution of the Polystomatidae is intimately associated with that of the tetrapod vertebrates. A molecular interpretation of host–parasite co-evolution (Verneau *et al.* 2002) calibrated the diversification of parasite lineages with earliest estimates of the major divisions within vertebrate phylogeny. There are about 200 polystomatid species currently recognized (Bentz *et al.* 2001) with an almost worldwide distribution. Most known species infect anuran amphibians and Verneau *et al.* (2002) estimated that the polystomatids

specific to this host lineage diverged at c.250 mya. Given the diversity of the Anura (over 3,500 species, almost equivalent to the mammals), a very rich polystomatid fauna might be predicted. However, Tinsley (2003) reviewed data indicating a remarkably poor exploitation of anuran amphibians. First, at the host species level, field data for relatively well-studied areas of the world show consistently low representation of polystomatid taxa. In North America, with nearly 1000 species of anurans, only three polystomatid species have been recorded. In Western Europe, also with a well-studied fauna of amphibians and their parasites, only four anuran species are infected by polystomatids. Relatively comprehensive studies of anurans in specific geographical areas of Africa have revealed very low parasite occurrence in the host species potentially available within the area. Second, at the host population level for those anuran species known to carry polystomatids, prevalence and intensity of infection are (with a few specialized exceptions) uniformly low. Data from extensive field surveys of *Polystoma* species in Africa show a prevalence of 10% or less in nearly half of the studies and intensity only very exceptionally exceeding 6 worms/host, most commonly only one or two. For the majority of African *Polystoma* species records, there is less than one adult parasite for every five host individuals in natural anuran populations (Tinsley 2003). Very low infection levels cannot be explained in terms of life cycle difficulties imposed by intermittent host links with water. Thus, *P. xenopodis*, transmitted throughout the year between aquatic hosts, does not achieve higher prevalence and intensity than *P. americanus*, transmitted during less than 24 h each year between desert hosts. Tinsley (2003) drew upon the field and laboratory data outlined above which demonstrate that polystomatid populations are subject to powerful regulation by the host immune response. For *P. xenopodis*, with the 'easiest' of all polystomatid life cycles, 60% of *X. laevis* in natural populations are free of patent infection and most of those infected carry only a single adult parasite (Tinsley 1995). For *P. americanus*, the attrition of successfully established worms post-infection is such that only 3% survive to the first opportunity for onward transmission. The exceedingly narrow margin between survival and complete failure suggests the possibility that relatively simple environmental perturbations could tip the balance towards extinction.

Experimental studies on these polystomatids, outlined above, demonstrate that increase in temperature significantly reduces infection levels. For *P. xenopodis* (with continuous transmission), an overall increase in temperature operating proportionally throughout the year, or a longer period of elevated summer temperatures each year, could have a critical negative effect on reproducing parasite population size. For *P. americanus* (with once-per-year transmission), if environmental temperatures were to remain high during winter, parasites would not survive from one transmission season to the next (Tocque and Tinsley 1994; Tinsley 1995). Rainfall variations interrupting parasite transmission represent an alternative environmental perturbation that could prejudice polystomatid survival. In the case study documented by Tinsley (1999*b*), three successive years of interrupted transmission led to extinction of *P. americanus* in local host populations (see above).

Based on these lines of evidence, Tinsley (2003) proposed that relatively simple environmental changes, such as those characteristic of past climatic oscillations and operating over longer time scales, could have led to extinction of polystomatid lineages. The critical factor in this process is the powerful attrition of parasite burdens by host immunity evident in present-day host–parasite associations, resulting in a very narrow margin between survival and extinction. In natural host species populations, the majority of individuals are parasite-free by virtue of acquired immunity. In natural anuran communities, a majority of species may be free of polystomatid exploitation because of past interactions leading to extinction of host-specific parasites. This scenario would reflect combined operation of the most hostile conditions for parasite survival: an interaction of host immunity that achieves very effective regulation of reproducing parasite populations with critical climatic changes that further reduce these infections and precipitate extinction.

6.10 Discussion and concluding remarks

The dominant theme of this review has concerned the environmental conditions experienced by parasites within the host, and this has been explored particularly with examples of natural host–parasite systems. Among the many fundamental effects of human activity on ecosystems, the management of societies and agriculture has typically created unnaturally close confinement of some animal populations (including humans themselves) (see Chapter 10). One effect has been reduction in the hazards of the external environment that regulate transmission of infection, producing conditions in which the hostile interactions between host and parasite are most intense. In attempts to control disease, humans have devised lethal factors to eliminate specific target organisms with anti-parasitic chemicals. However, these have become potent agents of natural selection, amplifying those parasite genotypes that have any capacity to blunt the population impact. The efficiency of the evolutionary response by parasites is exemplified by the rapid development of drug resistance. Other medical interventions have also been important in manipulating the hostility of the host environment. Immunosuppressive therapy removes the capacity to create a hostile environment and predisposes to uncontrolled infection. Vaccination creates hostile conditions before the experience (and the costs) of a primary infection (although it reflects the complexity of interactions with protozoan and helminth infections that effective vaccines to these parasites have proved difficult to develop).

While it is ironic that human social organization may have assisted the success of some parasite infections (see Chapter 10), studies of truly natural host–parasite systems emphasise an alternative view: despite the superb adaptations of parasites, the host constitutes a highly hostile environment. These systems provide relatively rare quantitative data on the extent of host-induced parasite mortality. Even the 'favourable environments', in which parasites can survive and reproduce, may be responsible for major mortality within parasite infrapopulations and this contributes to the regulation of host–parasite interactions. The consistent estimates of >95% mortality between successful invasion and first reproduction in polystomatid monogenean life cycles (above) testifies to this hostility.

Hostility of the environment cannot be considered a characteristic property of the conditions that define that environment (with respect to high and low temperature, pressure, availability of water, oxygen etc.). Adaptations of the organisms inhabiting the environment determine whether processes of growth and reproduction can occur or whether survival is in jeopardy. This applies both to the external macroenvironment and to the microenvironment provided by a host which, to a parasite specific to that host, may appear favourable whereas, to an unnatural infection, it may cause rapid destruction.

In many parasite life cycles, major external environmental constraints with the power to block transmission and population growth can be identified, but they sometimes do not actually represent conditions that could be described as hostile. In a series of unrelated examples, the temperature threshold below which parasite development is arrested is remarkably 'moderate': about 10–15 °C in different species. In temperate regions, this limitation precludes new contributions to transmission for up to 6 months each year (except for the pool of already-infective eggs, larvae, or encysted stages accumulated, in some life cycles, during the preceding warmer period). There should be significant selective advantage for parasite adaptations exploiting lower temperatures and reducing this long interruption. Given the evidence of rapid parasite adaptation in the evolution of drug resistance, for example, it seems a paradox that temperature effects should halt recruitment into parasite populations for such a major part of each year.

There are remarkable examples of parasite adaptations to extreme abiotic stress: survival mechanisms including anhydrobiosis in nematodes are important in maintaining life cycles despite normally lethal constraints. Nevertheless, in environments where macro-conditions are most hostile, there is evidence that parasites may exploit specific micro-environments where conditions are ameliorated. Thus, transmission of nematodes of reindeer and *Echinococcus multilocularis* in voles and foxes

may continue over winter on Svalbard when air temperatures may be as low as –40 °C. However, the environment is not so extreme under protective snow cover, where infective stages may occur. *Fasciola hepatica* must be assumed to exploit regular short periods when temperatures are above the minimum threshold permitting development in the high Andes. Parasites of ectotherms, that might be expected to experience external temperature conditions directly, may also benefit from amelioration of environmental conditions resulting from host behaviour. The behavioural responses of *S. couchii* in selecting optimum sites in the desert (including excavation of hibernation burrows) coincidentally provides its parasites with the most equable micro-environmental regime.

An important requirement in strategies involving a wait for the return of favourable conditions is a trigger factor and its recognition by the waiting parasite stage. With appropriate life cycle timing, the conditions experienced in hostile ecosystems may not be hostile at all. In its brief excursion into the desert environment, *P. americanus* encounters conditions that are entirely 'benign', both in terms of physical factors (transmission in water) and factors affecting invasion (host density). This is reflected in transmission success that is higher than in any other platyhelminth (see above). In this case, the conditions typically associated with a hot desert have relatively limited direct impact on parasite survival: the most important constraints on parasite life history are a product of the interaction with the host.

The extensive documentation of the complex interactions involving host immunity and parasite evasion provides evidence that the hostile environment created by the immune response has been a potent force selecting for parasite adaptation. The complex array of anti-parasite killing mechanisms brought into action by an infected host gives a vivid impression of an exceedingly hostile environment; but, the survival of *Schistosoma mansoni* in the human blood stream for over 20 years is testimony to the effectiveness of immune evasion mechanisms.

The hostile conditions created within the host have lethal effects, but this is a dynamic interaction: the severity of conditions is dependent on resource allocation by the host but can also be manipulated by the parasite. The parasite, too, incurs a cost in deployment of survival mechanisms.

The distinctive characteristic of parasitism is that the immediate environment is provided by another living organism. There are two effects. First, as argued by Combes and Morand (1999), parasites provoke the hostile conditions. There are no parallels in the ecology of free-living organisms that the environment reacts specifically to the presence of that organism, producing a lethal effect to eliminate it. Indeed, the environment may be more or less benign in this respect until invaded. The second interaction is also a special feature of parasitism: a capacity to modify the host environment. Thus, parasites can reduce or suppress the hostile conditions; they can also manipulate processes that should be hostile to benefit their survival. Free-living animals can exploit existing variations in external environmental conditions and they can moderate the microenvironment that they actually experience (by behavioural thermoregulation, for instance); but they rarely *change* the environment (*Homo sapiens* is, of course, a notable exception).

Combes and Morand (1999) considered that, once adapted to a particular environment, an organism's fitness is highest in that environment irrespective of anthropomorphic views regarding the severity of conditions experienced. They concluded that it is more comfortable to live inside a polar bear than outside on the pack ice: 'The polar bear does live in an extreme environment. Its parasites do not.' However, the principal indication of this review is that the most hostile of all environments experienced by parasites is that encountered *within* the body of the host. Thus, one of the parasites of polar bears, *T. nativa*, has the distinction that it may never encounter the external environment during its life cycle, transmitted directly from one endothermic host to the next. If this thermostatic regulation should fail (at host death), then encysted larvae can still survive prolonged periods in frozen carrion until eventual ingestion by a scavenger. For maturing parasites in the intestine, the inside of the host is exceedingly hostile: survival of adult worms is limited to only a few weeks or months. In immunocompromised hosts, this brief lifespan is

extended, demonstrating that it is the hostile immune environment that imposes the lethal effect. Within the host's body, *T. nativa* larvae provide superb illustration of the ability of parasites to modify their environment, taking over the developmental machinery of individual muscle cells to create a protected home. However, even in this state, the larvae are subject to attrition: calcification begins after 9–12 months. Larval mortality is more rapid in some host species (including pigs) than others (laboratory rodent models). Paradoxically, *T. nativa* larvae may survive longer within a dead host in the frozen environment of the Arctic (4 years at –18 °C) (Kapel *et al.* 1999) than when inside a living host buffered from the extreme environment outside!

For several independent host–parasite interactions systems, it has been suggested that increased lifespan represents an important adaptation for survival in a hostile macroenvironment. Thus, in studies of *F. hepatica* at high altitude in the Andes, Mas-Coma *et al.* (2001) noted that high infection levels were associated with increased longevity of infections within snail hosts. In the case of trichostrongyle nematode transmission, although the adaptations of these free-living stages are not related specifically to Arctic environments, the comparatively long life expectancy of adult worms in Svalbard reindeer might represent an adaptation to extreme conditions. The life cycle of *P. americanus* in a desert ecosystem demonstrates the strong selective advantage of increased lifespan when transmission events are strictly annual: body size is age-related and longer lifespan has an important influence on fecundity. Relatively long life expectancy may also have selective advantage in bridging gaps between transmission opportunities, and in compensating for limits on reproductive investment imposed by hostile external conditions. However, the *Protopolystoma / Xenopus* system shows that lifespan is not simply a characteristic of the parasite, it is actually strongly influenced by the interaction with the host. Parasites make the longest contribution to reproduction when in the most compatible combinations of host and parasite genotypes. This, too, was the explanation proposed by Mas-Coma *et al.* (2001) for hyperendemic fascioliasis in the Bolivian Altiplano: reproduction by selfing from the snail(s) originally introduced has generated a homogeneous intermediate host population highly susceptible to infection.

Within host and parasite populations, variations in parasite infectivity and host susceptibility may determine that, at a 'micro-level', some host environments are more hostile than others. This has been documented in many systems illustrating 'local adaptation'. Amongst the case studies emphasized in this review, quantitative evidence is provided by data from experimental infections of *P. xenopodis* in *X. laevis*. Using full-sibling groups of lab-raised hosts (with reduced genetic variation), survival to patency for different isolates of parasites (from both local and distant localities) showed major variation (Jackson and Tinsley 2003*b*). Development within the host at higher or lower temperatures (within the natural seasonal range experienced by host and parasite) may result in significant differences in survival (Jackson and Tinsley 2002). Studies of lifetime reproductive performance showed marked variation between infections in survivorship, egg output, etc. (Jackson and Tinsley 2001). All this combines to illustrate, for this natural host–parasite system, variations in the degree of hostility that may exist in nature.

In considering the hostility created by the immune response, there are fundamental differences between endothermic and ectothermic vertebrate hosts. The components of immune attack are broadly equivalent (for instance, *Xenopus* has most of the immune attributes better-documented in mammals), but the intensity of interactions is modulated in ectotherms by natural temperature variation. In ectotherms, host–parasite interactions experience periods of respite at low temperatures:

parasite feeding rate and host metabolic loss are reduced; the activity of the host immune response is, to a greater or lesser extent, suppressed. The outcome, demonstrated by the detailed quantitative studies on monogeneans such as *Gyrodactylus*, *Pseudodiplorchis*, and *Protopolystoma* species (above), shows that during periods of reduced parasite-induced pathology, host-mediated attack on parasite populations is also reduced. At higher temperatures, when there are rate increases in parasite activity (including feeding, reproduction and transmission), the host immune response is also correspondingly upregulated. In endotherms, by contrast, both pathogenic effects and immune attack are continuous at constant temperature—both are relentless in provoking and generating hostile conditions within the host micro-environment. It might be speculated that improved immune defence against infection was a major selective advantage in the evolution of endothermy, developing in just two classes of one animal phylum (and perhaps also in the now-extinct advanced reptiles that were ancestral to the mammals and birds). However, while the increased activity of immune defences will have been a major benefit of endothermy, the concomitant change in the thermal environment of parasites will have escalated the evolutionary arms race, so that parasite pathology and reproductive output, in an endotherm, is also relentless. The evolution of endothermy may therefore have been a critical factor in increasing the virulence of parasite infections, so that the capacity for creation of highly hostile environmental conditions has also selected for improved adaptation to survive in and circumvent these hostile conditions.

CHAPTER 7

Parasitism and environmental disturbances

Kevin D. Lafferty[1,2] and Armand M. Kuris[2,3]

Several new diseases have gained celebrity status in recent years, fostering a paradigm that links environmental stress to increased emergence of disease. Habitat alteration, biodiversity loss, pollution, climate change, and introduced species are increasing threats to the environment that are postulated to lead to emerging diseases. However, theoretical predictions and empirical evidence indicate environmental disturbances may increase some infectious diseases but will reduce others.

7.1 Introduction

To build a predictive framework for how environmental disturbances can affect parasitic diseases, we limit our scope to those environmental disturbances that result from human activities. Anthropogenic change that may affect parasite communities can be divided into five broad types: habitat alteration, biodiversity loss, pollution, climate change, and introduced species. We do not limit ourselves to the facile prediction that environmental change will lead to increases in parasitism. As we will make clear, there are substantial theoretical and empirical reasons to expect the opposite will also often result from such changes.

With the possible exception of invasive species, environmental disturbances can collectively be considered as stressors (Lafferty and Kuris 1999). Perhaps the first thing that comes to mind when one thinks about the effect of stress on disease is our own health. Studies link stress to reduced immune function and various associated maladies of the modern age (Yang and Glaser 2002). Immune systems are costly to maintain and stressed individuals

[1] USGS Western Ecological Research Center.
[2] Marine Science Institute, University of California, Santa Barbara, California.
[3] Department of Ecology, Evolution and Marine Biology.

may lack sufficient energy to mount an effective defence (Rigby and Moret 2000), making them more susceptible to opportunistic infections (Scott 1988; Holmes 1996). But stress is not just fretting about how to make an unreasonable deadline or frustration over being late for an appointment while sitting in stalled traffic. Toxic chemicals (Khan 1990), malnutrition (Beck and Levander 2000), and thermal stress (Harvell *et al.* 1999) are all examples of stressors hypothesized to increase individual susceptibility to infectious diseases. This line of thought suggests that environmental stress should aggravate infectious disease. An opposing prediction emerges if one considers population dynamics. Abundant species have more parasites (Arneberg *et al.* 1998a). The likelihood and impact of an epidemic increases with host density because density determines contact rates between infected and uninfected individuals (Stiven 1964; Anderson and May 1986). Infectious agents require a threshold host density for transmission (McKendrick 1940). Outside stressors that reduce host vital rates will depress host population density, thereby reducing the chance of an epidemic process, or even the ability of a parasite to persist at all in a declining or low density population.

Stressors may also induce a more negative impact on parasites than on their hosts. This should increase recovery rates of infected individuals and mitigate the population-level impacts of the disease. In addition, infected hosts might experience differentially high mortality when under stress. This would remove parasites more rapidly from the host population than would occur without the stressor. While this increases the impact of disease on infected individuals, it simultaneously decreases the spread of an epidemic through the host population. Such a relationship underscores the point that population effects of stress and infectious disease cannot necessarily be predicted from their effects on individuals.

It is more likely that stress will have multiple effects on hosts and parasites such as increasing host susceptibility to disease while impairing host vital rates. This makes it unclear how a particular stressor should affect disease in a host population. Although stressed individuals should be more susceptible to infection if exposed, the stressor will likely also reduce the contact rate between infected and uninfected individuals to the extent that the stressor reduces host density. Simulation models help resolve the opposing predictions stemming from these alternative effects. Stress is most likely to reduce the impact of closed system, host-specific infectious diseases, and increase the impact of other types of disease (Lafferty and Holt 2003).

7.2 Habitat alteration

Humans have altered nature in ways that can affect diseases (Lafferty and Kuris 1999) (see also Chapter 10). Conversion of forest to agricultural land dramatically changes the environment for parasites and their hosts; and this has raised concerns for human health (Patz *et al.* 2000). In particular, deforestation, damming, road construction, fish farming, and rice farming increase malaria transmission by creating mosquito breeding habitat (Smith 1981; Desowitz 1991). In addition, domestic animals may provide new food sources for mosquitoes, leading to increased malarial transmission in the associated human population (Giglioli 1963).

Habitat alteration has also created conditions conducive for the transmission of trematodes. For instance, dumps and fish farms attract seagulls which fuel trematode life cycles (Kristoffersen 1991;

Bustnes *et al.* 2000). Increases in trematodes are of particular concern for those trematode species that cause human disease. Deforestation reduces acidic leaf litter and increases algal growth in ponds and streams, creating conditions suitable for snails that serve as intermediate hosts for schistosomes (Southgate 1997). The Aswan Dam that created Lake Nasser also created excellent habitat for the snails serving as the intermediate host for the trematodes that cause human schistosomiasis (Heyneman 1979). Construction of other large impoundments throughout Africa (e.g. Paperna 1969) has substantially increased schistosome transmission, resulting in increased human morbidity and mortality (Gryseels *et al.* 1994).

Due to concerns for human health, the literature tends to focus on the types of habitat changes that increase disease. However, there are many ways that habitat alteration, through its effects on biodiversity loss, should decrease infectious disease (as discussed below). In particular, the wholesale draining and conversion of wetlands has dramatically reduced the transmission of various infectious diseases (Lafferty and Kuris 1999; Reiter 2000). Management of water sources for breeding mosquitoes, through drainage and controlled water levels, was instrumental in the successful malaria control campaigns in the southern United States and Israel/Palestine (Kitron 1987).

7.3 Biodiversity loss

Although authors disagree on the present rate of extinction associated with human induced environmental degradation, there is no denying that it is orders of magnitude above background levels (Regan 2001). None of these estimates considers extinctions of parasites which, for some host groups, may exceed the extinction rate of host species (Sprent 1992). Few will lose sleep over the notion of parasites going extinct but one only need imagine the diversity of now extinct parasites specializing on dinosaurs (Kuris 1996) to realize that parasite extinction has been a vast, but hidden, component of evolutionary history (see also Chapter 6). In addition, given the possible role of parasites in stabilizing ecosystems (Freeland and Boulton 1992) conservation biologists may one day come to appreciate the potential need to protect parasites (Combes 2001, see also Chapter 8). Two caveats are: (1) many parasites are not strictly host specific and the fates of these parasites are not tied to the extinction or persistence of single host species; and (2) parasites, due to the nature of density-dependent transmission dynamics, are likely to go extinct well before their hosts (Lyles and Dobson 1993). For this reason, host extinction may not be the key for understanding parasite extinction. Reduced host species densities and host species ranges are more likely to be good predictors of parasite losses.

As mentioned previously, a decline in host density below a transmission threshold can cause host specific infectious diseases to go locally extinct. The number of species put on endangered lists is a good example of cases where host densities have been reduced to such low levels that parasite extinction seems likely. Sometimes, we have enough evidence for wide-scale declines in whole groups of taxa or habitats. For instance, amphibians (Houlahan *et al.* 2001) and British birds (Balmford *et al.* 2003) are now thought to be at substantially lower densities than during prior decades. Populations of monitored species from a wide range of taxa have declined appreciably in marine, estuarine, and freshwater ecosystems; this is likely because aquatic habitats are particularly susceptible to the sort of degradation that leads to reductions in host densities (Balmford *et al.* 2003).

About half of the primordial terrestrial habitats have been cleared or converted to human use (Balmford *et al.* 2003). However, habitat loss does not necessarily translate into reductions in host density if the remaining habitats are not degraded. Under this condition, disease transmission would be maintained. Transmission might even increase, at least temporarily, if habitat loss leads to crowding in the fragmented remaining habitats (Holmes 1996). Despite this potential maintenance of transmission on a local scale, habitat contraction and fragmentation reduces the geographic range of host species. Since most parasite species exploit a host only over a subset of the host's range, we predict that host range contraction will eliminate a proportion of the parasite species from a host species. A better understanding of the rate at which parasite communities

change over the landscape would provide more insight into this potentially large effect.

Parasites, particularly those with complex life cycles, should generally decline with a decrease in biodiversity (Robson and Williams 1970; Pohley 1976; Hughes and Answer 1982; Hudson *et al.* 1998). Digenetic trematodes are a good example. Trematode communities can vary considerably within a wetland (Lafferty *et al.* 1994; Stevens 1996; unpublished thesis) and among wetlands (Lafferty *et al.* 1994; Huspeni 2000, unpublished thesis). This is likely a direct consequence of the biodiversity of final hosts that use a particular area. A healthy marsh ecosystem provides rich feeding grounds and habitat for dozens of species of birds that act as definitive hosts for 20+ species of trematodes. The primary first intermediate host for the trematodes, the California horn snail, occurs throughout the marshes (Lafferty 1993*a,b*). Huspeni and Lafferty (in press) found that degraded sites in an estuary had fewer trematode infections and lower species richness relative to undisturbed control sites. This seems most likely because estuarine birds, the definitive hosts, should be less abundant and diverse in disturbed areas (Kuris and Lafferty 1994; Lafferty 1997). Correlations between trematode species richness and bird species richness at different sites in an estuary (Hechinger and Lafferty in review) and demonstration that the addition of bird perches to an estuary leads to increased prevalence of trematodes in snails (Smith 2001), further support the hypothesis that functioning ecosystems facilitate parasite communities.

Deforestation, particularly clear cutting, along lakes and streams, also leads to a significant decrease in the prevalence of trematodes and other parasites with complex life cycles. In the most heavily cut watersheds, rates of fish parasitization declined to the extent that only unparasitized fishes were present in those lakes (Marcogliese *et al.* 2001). Since other parasites with direct life cycles (copepods and monogenes) actually increased in the most impacted lakes, this supports the hypothesis that biodiversity maintains parasites with complex life cycles in ecosystems.

These effects can be seen over time as well. A decline in trematode species richness at Douglas

Lake, Michigan, is postulated to result from half a century of increasing human disturbance, and an associated reduction in shorebirds (Cort *et al.* 1937; Keas and Blankespoor 1997). Habitat restoration can generate the same pattern, but in reverse. Restoration of degraded salt marsh was followed by an increase in trematode prevalence and species richness so that after 7 years trematode communities at restored sites were comparable to control sites (Huspeni and Lafferty, in press). In addition to providing substantial evidence for the link between biodiversity and parasites, these studies indicate how parasites can be used to monitor changes in the environment over time (Lafferty 1997).

The cessation of hunting and other protections has favoured many marine mammal species. In the United States and elsewhere, regulations such as the Marine Mammal Protection Act of 1972 fully protect pinniped populations and these have soared (Stewart and Yochem 2000). Not surprisingly, the prevalence and intensity of larval anasakid nematodes in fish that use marine mammals as final hosts increased when and where seals became common (Chandra and Khan 1988). The combination of increased susceptibility due to stressors, and increased population density due to marine mammal protection regulations suggests that marine mammals are one group in which host specific diseases will increase.

In contrast, fishing and hunting can reduce populations of targeted species, even to extinction. Reduction in seal populations that are still hunted is expected to reduce the intestinal nematode parasites of seals by reducing host-density thresholds (Des Clers and Wootten 1990). Recent studies show how fishing has dramatically reduced populations of many species across the globe (Jackson *et al.* 2001; Myers and Worm 2003). In depleting a stock, a fishery can 'fish out' a parasite. This is possible if the fishery takes the population below the host density threshold for the parasite and can even be profitable if the host threshold density is higher than the density for Maximum Sustainable Yield (Dobson and May 1987). Fishing out a parasite at a local scale is most probable for host-parasite interactions where the parasite has a recruitment system that is relatively closed compared to the recruitment of its host (Kuris and Lafferty 1992). For example, in the

Alaskan red king crab (*Paralithodes camtschatica*) fishery, nemertean worms can consume nearly all crab eggs in some areas. Nemerteans develop rapidly, are probably transmitted locally and king crab larvae disperse widely. Hence, fishing king crabs intensively (including females) in certain fjords has the potential to extirpate the nemertean locally in those fjords (Kuris *et al.* 1991).

Several examples illustrate the potential to fish out parasites. Fishing reduces the prevalence of the tapeworm, *Triaenophorus crassus*, in whitefish, *Coregonus lavaretus*, (Amundsen and Kristoffersen 1990) and has apparently extirpated a swim bladder nematode from native trout in the Great Lakes (Black 1983). Similarly, the prevalence of a bucephalid trematode in scallops declined from 50–70% (Sanders 1966) to 1–2% (Sanders and Lester 1981) following intensive fishing of scallops and of the final host, the leatherjacket filefish. These examples suggest that parasites of fished species should be declining over time. In contrast, a fishery may be inadvertently managed to increase parasite populations (Lafferty and Kuris 1993). In some cases, as happened with bitter crab disease, fisheries can spread parasites by releasing infected animals because they cannot be marketed (Petrushevski and Schulman 1958). Further, inadvertent management, by targeting unparasitized stocks, can also protect parasites in the unharvested infected stocks. This may be able to sustain parasite populations that might otherwise collapse as host abundance is greatly reduced in efficient fisheries. For example, some fishermen avoid areas where fish have high intensities of sealworm, because this reduces the value of the catch (Young 1972). Fishing practices may unintentionally protect parasites. Crab fisheries often protect reproductive output by releasing trapped females. This protects parasites of females or parasites that feminize males (nemertean worms, rhizocephalan barnacles) (Kuris and Lafferty 1992).

While removal of top predators may break transmission of parasites with complex life cycles it can also have indirect positive effects on some diseases of prey populations (Hochachka and Dhondt 2000; Jackson *et al.* 2001). At the California Channel Islands, lobsters historically kept urchin populations at low levels and kelp forests developed in a community-level trophic cascade (Tegner and Levin 1983). Where lobsters were fished, urchin populations increased and they overgrazed kelps (Lafferty and Kushner 2000). In 1992, an urchin-specific bacterial disease entered the area where urchin densities well exceeded the host-threshold density for epidemics (Lafferty in press). This study found that epidemics were more probable and led to higher mortality in dense urchin populations. Hence, this bacterial disease acted as a density-dependent mortality source. Another example may be the removal of sea otters and Native Americans as black abalone predators on the Channel Islands in the 1800s. This facilitated an increase in black abalone populations to great abundance which then enabled a previously unknown rickettsial disease to cause a catastrophic collapse of the black abalone populations (Lafferty and Kuris 1993). These examples show how fishing top predators can favour disease transmission in prey populations (Hochachka and Dhondt 2000). Indeed, this may be the major cause of increased diseases in marine organisms at lower trophic levels, rather than climate change (Jackson et al. 2001). Predator removal is a management strategy sometimes used to protect livestock or increase wild prey populations of conservation concern or (because they are endangered or hunted for sport) (Packer et al. 2003). Mathematical models find that this practice can inadvertently increase the incidence of parasitic infections, reduce the number of healthy individuals in the prey population and decrease the overall size of the prey population, particularly when the parasite is highly virulent, highly aggregated in the prey, hosts are long-lived, and predators formerly selected infected prey (Packer et al. 2003).

7.4 Pollution

Pollution interacts with parasitism in complex ways, making it difficult to generalize broadly about its effects on disease (Lafferty 1997). This is most clear in reviews of parasites of fishes (MacKenzie et al. 1995). Some pollutants are toxins and these can impair host immune systems and host vital rates. Pollutants may also impair parasite

vital rates and some may even preferentially concentrate in parasite tissues (Sures et al. 1997). However, sometimes parasites have reduced levels of toxicants in their tissues (Bergey et al. 2002). These possibilities lead to a diverse set of predictions about the effect of toxic pollutants on parasites (Overstreet and Howse 1977). However, specific predictions for some parasite–pollution pairs are possible.

Perhaps the best case for a link between toxic pollution and an increase in infectious disease is from parasitic gill ciliates and monogenes of fishes (Khan and Thulin 1991). Intensities and prevalences of ciliates increase with a wide range of pollutants (Lafferty 1997). This appears to be due to an increase in host susceptibility. Toxins somehow impair mucus production which is a fish's main defence against gill parasites (Khan 1990).

Marine mammals have the potential for interactions between pollutants and increased susceptibility to parasites. As top predators, marine mammals bioaccumulate lipophillic toxins that can be broadly pathogenic (O'Shea 1999). These contaminants can affect the mammalian immune system (Swart et al. 1994); for example, harbour seals fed fish from polluted areas have lower killer cell activity, decreased responses to T and B cell mitogens and depressed antibody responses (DeStewart et al. 1996). In seals, such immunosuppression may be a cofactor in the pathology associated with morbillivirus (Van Loveren et al. 2000), Phocine Distemper (Harder et al. 1992), Leptospirosis and calicivirus (Gilmartin et al. 1976). Similarly, marine contaminants may increase sea otter susceptibility to infectious diseases (see Lafferty and Gerber 2002).

Toxic chemicals have a consistent negative effect on helminths (Lafferty 1997). For example, selenium is more toxic to tapeworms than to fish hosts (Riggs et al. 1987). Free-living stages of parasites may be particularly sensitive to toxins (Evans 1982). Trace metals kill free-living trematode cercariae and miracidia, reducing infection rates of snails in polluted waters (Siddall and Clers 1994). This can help otherwise heavily infected snail species compete with other species, greatly altering snail communities (Lefcort et al. 2002). Additionally, if infected hosts are differentially killed by pollution, the parasite population will decline (Guth *et al.* 1977, Stadnichenko *et al.* 1995), further reducing prevalence. For instance, cadmium kills amphipods infected with larval acanthocephalans more readily than it kills uninfected amphipods (Brown and Pascoe 1989). In addition, pollution can negatively affect fish vital rates. For example, oil pollution causes liver disease and reduces reproduction and growth (Johnson 2001). Such effects should reduce density and contact rates, further reducing parasitism.

In contrast to toxic pollution, eutrophication and thermal effluent often raise rates of parasitism in aquatic systems because the associated increased productivity can increase the abundance of intermediate hosts. Parasites that increase under eutrophic conditions tend to be host generalists and have local recruitment; cestodes with short life cycles and trematodes seem to be particularly favoured (Marcogliese 2001). The most dramatic examples include parasites whose intermediate hosts favour enriched habitats. These include some species of tubificid oligochaetes and snails. Myxozoan parasites of fishes, which require oligochaete hosts, are frequently more prevalent at sites polluted by sewage (having high coliform counts) (Marcogliese and Cone 2001). Beer and German (1993) described how eutrophication improved conditions for snails that serve as first intermediate host for the digene, *Trichobilharzia ocellata*. Similarly, Valtonen *et al.* (1997) found that eutrophication correlates positively with greater overall parasite species richness in two fish species. An increase in frog deformities has been linked to eutrophication of ponds which increases the density of snails infected with *Ribeiroia ondatrae*, a trematode known to cause abnormal growth in second intermediate hosts (Johnson *et al.* 2002). The association between eutrophication and pollution is not likely to be linear. At high nutrient inputs, toxic effects may occur and parasitism can decline (Overstreet and Howse 1977). The influence of pollutant stressors, must be analysed in the context of natural history. Some tubificids require clean water and will not be present at enriched sites (Kalavati and Anuradha 1992).

Evaluating the changes in the fish parasitofauna of oligotrophic and eutrophic lakes in Michigan, Esch (1971) recognized that eutrophication opens up the scale of interactions in an aquatic ecosystem. As biomass increases due to increased productivity, birds and mammals increasingly feed at enriched sites. Hence, snails and fishes acquire increasing numbers of larval parasites that will be trophically transmitted to the non-piscine top predators. In oligotrophic lakes, some of these same fishes are the top trophic level and harbour mostly adult parasites. Since larval parasites are more pathogenic than adult parasites there will be a further cascade of disease effects on a eutrophic ecosystem.

Acid precipitation associated with air pollution can negatively effect parasites in waters with poor buffering capacity. Marcogliese and Cone (1996) found that yellow eels (*Anguilla rostrata*) from Nova Scotia have an average of 4 parasite species at buffered sites, about 2.5 parasite species at moderately acidified sites, and 2 parasite species at acidified sites. This decline in parasite richness with acidity is due to drops in the prevalence of monogenes and digenes. The latter require molluscs as intermediate hosts and these cannot survive in acidified conditions. Parasites that use freshwater crustaceans as intermediate hosts may be similarly impacted by reduced access to calcium ions.

7.5 Climate change

The most notable prediction of anthropogenic global change is widespread increases in average temperatures (Houghton *et al.* 1996). This is particularly troubling to most parasitologists from temperate climes because many of the most deadly human parasitic diseases we teach about, but are not at direct personal risk to, are tropical (Rogers and Randolph 2000). The fear is that if our world becomes more tropical, tropical diseases will go hand in hand with the more benign benefits of pleasant weather. This is a bit simplistic; forecasts of climate change do not predict that the weather in Milwaukee will necessarily resemble that in Manaus. Still, there is a general expectation that temperatures will rise and precipitation patterns will change. The distributions of parasites, as for all species, are bounded by suitable climatic conditions. Thus, climate change should alter the future distribution of parasitic disease (Marcogliese 2001).

Some parasites should be more sensitive than others to warming. Temperature would seem particularly important when hosts are ectotherms that do not actively regulate their temperature. In addition, parasites with free living stages should have more opportunity to interact with climatic conditions (Overstreet 1993). For example, trematodes of littorine snails that have free swimming cercarial stages are not able to persist in arctic regions, presumably due to the effect of harsh weather (Galaktionov 1993).

Moderate increases in temperature are likely to alter birth, death, and development rates in ways that could conceivably favour parasites or intermediate hosts. For example, if individuals are infectious for longer time periods under warmer conditions, then disease will increase with temperature. The impact of parasites on their hosts may increase with temperature if parasites are, as a result, able to grow more or mature more rapidly (Chubb 1980). More complicated situations arise in vector-borne diseases where increased temperature may simultaneously increase pathogen development and vector mortality rates (Dye 1992). Much of the research on the effects of temperature on disease concerns fungal pathogens of plants. In general, fungal pathogens induce most damage to their plant hosts at warm (but not too warm) temperatures and at high humidities.

Studies of seasonal variation in parasites provide insight into the effect of temperature. Direct life cycle parasites (such as some monogenes) may be able to reduce generation times in warm water, leading to increases in these parasites (Pojmanska et al. 1980). However, aquatic helminths vary in their optimal temperature (Chubb 1979), making it impossible to make a general prediction about the effect of warming. The cestode Cyathocephalus truncatus has poor establishment success in trout if the water is warmer than 10 °C (Awachie 1966), presumably because host resistance is stronger at warm temperatures (Leong 1975). Other parasites with complex life cycles may be favoured by warming. Trematode cercariae are released from snails only when the water is warm (Chubb 1979), suggesting that the season for completion of trematode life cycles will be prolonged under global warming scenarios, a prediction borne out by observations of parasite communities in a thermal effluent (Sankurathri and Holmes 1976). Nonetheless, it is hard to predict the effect of warming on the parasite community as a whole. In one case where this has been studied, parasite communities in turtles declined with increasing thermal pollution (Esch et al. 1979).

Most fitness traits for hosts and their parasites will exhibit a peak performance at a thermal optimum. If the relationship between performance and temperature differs between host and parasite, the resulting gene by gene by environment interaction will either increase or decrease disease at a given temperature, at least on the level of the individual host (Elliot et al. 2002). For example, the optimal temperature of a fungal pathogen is higher than the optimal temperature of its sea fan host, placing the sea fan at risk to global warming (Alker et al. 2001). But the evidence does not always suggest that warming will increase parasitism. Insect hosts gain several advantages with moderate increases in temperature. Haemocyte production increases and this promotes general defences (Ouedraogo et al. 2003). The ability to encapsulate parasitoid eggs (and presumably other foreign bodies) increases (Blumberg 1991). Pathogenic fungal cells lyse at high temperatures, enabling insect host recovery (Blanford et al. 2003). So, in contrast to the general assumption that parasitism should increase with temperature, there is a general trend for less parasitism at higher temperatures, at least for insect hosts (Thomas and Blanford 2003). Some hosts use this to their advantage by changing their behaviour to increase body temperature in an effort to fight infections (Elliot et al. 2002). This suggests that warming may release some insect pests from their parasitic natural enemies, potentially leading to a variety of economic and ecological impacts. If climate change increases the abundance of insects that transmit diseases, there may be a subsequent increase in the spread of diseases such as malaria (see below).

Precipitation is another aspect of climate that may change with environmental degradation. Increased precipitation should favour parasites (e.g. trematodes) that have an aquatic phase. Outbreaks of water-borne diseases may increase with climate change (Shope 1991), as these are linked to periods of increased rainfall (Curriero *et al.* 2001; Pascual *et al.* 2002). This should also result in increases in parasites that require vectoring by biting arthopods with juvenile aquatic stages (particularly mosquitoes, but also black flies). Despite these direct effects of precipitation, some scenarios do not predict increases in aquatic habitat with increased precipitation because increased temperature may increase evaporation even more than precipitation (Schindler 2001). Although, in some areas, humidity associated with increased precipitation should favour some parasites, especially nematodes transmitted by eggs or with free-living juvenile stages, elsewhere, higher temperatures will dessicate soil (Kattenberg *et al.* 1996). Increased aridity should impair the transmission of parasites with stages that live in soil.

There are important differences in the effect of climate change on aquatic and terrestrial systems. The first obvious difference is that atmospheric humidity is irrelevant in aquatic systems. This means that free-living stages of fully aquatic parasites are less likely to be affected by some aspects of climate change. The second difference has to do with respiration. Because the ability of gas to dissolve in liquid decreases with temperature, warmer water contains less oxygen. This, coupled with the fact that ectotherms have increased metabolic demands at high temperature, suggests that increases in temperature can place aquatic species under respiratory stress. The extent to which hosts or parasites are differentially sensitive to such stress has not been studied to our knowledge but we suspect that hosts, particularly infected hosts, will, on average, be at a greater disadvantage as temperatures rise and less oxygen is available. For example, high temperatures promote rapid reproduction of gill parasites that impair respiration at a time when oxygen is limited (Pojmanska *et al.* 1980). Also, marine snails, infected with larval trematodes, had elevated mortalities under reduced oxygen conditions (Sousa and Gleason 1989). Once again, the ecological consequence of this interaction may be to decrease or eliminate parasites from such populations by increasing parasite mortality.

Global warming could shift ranges of parasites poleward. For example, along the Atlantic coast of the United States, northward expansion of the protozoan *Perkinsus marinus*, which causes Dermo

disease in oysters, is associated with increases in winter water temperatures, greatly expanding the economic impact of this disease (Cook *et al.* 1998). One likely ramification of increased temperature and precipitation is a shift in the distribution, and a probable expansion of the geographic range of mosquitoes and other haematophagous insects that serve as vectors for infectious disease (Shope 1991; Dobson and Carper 1993). The potential for malaria to expand is probably the most feared health consequence of climate change (Patz *et al.* 2000). The present distribution of malaria in tropical areas and reports of increasing outbreaks of malaria (Mouchet and Manguin 1999; Guarda *et al.* 1999; Keystone 2001; Hay *et al.* 2003), in conjunction with concern over warming, has prompted fear that current and future warming will expand malaria's distribution. This hypothesis recognizes that variation in malaria transmission is associated with climate. In Venezuela and Colombia, malaria mortality and morbidity predictably increase following El Niño events (Bouma and Dye 1997; Poveda *et al.* 2001). Modellers have used the associations between climate and mosquito distributions along with predicted patterns of climate change to further predict that the potential for malaria transmission will greatly expand in the future (Martens *et al.* 1999). This concern has attracted widespread public attention. However, other models using multivariate approaches to consider a range of factors find that the distribution of malaria is unlikely to expand as a result of global climate change (Rogers and Randolph 2000). In this regard, recall that malaria was once endemic in relatively temperate areas of the Americas and Europe (Reiter 2000), suggesting that climate, *per se*, is not the best predictor of future malaria distribution (Dye and Reiter 2000). Instead, the abandonment of vector control programmes coupled with the evolution of drug resistance by the parasite and pesticide resistance by the vectors are much more likely reasons for the current and future spread of malaria (Hay *et al.* 2002).

7.6 Introduced species

Humans import animals and plants for pets and agriculture. Many of these are raised near wild species or have escaped to form feral populations. In addition, humans intentionally release species for hunting and fishing, plant them for dune stabilization and use them for biological control. Global trade and travel accidentally introduced many additional species (Ruiz *et al.* 2000). When exotics bring infectious agents with them, they may expose similar native hosts that have no evolved defences to new diseases. Some species have invaded or were introduced without their parasites and are apparently not susceptible to local parasites (Torchin *et al.* 2002, 2003), while others may bring with them a subset of their native parasite fauna (e.g. Lyles and Dobson 1993; Lafferty and Page 1997) (see also Chapter 3). Lafferty and Gerber (2002) recently reviewed published records of infectious diseases of conservation concern. For common native species that were decimated by an epidemic, the source of the disease was usually novel and was first recognized as a pathogen of the species during the epidemic. Sources for these epidemics were usually intentionally introduced species. Most of these diseases have broad host specificity and are less severely pathogenic in their original and abundant (exotic) hosts (McCallum and Dobson 1995; Woodroffe 1999; Gog *et al.* 2002). Relatively low virulence in their coevolved hosts has contributed to poor management decisions concerning the spread of an avian malaria with introduced wild turkeys (Castle and Christensen 1990). Chestnut blight (a fungus introduced with Chinese chestnut trees) is infamous for killing nearly every American chestnut tree. Infectious diseases from domestic sheep have extirpated populations of bighorn sheep (Goodson 1982) and rinderpest (brought to East Africa with cattle) has devastated native ungulates (Dobson 1995*a,b*; see also Chapter 8). A monogene was introduced into the Aral sea along with the Caspian stellate sturgeon; this parasite infected the gills of the native spiny sturgeon, leading to mass mortalities of this naïve host (Dogiel and Lutta 1937). Whirling disease, presumed to have originated with introduced European trout, has spread from stocked trout to native trout in North America, with severe consequences for native populations (Bergersen and Anderson 1997; Gilbert and Granath 2003). Canine

distemper virus (originating from domestic dogs) led to the death of 35% of the lions in the Serengeti (Roelke-Parker *et al.* 1996) and has created problems for several other species at risk (Lafferty and Gerber 2002). Similarly, parapox virus may play a crucial role in the replacement of red squirrels by grey squirrels in Great Britain (Tompkins *et al.* 2003). Perhaps the most tragic example of an introduced vector is the night mosquito in Hawaii which permitted avian malaria to exterminate several malaria-sensitive endemic bird species in the lower altitudes where the mosquito lives (Warner 1968). Finally, an introduced tachinid parasitoid uses abundant exotic gypsy moths as hosts without sufficiently controlling those forest pests. Spillover from the gypsy moth reservoir has led to substantial declines of native North American moths (Boettner *et al.* 2000).

Non-indigenous species are an increasingly common component of estuarine systems (Cohen and Carlton 1998). One of these invaders, the European green crab, *Carcinus maenas*, and its parasites have been well studied. Torchin *et al.* (2001) found that the catch per unit effort of green crabs in their native range (Norway to Gibraltar), decreases with the prevalence of parasitic castrators (rhizocephalan barnacles and entoniscid isopods), supporting the hypothesis that these infectious agents control green crab populations. In addition, samples from introduced regions indicated that parasites are strikingly less common or absent where *C. maenas* is introduced compared to where it is native. Additional analyses indicate that reduced parasitism is a principle reason that green crabs perform better in introduced locations. This is not to say that introduced species remain completely unparasitized. As an example, a native nemertean worm was able to colonize *C. maenas* in California by transferring from the native shore crab, *Hemigrapsus oregonensis* (Torchin *et al.* 1996).

Averaging across several taxa, introduced animals leave an average of 84% of their parasite species behind; in addition, native parasites do not sufficiently colonize introduced species to make up for this release from natural enemies, leaving introduced animals with fewer than half the parasites species they have in their native range (Torchin *et al.* 2003).

The same pattern is true for plant pathogens (Mitchell and Power 2003). Such a release from natural enemies could greatly facilitate subsequent impacts of an introduced species.

Introduced species may indirectly impact native species if they help maintain transmission of native diseases (Daszak *et al.* 2000). On average, about four species of native parasites occur in introduced hosts (Torchin *et al.* 2003) and these, by gaining a wider host base, could increase in prevalence, intensity, and geographic range. This is particularly problematic if the disease has little impact on the invader and a big impact on native species.

7.7 Pollutogens

A distinctive class of infectious agents appears to be increasing in prevalence and ecological impact. We define pollutogens as infective agents that have a source exogenous to the ecosystem, but are able to develop within a host in that ecosystem yet do not require that host for reproduction. Two diseases of California sea otters are good examples of pollutogens; Valley Fever is caused by a fungus that enters the marine environment from eroded soil and Toxoplasmosis is caused by a protozoan that enters the ocean along with faces from domestic cats (see Lafferty and Gerber 2002). Another example under extensive investigation is *Aspergillus sydowii*. This is a terrestrial fungus that has appeared across the Caribbean Sea as a severely pathogenic parasite of several species of sea fans (Garzón-Ferreira and Zea 1992). It is believed to have arrived in the Caribbean from a terrestrial source and that secondary infection occurs only when prevalence is high (Jolles *et al.* 2002). Like other classes of infectious agents, pollutogens have an internal physiological dynamic within their hosts. They may also elicit defensive responses. However, unlike other parasites, they have little or no infectious dynamics within the host population. Hence, neither macroparasite nor microparasite models are relevant (no feedback occurs). Pollutogens have no threshold for transmission, no virulence tradeoff consequences, and no coevolution (the host can evolve resistance, but the pollutogen cannot selectively respond because its reproductive success is very low or nil in those hosts). In a sense,

this new class of emerging infectious disease is an extreme form of spillover from a reservoir host (even if they are not actually or primarily parasitic in their evolved habitat).

7.8 Concluding remarks

Given the diversity of interactions between environmental disturbance and infectious disease, is it possible to generalize about whether these diseases are increasing or decreasing in association with environmental degradation? Recent attention has been given to mass mortalities in marine systems (e.g. Caribbean sea urchins, Lessios 1988), phocine distemper virus (Heide-Jorgensen *et al.* 1992), pilchard mortalities (Jones *et al.* 1997), and infectious coral bleaching (Hoegh-Guldberg 1999). This has led several authors to speculate that disease outbreaks in marine organisms have increased in recent years (Williams Jr. and Bunkley-Willimas 1990; Epstein *et al.* 1998; Harvell *et al.* 1999; Hayes *et al.* 2001). Unfortunately, a lack of baseline data precludes a direct evaluation of this hypothesis.

Ward and Lafferty (2004) developed a proxy method to evaluate a prediction of the increasing disease hypothesis: that the proportion of scientific publications reporting marine disease has increased in recent decades. Reports of parasites and disease, normalized for research effort, have increased in turtles, corals, mammals, sea urchins, and molluscs. There are no significant trends for reports of disease in sea grasses, decapods, and sharks/rays (though disease occurs in these groups). Consistent with the expectation that fishing reduces parasites, disease reports have significantly decreased in teleost fishes. The increase in reports of coral disease is notable, but this is driven by reports of non-infectious coral bleaching, not reports of infectious disease. These latter results are consistent with the general theory that environmental degradation should increase non-infectious and generalist diseases and parasites (Lafferty and Holt 2003). Increasing host populations, such as seen in many marine mammals, should see increases in most types of infectious disease, while decreasing populations, such as recently experienced by many commercially fished species of fin fish, crabs, lobsters, and shrimps, should result in decreased prevalences and intensities, and may even prevent transmission of inefficiently transmitted infectious diseases with high host-threshold densities. So, although environmental degradation is occurring at an alarming rate, an increase in infectious disease is not a necessary outcome of these changes. Some parasites will increase, but we expect that many more will decrease, even to the point of extinction. This may seem a blessing amidst otherwise sobering expectations for the future. However, before we count loss of parasites as something to look forward to, we should consider that parasites play important roles in ecosystems. Fungal pathogens (Gilbert *et al.* 1994) and specialized herbivorous insects (Barone 1998), for example, are thought to be responsible for maintaining the high diversity of tropical forest trees through density dependent mortality of seedlings close to parents (Janzen 1970; Connell 1971, see Chapter 8). Although their roles are generally unseen and little appreciated, the loss of parasites may create more problems for us than it solves.

While the evidence for global warming is strong, its ecological effects are not obvious. We are faced with a difficult confound. Other major factors with strong effects on infectious disease dynamics are changing in temporal concert. These certainly include population increases of humans and some other anthropophillic species, invasive species that are now so pervasive in some regions that a parasite-diminished homogicene has been established; economic pressures reducing or eliminating programmes to decrease transmission of diseases; loss of top predators—mostly long gone from terrestrial systems and now severely depleted in aquatic ecosystems; eutrophication; expanded use of pesticides, antibiotics, anthelminthics; and herbicides in agriculture; and evolution of drug resistance by malaria, tuberculosis, and other important infectious diseases. Interpreting changes over time simply with climate change will hinder comprehension of the interactions between disease and the environment. Analysing these specific effects is now an important task for ecologists, parasitologists, and public health investigators.

Given that there are no simple answers to the questions about how environmental disturbances will affect parasitic diseases, substantial research effort will be needed to unravel the complex linkages between these two forces. Until recently, this has been sparsely supported. The US National Institutes of Health (NIH) has traditionally funded few studies that consider the relationship between environmental degradation and infectious disease because its mission focuses on human health. Ironically, the National Science Foundation, which traditionally funds ecological research, has shied away from issues related to infectious disease (as these are perceived to be within the mission of the NIH). Emerging diseases such as Lyme Disease, West Nile Virus, and SARS have forced health professionals to consider the ecological context of infectious disease in a changing world (Aguirre *et al.* 2002) (see Chapter 10). Now, both agencies are aware that an ecological perspective seems necessary to meet these challenges and have recently combined to fund research through their joint Ecology of Infectious Diseases programme in the context of anthropogenic changes. These new research efforts should considerably expand our understanding of how environmental disturbances interact with infectious diseases.

CHAPTER 8

Parasitism, biodiversity, and conservation

Frédéric Thomas,[1] Michael B. Bonsall,[2] and Andy P. Dobson[3]

Sharing parasites has wide range of implications for the structure and diversity of ecological assemblages. Parasites have the potential to affect ecosystem structure and function by mediating competition between different host species, influencing trophic cascades, manipulating coexistence through life history tradeoffs and impacting on persistence of host species. Parasites impact on the diversity of ecosystems, and the conservation of threatened species. Understanding the ecology of shared parasites is essential for protecting endangered species or ecosystems.

8.1 Introduction

One of the major goals of community ecology and conservation biology is to identify the ecological and evolutionary processes that generate, maintain, and erode biological diversity in ecosystems (Begon *et al.* 1990; Tokeshi 1999). Ecologists have long realized that interactions between organisms largely influence the distribution and abundance of species. However, while competition and predation have been traditionally considered as major biotic determinants of community structure, parasites have been virtually ignored during most of the history of community ecology.

From the pioneering work of Park (1948) showing that one parasite could change the outcome of competition between two beetle species, ecologists now acknowledge the importance of parasites not only on individual hosts, but also on the population

dynamics and community interactions between species. Over the past 15–20 years, considerable progress has been made in understanding the functional importance of parasites in ecosystems. Numerous theoretical and empirical studies have shown that parasites, in spite of their small size, are biologically and ecologically important in ecosystems. As a result, most ecologists and conservation biologists are, for instance, aware that the introduction or elimination of a parasite in an ecosystem can strongly affect the interactions between a diverse range of species in the community, and hence affect biodiversity. However, the link between parasitology and community ecology is still under construction (see Chapter 1). Given the ubiquity of parasites and the large spectrum of their effects, understanding how they affect species assemblages and ecosystem dynamics is a central question in conservation biology. The aim of this chapter is to highlight the different ways in which parasites have been shown, or are suspected, to influence biodiversity in ecosystems by influencing different ecological processes. We begin by considering the effects of species sharing parasites, before examining how parasites might affect the structure of ecosystems through trophic interactions and as ecosystem engineers. We argue that the understanding of life history trait tradeoffs

[1] Génétique et Evolution de Maladies Infectieuses GEMI/UMR CNRS-IRD 2724, Equipe: 'Evolution des Systèmes Symbiotiques', IRD, 911 Avenue Agropolis, B.P. 5045 34032 Montpellier Cedex 1, France.
[2] Department of Biological Sciences, Imperial College London, Silwood Park Campus, Ascot Berks, SL5 7PY, UK.
[3] Department of Ecology and Evolutionary Biology, Eno Hall, Princeton University, Princeton, NJ 08544-1003.

is fundamental to the influence of parasites on species coexistence. We conclude the chapter with a discussion of the diversity and conservation implication of parasites.

8.2 Parasites and apparent competition

8.2.1 Basic ideas and general processes

The interaction between two species in an ecosystem can often be influenced by a third species. It is for instance well known that a predator attacking a highly abundant prey species can facilitate the coexistence of two or more prey species (Paine 1966). Parasites can also play this role of mediator by harming some host species more than others. Parasites indeed do not usually infect host species at statistically similar frequencies, and virulence can differ from one host species to another. As a consequence, the host species whose fitness is impaired by parasitism is at a selective disadvantage in competition with a relatively unaffected species (Fig. 8.1). This influence of parasites is expected to be more pronounced for closely related host species than for distantly related species because the overlap in susceptibility to the parasite community increases with increasing phylogenetic relatedness (Freeland 1983; Holt and Pickering 1985).

Whether some of the host species involved in the competition are totally resistant to infection (cannot

be infected) or tolerant (suffer few fitness reduction from the infection) is of importance to understand the different implications of competition. In the first case, the differential regulation induced by the parasite (effect on survival and/or fecundity) directly gives an advantage to the resistant species in exploitation competition. In the second case, competition also occurs through interference as the tolerant host species negatively interferes with other species either by directly transmitting the disease, or by locally favouring the demography of the parasite.

Interactions between species mediated by parasites through these basic processes is of considerable importance in nature and must even be regarded as one of the major types of interaction in ecological systems, comparable in importance to direct competition and predation (Price *et al.* 1986, 1988). Not surprisingly, in addition to 'apparent competition' (Holt 1977), this phenomenon has also been called 'germ warfare' (Barbehenn 1969), 'biological warfare' (Price 1980), 'weapons of competition' (Holmes 1982), 'agents of interference competition' (Rice and Westoby 1982) and also 'parasitic arbitration' (Combes 1995). Both mathematical models and empirical evidence show that apparent competition mediated by parasites (and/or its cascading effects) clearly has the potential to influence the structure of ecological communities, the viability of each species within a community and the potential for new species to invade a community. Net effects for biodiversity can, however, be positive or negative. Basically, a non-specific parasite infecting related host species in a frequency dependent manner, or a more host specific parasite with a preference for the competitively superior host species can help maintain a high host species diversity (Fig. 8.1). Endemic pathogens and parasites may through this process truly operate as keystone species, playing a crucial role in maintaining the diversity of ecological communities and ecosystems. However, a host-specific parasite can also have a preference for the competitively inferior species. In this case the parasite will be detrimental for biodiversity driving one or several species to local extinction (Combes 1996).

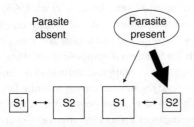

Figure 8.1 Basic process through which parasites are known to influence communities of free-living species. When the parasite is absent, the species S1 is dominated by the species S2. However, because of S2 display a higher susceptibility to the infection or its consequences, the outcome of the competition is changed when the parasite is present. The thickness of lines represents the relative strength of interspecific effects. The size of squares represents the relative abundance of species in the communities.

8.2.2 Parasites, apparent competition, and mathematical models

Williamson (1957) first argued that, in the absence of other limiting factors, the interaction between two species sharing a natural enemy leads to a variety of outcomes and often to the exclusion of one of the species. However, it was a further twenty years before Holt (1977) first formalized the concept that prey that share a common natural enemy can be excluded through these indirect competitive effects (colloquially termed 'apparent competition'). For apparent competition to be manifest, the natural enemy must be primarily food-limited and have a numerical response to each prey species. Formulating this biology leads simply to a set of coupled ordinary differential equations that are straightforward extensions of the Lotka–Volterra predation model. It can be shown that the prey species that wins out in apparent competition can be predicted from the population growth rate of the prey species and the transmission rate of the parasite. In the absence of any other limiting factors, an apparent competition between two prey species that share a natural enemy can be expressed as:

$$\frac{dH_1}{dt} = r_1 H_1 - \alpha H_1 P,$$

$$\frac{dH_2}{dt} = r_2 H_2 - \beta H_2 P,$$

$$\frac{dP}{dt} = \alpha H_1 P + \beta H_2 P - d_p P$$

where r_1 and r_2 are the growth rates of prey 1 and 2 respectively, α and β are the parasite transmission rates on prey species 1 and 2, respectively, and d_p is the mortality rate of the natural enemy. If it is assumed that H_2 is rare, then the invasion criteria for H_2 to invade the interaction where H_1 and the parasite are present is simply $r_2 - \beta P_1^* > 0$ where P_1^* is the equilibrium value of the parasites in the presence of H_1 only and is given by $P_1^* = r_1/\alpha$. Similarly, if H_1 is rare then the invasion criteria is $r_1 - \alpha P_2^* > 0$ where, similarly P_2^* is the equilibrium value of the parasites in the presence of H_1 only and is given by $P_2^* = r_2/\beta$. Coexistence requires that both these invasion criteria be satisfied. However, it can be seen that if the resident host–parasite interaction is at an equilibrium (H^*, P^*) then the invasion criteria can be expressed as $r_1/\alpha > r_2/\beta$ and $r_2/\beta > r_1/\alpha$.

It is clear that these two invasion criteria are mutually exclusive and there is no possibility for mutual coexistence (provided there are demographic differences between the host species). This is known as the P^* rule: the host species expected to win out in apparent competition is the one that can withstand the higher parasite attack and maintain a positive population growth rate (Holt *et al.* 1994). Apparent competition is therefore most likely in interactions where the predator limits host numbers to levels at which they experience little density dependence. As such any increases in the host's carrying capacity is likely to magnify any effects of apparent competition. Further Holt and Pickering (1985) argue that sustained coexistence of alternative hosts limited principally by a shared parasite requires that the within-species transmission of the disease be greater than the between-species disease transmission. If parasites are capable of excluding alternative hosts then it is possible (and predicted from simple host–parasite models) that the outcome of shared parasitism might result in habitat partitioning or segregation of geographical ranges of hosts.

Relaxing the assumption that only parasitism limits host growth rate or that the natural enemy has a positive numerical response weakens the effects of apparent competition and may introduced ecological processes that foster species coexistence. In particular, under certain circumstances apparent competition can give way to apparent mutualism (Abrams *et al.* 1998). Hosts that share a common natural enemy can interact not only through the numerical response, but also through the functional response of a natural enemy. If the functional response saturates at high prey densities (and the natural enemy is limited by factors other than host consumption) then there may be positive indirect effects or apparent mutualisms between hosts. More recently, it has been illustrated that understanding not only the mean response, but also the variance in the response of natural enemies to distributions of hosts can affect the probability of coexistence. In particular, natural enemy responses can be either dependent or independent of host densities. In single pairwise interactions, it is the variability in these two responses that is sufficient to promote persistence. In contrast under apparent competitive interactions, it is only the host density dependent responses that generate sufficient positive covariances in attack or transmission rates between alternative prey that are sufficient to promote coexistence. Under host density independent responses the multiplicative variation in density of the two hosts between patches is insufficient to overcome the interspecific apparent competitive effects generated by the natural enemy, and coexistence is not promoted (Bonsall 2003). Such mathematical models of apparent competition predict a number of different outcomes that reflect the nature of the diversity of shared enemy interactions. It is important to emphasize that changes in prey abundance through the actions of a non-specific parasite can lead to a range of indirect effects that are predicted to enhance or destroy host species diversity. Understanding these effects is the challenge of empirical parasite community ecology.

8.2.3 Parasites and apparent competition: empirical evidence

As mentioned, Park (1948) was the first to show experimentally that parasites could maintain biodiversity through the effects of shared enemies and apparent competition. Park examined the competitive interaction competition between two flour beetles (*Tribolium castaneum* and *T. confusum*). When the two beetle species were kept together in the same containers, *T. castaneum* usually drove *T. confusum* to extinction, suggesting that *T. castaneum* was the superior competitor. However, when the sporozoan parasite *Adelina tribolii* was also present in the mixed cultures, the reverse tendency was observed. In fact, the parasite *A. tribolii* was simply more deleterious to *T. castaneum* than *T. confusum* and its presence shifted the outcome of competitive interactions between the two beetle species. Fifty years later, Yan *et al.* (1998) showed in the same system that when another parasite species was considered (the cestode *Hymenolepis diminuta*), a new scenario occurs. Indeed, the parasite this time benefits the superior competitor *T. castaneum* so that its net effect was to reduce the biodiversity by accelerating the extinction

of *T. confusum*. These studies by Park and Yan clearly illustrate that, depending on which species is the most affected by the infection, parasites can be beneficial or conversely detrimental to species diversity.

Although parasitism, by decreasing survival, may significantly reduce the host population, its effect on the host reproductive potential can also affect the population dynamic interactions between species. Not surprisingly, Jaenike (1995) showed that apparent competition mediated by parasites is also possible when the mechanism of differential regulation parasite-induced affect host fecundity rather than survival. While the castration induced by the nematode *Howardula aoronymphium* is total on *Drosophila putrida*, it is only partial (50%) in *D. falleni*. As in Park's experiment, the outcome of the competition in mixed cultures is inversed in the presence of the parasite (Jaenike 1995).

Indirect methods traditionally employed to detect parasite-induced host mortality (Anderson and Gordon 1982; Rousset *et al.* 1996) can be useful tools to study apparent competition in the field. The two congeneric and syntopic invertebrates *Gammarus insensibilis* and *G. aequicauda* (Crustacea, Amphipoda) can both be infected with metacercariae of the trematode *Microphallus papillorobustus* (Helluy 1981). In the field, however, two distinct infection patterns are observed between the two amphipod species (Thomas *et al.* 1995; Fig. 8.2), suggesting that *M. papillorobustus* can act as an important mechanism regulating the density of *G. insensibilis* populations and *G. aequicauda*. Because the higher reproductive success of *G. insensibilis* (Janssen *et al.* 1979) is offset by its lower tolerance to *M. papillorobustus*, the sympatric coexistence of the two amphipod species is mediated by the parasite (Thomas *et al.* 1995). A similar scenario of parasitism affecting the competitive interaction between species occurs with the association between the trematode *M. clavifomis* and its two possible intermediate hosts *Corophium volutator* and *C. arenarium*. By having more harmful effects on the competitively superior species (*C. volutator*), this trematode apparently permits the inferior competitor to persist and hence, help to maintain biodiversity (Jensen *et al.* 1998).

Parasitoids (insects lay their eggs and in develop in, on or near other arthropods), like parasites have

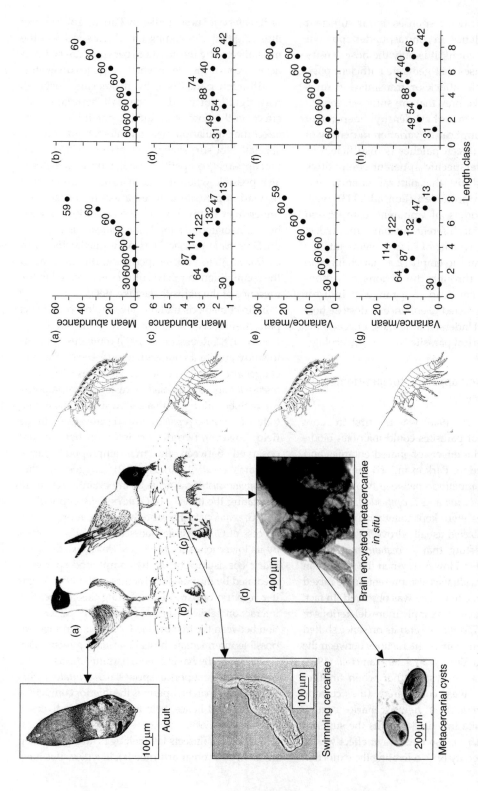

Figure 8.2 Life cycle of the trematode *Microphallus papillorobustus* and infection patterns (mean abundance and the variance to the mean abundance ratio) in relation to host size in its two possible intermediate hosts *Gammarus aequicauda* (a, b, e, f) and *G. insensibilis* (c, d, g, h). Life cycle: Adults *M. papillorobustus* reproduce in an aquatic bird intestine (a). Once a miracidium successfully infects a snail (b), it reproduces asexually to produce cercariae (c). Infected snails are generally castrated by the parasite. Cercariae, once outside the snail, swim, and crawl on the surface of the mud; then cercariae infect a gammarid by penetrating its branchial cuticle. Once inside *G. insensibilis*, the cercaria drops its tail, crawls toward the amphipod's brain and encysts in cerebroid ganglions to form a metacercarial cyst (d).

Notes: Inside this cyst is a miniature adult worm that will be released when the enzymes in a bird's gut digest away the cyst wall. Things are similar when cercariae enter a young *G. aequicauda*. However, for unclear reasons, cercariae infecting an older individual of *G. aequicauda*, encyst in the abdomen instead of the brain. Adult *M. papillorobustus*, like many adult helminths, live only a few days in the bird intestine, where they seem not to be associated with any particular disease.

Source: From Thomas *et al.* (1995). Drawing from Armelle Dragesco.

been shown to intervene in apparent competition processes. For instance, Boulétreau *et al.* (1991) showed that the coexistence between *Drosophila melanogaster* and *D. simulans*, normally impossible (*D. melanogaster* eliminates *D. simulans*) become possible and stable when the parasitoid *Leptopilina boulardi* (Cynipide) is present. A slight preference of the parasitoid for *D. melanogaster* is apparently responsible for this phenomenon.

Although there is considerable anecdotal evidence on the role of shared parasites, very few manipulative experiments have evaluated the population dynamic impact of apparent competition in structuring ecological assemblages. Using a simple laboratory system involving the stored product moth hosts, *Plodia interpunctella* and *Ephestia kuehniella* and their shared parasitoid, *Venturia canescens*, Bonsall and Hassell (1997, 1998) showed how one of the host species is rapidly eliminated by the action of the parasitoid. In the pairwise host–parasitoid interactions, the interaction between each of the moths and the parasitoid (*P. interpunctella–V. canescens*, *E. kuehniella–V. canescens*) were persistence with the populations showing dynamics to a stable equilibrium. However, in the three species interaction (*P. interpunctella–E. kuehniella–V. canescens*) when the only feasible interaction between the two moths was via the numerical response in the parasitoid, one of the hosts (*E. kuehniella*) was rapidly eliminated. The parasitoid is classed as a dynamic monophage (Holt and Lawton 1993). Through its population dynamic interactions with alternative hosts, the parasitoid is monophagous as it drives all but one of its hosts to extinction. More recently it has been shown, theoretically, that the spatial scale and the foraging decisions by *V. canescens* were too marginal to allow the parasitoid to discriminate differences in the heterogeneity of host distributions and promote persistence of the three species interaction (Bonsall 2003).

8.2.4 Parasites, apparent competition, and migration

Given the dynamical consequences of shared parasites, it is entirely feasible that parasites could constitute a major obstacle to migrations and hence

affect the success of biological invasions (Freeland 1983). Ecological theory predicts that a new species will be able to invade an ecosystem only if its susceptibility to local parasites is lower than that of the related resident species. Because in the novel habitat, parasite species (at least some of them) are likely to be new for immigrant individuals, the behavioural and/or immune defences are also likely to be non efficient for avoiding infection and/or limiting its detrimental fitness consequences. Selection for resistance in the invading species may be not possible if the constant arrival of new individuals dilute the gene pool (Combes 1996). Through these processes, parasites undoubtedly protect many ecosystems from alien invasions. Although it seems impossible to know exactly how many invasions failed because of parasites, the protective role of parasites is clearly established. For instance, areas of North America occupied by white-tailed deer *Odacoileus virginiatus* cannot be colonised by other ungulates largely because of the presence of the meningeal nematode *Parelophostrongulus tenuis*. While *O. viginiatus*, the usual host of the parasite, is tolerant to the infection, other ungulate species developed severe neurological disorders when infected (Anderson 1972). As a consequence, repeated attempts to colonize cervids (e.g. caribou, elk, red deer, reindeer) systematically failed in the presence of meningeal worms (Anderson 1972).

The protective role of parasites against invasions can also be manifest through the ethology of resident species. Natural selection has indeed favoured various behaviours aimed at limiting or reducing the risks of infection (Hart 1994). Löehle (1995) suggested that infection risks could explain certain aggressive behaviours expressed by resident species toward migrants. Because such aggression increases the level of environmental adversity for migrants, it increases their probability that colonizations will fail.

Parasites, however, do not always systematically protect ecosystems from alien invasions. Instead, these natural enemies can, under certain circumstances, have dramatic consequences for biodiversity. If infected hosts invade a new area and their parasites become established, these invasive parasites may impact native species if they can recruit to novel hosts (see also Chapter 7). The introduction of avian malaria to the Hawaiian islands remains one of the most famous disaster of this kind. The mosquito *Culex pipiens fatigans* was accidentally introduced into Hawaii in 1826 and transmitted bird malaria from migrating shore birds to indigenous birds. Because indigenous birds were highly susceptible to this new disease, numerous land species which were directly in contact with the mosquito were also driven to extinction (Warner 1968). Another dramatic example of these effects that parasites can have on biodiversity is the introduction of the Caspian sturgeon *Acipenser stellatus* and its gill monogean *Nitzchia sturionis* into the Aral sea during the 1930s. The populations of the local sturgeon *Acipenser nudiventris* were almost decimated by this parasite (Zholdasova 1997).

Parasites can also indirectly influence the impact of an introduced species in a novel environment. Settle and Wilson (1990) reported an interesting case of invasion facilitated by a parasitoid. In 1980, the variegated leafhopper *Erythroneura variabilis* invaded California's San Joaquin Valley. Correlated with this invasion, are the declining populations of the endemic grape leafhopper *E. elegantula*. The parasitoid *Anagrus epos* (Mymaridae) was known to be the principal mortality agent in the endemic species and was also known to parasitize the invading species but at a lower rate. Reasons of this differential susceptibility relies on the way host eggs are laid: while the eggs of the endemic species are inserted just under the epidermal layer of the leaf, those of the invading species are injected deep within the leaf and are consequently less accessible to the parasitoid. During the initial phase of the invasion, the parasitoid reduced populations of the endemic host to low levels, allowing invading populations to advance and increase unencumbered by interspecific competition with the endemic host species. As the invading species became dominant, it also contributed to an increasing proportion of the overall parasitoid population, which in turn accelerated the decline of the endemic species. In this case, the parasitoid did not drive the initial invasion but it did contributed to the final outcome and the relative abundances of the two host species.

8.3 Parasites, ecosystem stability, and cascade effect

When a parasite is responsible for the local decline of one or several species (either through reducing their survival or their fecundity), we expect some consequences not only for competitors, but also for its prey and its predators. If the species affected by parasitism is a predator, we may for instance expect the population density of its prey to suddenly increase at the expense of others with the outcome being a reduction in local diversity. A spectacular example is that of an epizootic of sarcoptic mange (Lindstöm *et al.* 1994) that decimated the populations of red foxes in Scandinavia and strongly altered the demography of the small mammals that are their usual prey. Alternatively, if the species whose survival (or activity) is impaired by parasitism is a prey species, predators may be expected to switch to other prey species that would otherwise not be selected.

More generally, the more a species is functionally important in an ecosystem, the more likely it is that a parasite that impairs this species is responsible for major changes in the ecosystem through cascade effects. For instance, following the introduction of the myxoma virus into England, there was a rapid decline in the rabbit (*Oryctologus cuniculus*) population, which in turn changed the vegetation patterns and both invertebrate and vertebrate populations (Minchella and Scott 1991). Similarly the arrival of Dutch Elm disease in England had a variety of consequences for plant and animals communities. As numerous trees died, the availability of habitats for number of bird species was also decreased. However, at the same time, the increase in the number of beetle larvae in the dead trees resulted in an increase availability of food for other bird species (Osborne 1985). Similarly, when a parasite normally present disappears, the regulatory roles it had on one or several species also disappear (see Section 8.6.4 of this chapter for a detailed example).

All these examples illustrate how the interactions between dominant species in a community may be largely mediated by parasites and pathogens. However, Torchin *et al.* (2002) suggested other subtle ways through which parasites could indirectly influence the impact of an introduced species on the function and persistence of an ecosystem. When exotic species exclude particular native species, the parasitic community linked to that native species may also be excluded (see also Bartoli and Boudouresque 1997). Cascading effects resulting from this phenomenon could in theory be important for a number of ecosystem processes. For instance in several marshes of the West Coast of North America, the introduced Japanese snail *Batillaria attramentaria* competitively excludes the native snail *Cerithidia californica*. Knowing that *C. californica* serves as first intermediate host for at least 18 native trematode species, the local extinction of *C. californica* means the concomitant extinction of all of these trematodes. As trematodes frequently have major effects (fecundity, behaviour, demography, etc.) on the population biology of their second intermediate hosts (numerous molluscs, crustaceans, fishes), their sudden disappearance might have important repercussions on the entire ecosystem functioning and stability. The ecosystem consequences of the loss of a single species (i.e. one snail species excludes another snail species) can potentially be strongly amplified by the interactions of shared parasites.

Parasites, through their debilitating effects on host species, could in theory influence the foraging strategy and local abundance of predators (Lafferty 1992; Thomas and Renaud 2001). Indeed, although predators risk infection when feeding on infected prey, they also often benefit from enhanced prey capture (Lafferty 1992; Norris 1999; Hutchings *et al.* 2000). Parasitized hosts are usually in poorer conditions compared to uninfected conspecifics and are consequently easier to capture. In addition, many trophically transmitted parasites adaptively change the phenotype of their hosts in a way that increases their probability of being captured by predators (definitive hosts), making them for instance more conspicuous or less able to escape (Combes 1991; Lafferty 1999, see also Chapter 9). As such these parasites usually cause little harm to definitive hosts (Lafferty 1992, 1999), predators should show a preference for foraging on such prey (see Lafferty 1992). For example, Aeby (2002) has demonstrated that the coral-feeding butterflyfish *Chaetodon multicinctus*

in Hawaii reefs prefer foraging of polyps (*Porites sp*) that are infected by the trematode *Podocotyloides stenometra*. Infected polyps are easier to capture as they are no longer able to retract into their protective coral skeletons. In addition, because costs of infection are low for *C. multicinctus*, the benefits of feeding on infected coral outweigh the costs associated with parasitic infection. By increasing the accessibility to prey species that are normally difficult to capture, the net effect of manipulative parasites may be to enhance the strength of trophic interactions in ecosystems. Whether there is a positive relationship between the local abundance of manipulative parasites, and the richness/diversity of predators frequenting these habitats is poorly documented but is undoubtedly an interesting question (Hechinger and Lafferty, in press).

8.4 Parasites, host life history traits, and species coexistence

Parasites could in theory play an important ecological and evolutionary role in community ecology beyond the effects of shared parasitism and 'apparent competition' through their influence on host life history traits and evolutionary trait tradeoffs (Thomas *et al.* 2000a). Ecologists readily acknowledge that not only ecosystem traits (e.g. complexity, stability, productivity) are relevant to understand species coexistence but that organismal traits (e.g. body size, dispersal ability, fecundity, timing of reproduction, etc.) are also highly important (Tokeshi 1999). For instance coexistence based on resource partitioning is more likely to occur when species display different life history traits allowing resource specialization. Temporal segregation of reproductive periods within a group of species has often been favoured by selection as it reduces the possibility and the magnitude of resource competition. Different sizes or morphologies allowing the use of different types of resources between closely related species is often a necessary condition for their coexistence. Because parasites frequently alter life-history traits in their hosts, they can, in theory, also have the potential to play an important ecological and evolutionary role in community ecology.

Parasites are responsible for changes in their host life history traits by directly exploiting them and/or by inducing adaptive response from their host. Parasitic exploitation is *per se* an important cause of between-individual or between-population variation in the life history traits such as fecundity, growth, or survival. Changes in host life history traits can also be an adaptive response to parasitism in order to compensate for the negative effects of parasitism on fitness (Minchella 1985; Michalakis and Hochberg 1994). For instance, there are several examples that illustrate that by reproducing earlier when infected by a harmful parasite (e.g. killer or castrator), hosts may partly compensate for the losses due to the parasite (see, for instance, Minchella and Loverde 1981; Hochberg *et al.* 1992; Forbes 1993; Sorci *et al.* 1996; Polak and Starmer 1998; Adamo 1999). Parasites also have the potential to impose selective pressure on other life history traits such as reproductive effort (e.g. Richner and Triplet 1999), dispersal (e.g. Sorci *et al.* 1994; Heeb *et al.* 1999), or growth (e.g. Agnew *et al.* 1999). Under particular circumstances the parental parasite load can even influence the life history traits of offspring (i.e. intergenerational effects) (Sorci *et al.* 1994).

When traits altered by parasites correspond directly, or are related, to traits involved in the coexistence of species, parasites are likely to interfere with community processes. In fact, several situations previously assigned to 'apparent competition' fall within the scope of this idea (e.g. when the traits altered in hosts are fecundity or survival). However, other relevant examples do not correspond to what is classically considered as 'apparent competition'. For instance, when a parasite selects for early investment in reproduction in a given host species (e.g. Lafferty 1993*a*), this parasite also has the potential to alter positively or negatively the magnitude of the temporal segregation during a breeding season. Depending on which species is predominantly affected by the parasite, the resulting competitive interactions may contribute positively or negatively to species coexistence (Thomas *et al.* 2000*a*; Fig. 8.3). Similarly, when temporal segregation between species in a community is maintained by a parasite, the disappearance of such

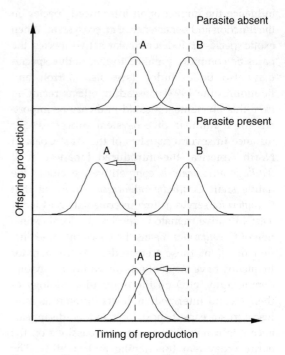

Figure 8.3 Effect of parasites selecting for early reproductive investment in their hosts and consequences for species (A and B) coexistence.

Notes: Top-most graph: parasite absent, coexistence between the two species; Middle graph: parasite present, selective pressure on species A, coexistence is favoured; bottom graph: selective pressure on species B, coexistence is compromised.
Source: From Thomas *et al.* (2000a).

parasites from the ecosystem could result in an increase of the magnitude of the competition between coexisting species.

Dispersal has often been identified as an important factor that influences the genetic diversity and structure of populations, as well as the probabilities of regional extinction/colonization. Parasites can influence dispersal in their hosts in different ways. First, given that a classical consequence of infection is a reduced activity, parasites probably often impair the dispersal potential of their host (see, for instance, McNeil *et al.* 1994; Thomas *et al.* 1999*b*). At a metapopulation level, gene flow is likely to be influenced by the mean infection rate of the different populations. Parasites that influence dispersal ability and behaviour undoubtedly control, at least partially, the level of geographical isolation of populations, and over evolutionary time, could potentially influence taxonomic diversification. In particular situations, parasites have been conversely shown to favour dispersal in host species. A high risk of infection in a given habitat may indeed select for increased dispersal to avoid future infection (Sorci *et al.* 1994). In such a case, parasites would favour gene flow between host populations.

The net effect for diversity of the selective pressures exerted by parasites on host life history traits is far from fully understood at the moment. The examples and ideas presented here however suggest that this area deserves to be examined more carefully in the future, from both ecological and evolutionary perspectives.

8.5 Parasites and ecosystem engineering

Ecosystem engineers are organisms, plants, or animals that directly or indirectly modulate the availability of resources to other species, by causing physical state changes in biotic or abiotic materials (Jones *et al.* 1994, 1997). Certain species change the environment via their own physical structures (autogenic engineers, for example, trees), while other transform living or non-living materials from one physical state to another (allogenic engineers, for example, beavers). Both kinds of engineers modify, maintain or create habitats within ecosystems (Jones *et al.* 1994). Acts of engineering from free-living species result from traits associated with their phenotype (e.g. morphology, behaviour). Parasites can interfere with these processes as they alter the phenotype of their host. In fact, parasites can either impact on existing ecosystem engineers, or act as engineers themselves (Thomas *et al.* 1999a). These different cases of interactions between host traits that are altered by parasites and those that are involved in engineering acts are illustrated in Fig. 8.4.

8.5.1 Infection of ecosystem engineers

Infection of ecosystem engineers does not necessarily have community consequences because host traits modified by parasites may have no link with engineering functions (Fig. 8.4(a)). In other situations, parasites alter host traits that are directly

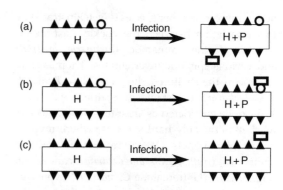

▲ : Phenotypic trait having no engineering function;
O : Phenotypic trait having an engineering function;
▢ : Trait altered by parasites.

Figure 8.4 Interactions between traits altered by parasites and traits involved in engineering processes.
Source: From Thomas *et al.* (1999a).

related to engineering functions present in the host (Fig. 8.4(b)). By altering key phenotypes in ecosystem engineers, parasites interfere with their engineering function and hence on community processes. There are numerous examples of parasites interfering with engineering function. For instance, numerous free-living species are autogenic engineers because they directly provide through their growth space for other organisms. Any parasites that reduce or enhance the growth of these hosts will affect the trait (host size) involved in the engineering process. For instance, gammarids *G. insensibilis*, infected with the trematode *M. papillorobustus* (Fig. 8.2) have a longer intermoult duration than uninfected ones, and usually harbour higher numbers of epibionts on their cuticle (F. Thomas, unpublished data). Furthermore, numerous acts of engineering from free-living species result from their activity level (e.g. woodpeckers and beavers, Jones *et al.* 1997). As activity is often reduced in parasitized hosts, parasites are likely through this simple effect to interfere with engineering processes. We could for instance imagine that gastrointestinal nematodes which frequently alter the appetite of herbivores (e.g. Arneberg *et al.* 1996) will have an (indirect) impact on plant communities.

The idea that by shifting the phenotype of their hosts from one state to another, parasites could create new resources for other species is well illustrated by the association between the cockle *Austrovenus stutchburyi*, the trematode *Curtuteria australis* and various epibiotic invertebrates (Thomas *et al.* 1998*a*). In the sheltered shores of New Zealand, *A. stutchburyi* lives just under the surface of the mud and can be considered as an autogenic engineer as its shell is the only hard substrate where invertebrates like limpets (*Notoacmae helmsi*) and sea anemones (*Anthopleura aureoradiata*) can attach (Fig. 8.5(a)). The trematode *C. australis* favours its transmission to oystercatchers by altering the behaviour of cockles, making it unable to burrow under the surface of the mud like unparasitized conspecifics. Given the important ecological differences (e.g. humidity, temperature, light, etc.) between living under or above the surface of the mud, it is realistic to consider that manipulated cockles could correspond to a new kind of substrate for invertebrates. Limpets which are normally outcompeted

for space on burrowed cockles by sea anemones, significantly prefer surface cockles (Thomas *et al.* 1998*a*) (Fig. 8.5(b)). Infected cockles not only provide an exposed surface suitable for grazing, they are also less occupied by anemones because of their lower resistance to desiccation at low tide. Using the terminology of Jones *et al.* (1994), the trematode *C. australis* is an allogenic engineer as it turns living material (the cockle) from one physical state (buried) into a second physical state (surface). This act of engineering alters both the availability and the quality of habitats for invertebrates. The net effect here is to reduce competition for space between invertebrates and because of this, the parasite probably facilitates the local coexistence of limpets and anemones. Many trematodes have been shown to impair the burrowing ability of their molluscan hosts (see Lauckner 1987). It is not known, however, whether similar behavioural changes induced by parasites in comparable ecological contexts yield the same ecological consequences. Beyond this specific study, such questions remain essential before generalizations can be made.

8.5.2 Parasites as ecosystem engineers

By altering particular traits of their hosts, parasites can be responsible for habitat creation (i.e. they give rise to an engineering function that did not previously exist) (Fig. 8.4 (c)). For instance, in the association between the crab *Carcinus maenas* and its crustacean parasite *Sacculina carcini* (Rhizocephala), the parasite usually stops the moulting process of its host. Because of this important physiological change, the carapace of crabs becomes a more permanent substrate for fouling organisms (serpulid polychaetes, barnacles, etc.) compared to that of non-infected crabs that moult regularly. More generally, because free-living organisms constitute themselves as an ecosystem for numerous parasite species, any parasite modifying the host is potentially an ecosystem engineer with a functionally important role in the parasite community.

Engineering processes can typically have both positive and negative effects on species richness (Jones *et al.* 1997). This is nicely illustrated in the association between the crustacean gammarid *G. insensibilis*, the

(a)

(b)

Infected cockle

Uninfected cockle

Figure 8.5 The cockle *Austrovenus stutchburyi* with the two most common invertebrate species living on its shell, the limpet *Notoacmea helmsi* and the anemone *Anthopleura aureoradiata* (a)—illustration of the effect of trematode infection on the fouling community (b).

trematodes *M. papillorobustus, Maritrema subdolum,* and the nematode *Gammarinema gammari.* As seen before, the trematode *M. papillorobustus* increases the vulnerability of gammarids to predation by aquatic birds (definitive hosts of the parasite) by strongly altering their behaviour (positive phototactism, negative geotactism, and an aberrant escape behaviour, Fig. 8.2). As this change is major, it is realistic to consider that this parasite turns gammarids from a phenotype A (normal behaviour) to a phenotype B (altered behaviour). As these new properties of manipulated gammarids affect infection risk, other parasites can have common or conversely conflicting interests in exploiting phenotype A or phenotype B gammarids. The second trematode species *M. subdolum* which also finishes its life cycle in aquatic birds prefers to infect phenotype B gammarids (Thomas *et al.* 1997). The cercariae (infective stage) of *M. subdolum* actively swim in the water column where they are more likely to encounter phenotype B gammarids (Thomas *et al.* 1997). If shared interests exist between the manipulator (*M. papillorobustus*) and this 'hitch-hiker' parasite (*M. subdolum*), there is conversely a clear conflict of interest between the manipulator *M. papillorobustus* and the nematode *G. gammari.* Indeed, the nematode uses amphipods as a habitat and source of nutrition, not as a 'vehicle' to be transmitted to birds. As expected, there is evidence that the nematode prefers phenotype A gammarids (Thomas *et al.* 2002). There is also the suggestion that the nematode 'sabotages' the manipulation exerted by *M. papillorobustus,* turning back gammarids from a phenotype B to a phenotype A (Thomas *et al.* 2002).

A final example serves to illustrate that, in natural conditions, the effects of parasites as ecological engineers leads to a complex set of effects as several processes can occur simultaneously. Mouritsen and Poulin (2002) reported an example of dramatic change in an intertidal community resulting from the effects of shared parasitism and host mortality, ecosystem engineering and cascade effects. Due to trematode infections, the amphipod *Corophium volutator* disappeared from Danish mud flats. This crustacean, because of its tube-building activity, played a major role in stabilizing the substrate. The local extinction of this allogenic engineers allowed rapid sediment erosion and important changes in the particle size composition of the mud flat, which in turn induced major changes in the diversity of infaunal invertebrates and finally the macrofaunal species.

Although there are a range of systems in which either common interests or conflicting interests exist between parasite species sharing the same invertebrate host population, studies on the community ecology consequences of this phenomenon remain relatively rare (see also Poulin *et al.* 1998, Lafferty *et al.* 2000, Outreman *et al.* 2002). Parasites can clearly have important effects on communities through their roles in engineering process. Knowing that most, if not all, free-living species organisms harbour parasites and knowing that a variety of host traits are altered by parasites, this new area of research at the junction of ecology and parasitology appears promising. Clearly, further empirical and theoretical studies on this subject are needed.

8.6 Parasite diversity and conservation biology

8.6.1 The abundance of parasites

The diversity and abundance of parasites is remarkable: for example, at least 2000 species of nematodes have been named but there may well exist 10 or 100 times this many species of nematodes, most of which still have yet to be classified (Poulin 1996*b*). Evidence from the phylogeny of nematodes illustrates that parasitism is something that has evolved many times even within this large order of organisms (Anderson 1988; Adamson and Caira 1995). Even within the major families of nematodes, we find that parasitism has evolved multiple times. Parasitism is a common way of life among other phyla of worms: all species in the trematodes, monogeneans, cestodes, and acanthocephala are parasitic. With the exception of the monogeneans, all species in these taxa have complex life cycles in which two different hosts species are parasitized sequentially. Parasitism is also a very common life form in many of the protozoans, the viruses, bacteria, and fungi. Peter Price was one of the first people to make a conservative estimate of the proportion of all known species that are parasitic; by including sucking arthropods (an important analogy which we will return to later), he concluded that around 50% of species are parasitic (Price 1980). This work was more vigorously formalized by Cathy Toft, who formally counted numbers of parasitic and non-parasitic species in all known phyla (Toft 1986). This estimate also suggests that as many as 50% of species are parasitic. Some of the major advantages of a parasitic mode of life are emerging from life history studies of parasites (Calow 1979; Skorping *et al.* 1991). Comparative studies of nematodes illustrate that the normal rules of allometric scaling and assimilation efficiency have to be reconsidered when we consider parasites (Skorping *et al.* 1991; Arneberg *et al.* 1998*b*; Morand and Sorci 1998). In contrast to most free-living animals, selection for increased body size in parasitic nematodes leads to both increased longevity and fecundity. This creates the potential for parasites to become 'Darwinian demons'.

We can obtain some estimate of the ubiquity of parasites by trying to estimate the number of parasitic worms associated with any population of free-living vertebrates. If we briefly survey published data for 50 different species of North American mammals we find that on average a mammal will contain 400 worms from four different taxa, trematodes, cestodes, nematodes, and acanthocephalans. At the population level there are on average 10 species of parasitic worms associated with any one population of vertebrates (Dobson *et al.* 1992*b*). This parasitic diversity is roughly evenly distributed among the trematodes, cestodes, and nematodes, with about 100–150 individuals of each, divided between two species of trematodes and cestodes, and often five species of nematodes. Acanthocephalans are less ubiquitous, but at least one species is associated with every other host population. This suggests that most free-living vertebrates will be parasitized by at least one parasitic worm species during the course of their life, many will harbour a community of parasitic organisms.

More detail can be added to this picture by looking at some data put together by Clive Kennedy, Al Bush, and John Aho (Kennedy *et al.* 1986*b*), who compared the species richness of helminth communities in fish, birds, and mammals. Their work suggests that

fish have on average one or two species of parasitic helminth in them, birds may have somewhere from 2 to 10 species of parasitic helminths, and mammals may have somewhere between 8 and 19 species of parasitic helminth. If we look at the actual numbers, then fish will contain anywhere between 2 and 100 worms per host, birds may contain anywhere between 20 and 1000 worms per host, and mammals may have anywhere between 10 and 50,000 worms per host. This implies that most populations of vertebrates, are effectively operating as a metapopulation of patches that serve as resources for parasitic helminths, a theme we will explore further below.

8.6.2 Host–parasite models as metapopulation models

Models for parasite–host communities are related to those for metapopulations of organisms in patchy environments. Indeed the simplest host–parasite model we could write would have a host population of constant size divided into susceptible and infected hosts. This model is identical to the classic Levin's population model (Levins and Culver 1971). The classic (SIR susceptible, infected, and recovered) epidemiological models (Kermack and McKendrick 1927) are a simple extension of this that include an extra category of patch, those that were previously occupied, but are now resistant to further occupation. The other major modification of host–parasite models is that they consider the vital dynamics of the hosts that are patches of habitat for the parasites and the impact that the presence of the pathogen has on these birth and death rates. The models used to describe parasitic helminthes add considerably more detail in that they consider, not whether a patch is occupied, but the numbers of individuals of each species occupying each patch, these are usually described by a series of frequency distributions. It is possible to compare the structure of a range of different host–pathogen models with those for metapopulation models. This suggests that all these models describe a spectrum of complexity and detail that starts out with the basic models first described by Levins and Culver (1971), and then passing through different types of models for host–parasite systems (Dobson 2003).

It is important to notice here that a large number of 'free-living' organisms will have population structure and dynamics that correspond to one of the frameworks previously described. For example the fish and invertebrates that occupy different parts of a coral can be considered as organisms that use patches of coral as habitat. Like parasitic helminths they may reduce the fitness of the coral, but their fitness will in turn also be dependent upon the dynamics of the coral and the ability of their offspring to locate a new patch of coral to colonize. Many insects that feed on plants, or even carrion, will also have a similar population dynamic structure (Ives and May 1985; Anderson 1989). The factors that determine community composition and relative abundance in these communities will be similar to those in the host–pathogen models: the vital dynamics of the plant hosts, the statistical distribution of each insect species across the plant population, and the birth, death, and dispersal rates of the different insect species.

8.6.3 Fungal pathogens and forest diversity

If parasitism is so ubiquitous in the natural world, and parasites have the potential to regulate or alter the dynamics of their hosts, then this implies they may play subtle and important roles in food webs (see also Chapter 4) and may even make significant contributions to indirect competitive interactions between species (see Section 8.2 of this chapter). Work by Augspurger's on the role of fungal pathogens in tropical forests illustrated that fungal pathogens have the potential to create spatially local patterns of frequency dependent recruitment (Augspurger 1983, 1984a). Janzen and Connell had posited that a mechanism of this form might be important in mediating the coexistence of different tree species in tropical forests (Janzen 1970; Connell 1978). Augsburger's work showed that rates of seedling death were significantly higher in the vicinity of the parent tree than they were at a distance removed from the tree (Augspurger 1984b). The majority of the deaths occurred in shaded regions where 'damping-off' diseases caused by fungal pathogens were the main source of mortality. Recent work on black cherry has illustrated that

similar effects may be occurring in temperate forests (Packer and Clay 2000).

The spatial dynamics of these systems can readily be modelled in one dimension. First assume that the seeds produced from a tree decline exponentially in abundance with increasing distance from the tree. When these tree seeds germinate their mortality is determined by whether they are infected with a fungal pathogen. Their probability of acquiring the pathogen is a simple function of the density of infected individuals in their immediate vicinity. If we assume that we can divide the area around the parent tree into an array of 'patches' within which pathogen transmission occurs, then we can describe the dynamics of the fungal pathogen in each patch by the magnitude of its basic reproductive number, R_0.

$$R_0 = \frac{\beta S}{(\mu + \alpha)}.$$

Here S is the density of seeds in a patch, β is the transmission rate of the pathogen, μ is the intrinsic rate of seedling mortality, and α is the additional mortality due to infection with the fungal pathogen. Within any patch the proportion of seeds that survive infection (SS) is approximately given by the following expression

$$SS = e^{-R_0}.$$

It is then relatively trivial to determine the pattern of recruitment of surviving seedlings at different distances from the parent. This mechanism produces a classic 'Janzen–Connell' recruitment curve. The high abundance of seedlings close to the tree produces high levels of disease that minimize their chances of recruiting. Seedlings are only able to survive at distances where seed abundance has fallen to sufficiently low levels that the pathogen dies out. Providing the fungal pathogen is specific to one host species then individuals of other species can recruit into these areas. It is important to notice that we are again dealing with a threshold phenomena, recruitment of seedlings can only occur at densities where the basic reproduction rate of the pathogen is less than unity. If seed production varies from year to year, then seedlings will tend to recruit closer to the parent tree in years of low seed abundance, and further from the tree in years when seed crop is high. Here, it is interesting to speculate about how the recruitment patterns of tropical trees might be affected by changes in the abundance of species that feed on seeds. In areas where these species have been lost due to hunting, then seed abundance on the ground will increase leading to recruitment at greater distances from the parent tree. Alternatively, if humans harvest seeds at high levels, then seedling recruitment will occur closer to each tree and the potential for other trees species to establish will decline.

One further speculative idea concerns the role that mycorrhizal fungi may play in controlling plant pathogens. Although most studies emphasize the role that mycorrhizae play in helping plants to assimilate nitrogen, there is also some evidence that they may also help suppress pathogens (Marx 1972; Allen 1991). If mycorrhizae evolved this ability in situations where pathogens caused a Janzen–Connel effect, then it should be possible for a single species of tree to dominate a community. As this would also lead to a tree's fitness being more directly associated with its net rate of seed production, then it may also allow large-scale mass seeding events such as masting to evolve. Plainly there is much still to be explored in this area.

There is also increasing evidence that parasitic nematodes play an important role in plant communities. Recent work by Van der Putten's succession in dune grass systems has shown that plant specific nematodes play an important role in reducing the fitness of different successional species (Van der Putten et al. 1993; Van der Putten and Van der Stoel 1998). A particularly nice twist to this tale is that species that enter the succession at a later stage are resistant to the nematode pathogens of species that have become more abundant at earlier stages of the succession. The models described above could be readily modified to examine this effect in more detail.

8.6.4 Rinderpest in East Africa

There have been relatively few studies of the impact that pathogens have at the ecosystem level. One enticing example comes from studies of rinderpest virus in East Africa (Sinclair 1979b; Plowright 1982; Dobson 1995a,b). Rinderpest is a morbillivirus that is closely related to canine distemper and human measles (Plowright 1968). The evolutionary split between the three pathogens occurred between 3000 and 5000 years ago, most likely as a consequence of the domestication of cattle and dogs (Norrby et al. 1985). Rinderpest has caused cattle plagues in India and Europe since at least the dark ages, there are no records of its presence in sub-Saharan Africa until the great pandemic of 1890–98. This was initiated when infected cattle were accidentally shipped into Africa during the Italian Mesopotanian military campaign (Plowright 1982). The virus quickly established in the local cattle population and spread from here into the wild artiodactyls. Historic records suggest that 80% of the populations of some wild ungulates perished in the pandemic that took 10 years to spread from the Horn to the Cape of Africa. Throughout the first half of the last century the presence of rinderpest prevented the development of a viable cattle industry in Africa and continued to cause epidemics in both wild species and domestic herds (Simon 1962; Branagan and Hammond 1965). The pioneering Serengeti wildebeest studies of Lee and Martha Talbot found that more than 40% of wildebeest calves died of rinderpest during their first months of life (Talbot and Talbot 1963), those calves that survived were then immune from further infection. This situation is directly analogous to measles in malnourished human populations.

A viable rinderpest vaccine was developed in the late 1950s and it was hoped that this would at least allow the cattle population to be protected from outbreaks that were maintained in the wildlife reservoir. Wide-scale vaccination of cattle produced a remarkable response in the wildlife populations of East Africa (Plowright and McCullough 1967; Plowright and Taylor 1967). Although only cattle were vaccinated the disease disappeared from the wild species, implying that cattle were in fact the main reservoir! The removal of rinderpest led to a massive eruption of both wildebeest and Cape buffalo populations, the two wild species that were most susceptible to the virus. In the Serengeti, wildebeest numbers increased from around

200,000 to 1.5 million (Sinclair 1979*b*). Buffalo populations increased by a factor of four to five and their range expanded into areas such as Ngorongoro crater, where no previous knowledge of their presence existed (Fosbrooke 1972).

Increases in the abundance of ungulate species led to an increase in the density of carnivores, particularly lions and hyenas. These increases in abundance were matched by decreases in the abundance of gazelles, most likely due to increased predation pressure (Dublin *et al.* 1990; McNaughton 1992). The most dramatic declines occurred in the wild dog population whose numbers declined from around 500 to eventual local extinction (Burrows *et al.* 1994). While competition with hyenas is the most likely mechanism producing this effect (Creel and Creel 1996), it is interesting to speculate upon the role that disease may have played. A significant crash in the wild dogs population was caused by a distemper outbreak. Plowright recorded that when he developed the rinderpest vaccine in Nairobi they disposed of dead cattle by supplying them to the local dog owners (Plowright 1968). Distemper effectively disappeared from the domestic dog population at this time! This implies that exposure to rinderpest in infected carcasses may cause cross-immunity to rinderpest in canids. It may be that loss of rinderpest from wildebeest increased the susceptibility of wild dogs to distemper (Dobson and Hudson 1986). Further evidence in support of this comes from the recent outbreak of distemper in the Serengeti lion population (Roelke-Parker *et al.* 1996). The outbreak was initiated by an epidemic in the domestic dog population in the lands that surrounds the Serengeti, it caused the lion population to decline by almost a third before the virus died out. It has led to a large-scale programme to vaccinate the domestic dogs around the park in order to prevent a future outbreak.

All of this suggests that rinderpest can act as a keystone virus in the Serengeti ecosystem. Although there was never more than a kilogram of rinderpest in the entire ecosystem, it had a major direct, or indirect impact on most of the large vertebrates (McNaughton 1992; Dobson and Crawley 1994). Furthermore, the changes in the numbers of artiodactyls would certainly have had an impact on grass that is the primary food of these species. There is even evidence to suggest that the decline in browsers, particularly impala during the first pandemic, may have allowed a large recruitment pulse in many tree species (Prins and Weyerhaeuser 1987; Prins and van der Jeugd 1993); the acacia stands in many parts of the ecosystem are remarkably even in their age and size distribution.

8.7 Concluding remarks

In conclusion, it is hard to imagine what natural communities and ecosystems would be like without shared parasites. Even if diseases and parasites have been undoubtedly responsible for a number of extinction (especially on islands), they also play in other situations a crucial role in maintaining biodiversity. How should conservationists manage parasites and pathogens? Although wildlife managers are increasingly aware of the roles played by parasites in ecosystems, at the moment we are far from an answer to this question. In particular situations, concrete actions have been attempted for instance in order to minimize disease risks in endangered species through reducing the population size of a reservoir population or through vaccination programmes (see Cleaveland *et al.* 2002 for a review). Some attention has also been devoted to the problem of disease risks associated to wildlife translocations (Cunningham 1996). These examples provided useful information but they remain sparse. In addition, generalizations are difficult to make at the moment because despite broad similarities, a close look at each situation often shows that the ecological or political context differ substantially. Finally, the problem of protecting endangered species from infection is only a specific and an extreme case of conservation problem. The challenge for parasitologists and conservationists is also to understand the ecological roles played by parasites, so as to understand which ecosystem functions would be lost if certain parasites would disappear. Sooner or later, such kind of research should logically yield to the conclusion that some parasites might need conserving!

CHAPTER 9

Subverting hosts and diverting ecosystems: an evolutionary modelling perspective

Sam P. Brown,[1] **Jean-Baptiste André,**[2] **Jean-Baptiste Ferdy,**[2] **and Bernard Godelle**[2]

Through a focus on the behavioural and physiological interactions among parasites and their hosts, we consider how within-host and among-host ecosystems are shaped by parasite strategies. Furthermore, we consider how these resultant ecosystem structures can feed back on parasite strategies via natural selection. We link our discussion to a diverse range of empirical systems and theoretical approaches.

9.1 Introduction

Most organisms are parasites. If an organism is not a parasite, then it harbours parasites, as do many parasites themselves. To make such broad claims, it is necessary to take an inclusive view of the concept of parasitism. In a hangover from the taxonomic focus of much nineteenth-century zoology, the term 'parasite' has come to be associated with the protozoan and metazoan occupants of free-living animal hosts, with a particular focus on helminth worms (e.g. Cox 1993; Wakelin 1996; Poulin 1998*a*). In the light of this continued taxonomic bias, it is of interest to note that the word parasite stems from the Greek *parasitos*, literally 'to eat at another's table' (*Oxford English Dictionary*, 2nd edition). This definition has several important merits: first, it views parasitism as a relationship, rather than as a taxonomic category. Second, it illustrates that this relationship takes place in a social context. Furthermore, it suggests that parasitism can occur within—as well as between—species.

The significance of the social context of parasitism is easily overlooked, yet is central to its origins and continued existence. Cooperative synergies among distinct biological individuals underlie all the major increases in complexity in evolution (e.g. genes to cells, protozoans to metazoans, individuals to societies; Maynard-Smith and Szathmáry 1995; Keller 1999). Yet simultaneously these cooperatively driven increases in complexity create increased opportunities for parasites, or cheats. The image of 'cheats' provides an interesting working definition of parasites—organisms that give less than they take in interactions with other organisms.

These social interactions with other organisms—both hosts and other parasites—can have great impacts on the ecosystems that these organisms comprise. Parasites divert some of the energy that flows through ecosystems. We shall consider in this chapter that the body of an individual host is the first ecosystem parasites live in, and that this host is itself part of a larger ecosystem. At the first level, parasites will distort or even completely hijack the body of their host (by manipulating host physiology, behaviour, life-history traits, etc. see Odling-Smee *et al.* 2003 'niche construction') in order to

[1] Department of Zoology, University of Cambridge, Downing Street, Cambridge CB2 3EJ.

[2] Laboratoire Génome, Populations, Interactions, Adaptation, UMR 5171, USTL-CC 105, Bât. 24, Place Eugène Bataillon, 34095 Montpellier Cedex 5, France.

fulfil their own ends. At the second level, parasite manipulation of the host physiology can have profound consequences on ecosystem structure at large. We shall then survey the conditions promoting adaptive manipulations of this global ecosystem. Finally, at this larger scale, the community of parasites constitute an ecosystem, composed of distinct ecological niches, and within which competition occurs. We shall then survey how the existence of distinct niches for parasites at this scale might alter the evolution of transmission modes and virulence in parasites.

Studies at these three scales are mostly theoretical. Giving an exhaustive review of the mathematical techniques these surveys require is beyond the scope of this chapter. The interested reader should find useful references within the text; boxes should provide some clues to understand important ways of approaching parasitism with an evolutionary–ecological approach.

9.2 Parasite manipulation of host phenotype

The first ecosystem that is transformed by parasites is the body of their host. The most illustrative examples of such effects are the so-called 'manipulators', exemplified by dramatic modifications of host behaviour (see below). However, in this chapter we wish to give a wider definition of the manipulation concept to include all host–parasite systems where the parasite adaptively brings about physiological, morphological, or behavioural changes in the host to promote its own fitness, either through an increase in transmission rate and/or persistence time (Brown 1999). Table 9.1 summarizes with examples the scope of our manipulation concept; the various effects are classified according to two poles. The first corresponds to the cases where the parasites control the overall properties of their infection (the infection being the product of the interaction host × parasite): the parasites manage the infection's physiology in order to turn the host's body into the best parasite transmitter. The second pole corresponds to the more spectacular cases where the parasites affect some traits specific to the host (behaviour, life history traits), these last cases being traditionally

Table 9.1 A broad classification of host manipulation: from management of infection physiology toward manipulation of host phenotype

Management of infection physiology
1. Resource intake from host tissues and dispersal (e.g. via tissue lesions caused by microparasite secretions; Robson *et al.* 1997; Williams *et al.* 2000).
2. Immune escape by antigenic variation (genetic or plastic; Moxon *et al.* 1994).
3. Immune escape by suppression of host immunity (e.g. virus involved in chemokine mimicry; Murphy 2001).
4. Resource defence against competitors by concomitant immunity; Brown and Grenfell 2001; Brown *et al.* in prep.)
5. Manipulation of host life history traits (size, reproduction; references in Kuris 1997).
6. Manipulation of the host's relationship with other organisms (e.g. increased susceptibility to predation through behavioural alterations caused by macroparasites in intermediate hosts; references in Poulin 1994a, 2000; Thomas *et al.* 1998a)
Manipulation of host phenotype

referred to as 'host manipulation'. Of course there is no clear-cut separation between these two poles but only a continuum of manipulations; for convenience we however separate their descriptions below into two subsections.

9.2.1 Parasite administration of infection physiology

In simple epidemiological terms, the adaptive rationale for manipulation is to increase either the transmissibility or the expected length of infection: the aim of parasites being to transform the body of their host into the best parasite transmitter (see Box 9.1). In microparasites (parasites that replicate for several generations inside their hosts and attain large population sizes, that is, bacteria, viruses, and protozoa, see Anderson and May 1979), the mechanism of this 'transformation' implies the replication, mutation, and death of a large number of microscopic entities: the individual microparasites and immune cells. The management of host physiology is therefore a matter of rates and numbers, which is widely open to mathematical modelling. In macroparasites, this transformation implies different mechanisms, but the same questions are nevertheless raised.

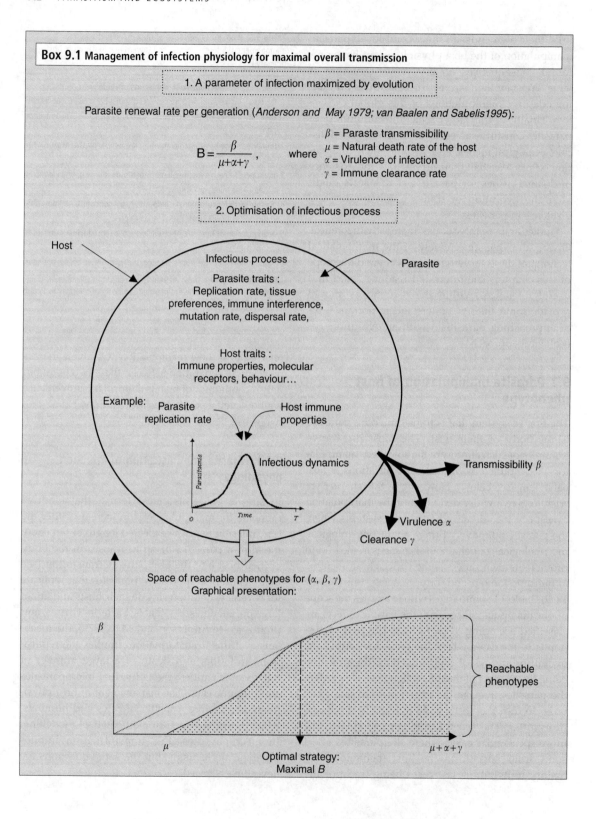

Box 9.1 Management of infection physiology for maximal overall transmission

1. A parameter of infection maximized by evolution

Parasite renewal rate per generation (*Anderson and May 1979; van Baalen and Sabelis 1995*):

$$B = \frac{\beta}{\mu + \alpha + \gamma},$$

where

β = Paraste transmissibility
μ = Natural death rate of the host
α = Virulence of infection
γ = Immune clearance rate

2. Optimisation of infectious process

Host

Infectious process

Parasite traits :
Replication rate, tissue
preferences, immune interference,
mutation rate, dispersal rate,

Parasite

Host traits :
Immune properties, molecular
receptors, behaviour…

Example: Parasite replication rate | Host immune properties

Parasitaemia

Infectious dynamics

Transmissibility β

o | *Time* | *T*

Virulence α

Clearance γ

Space of reachable phenotypes for (α, β, γ)
Graphical presentation:

β

Reachable
phenotypes

μ

$\mu + \alpha + \gamma$

Optimal strategy:
Maximal *B*

In a first approach, the infectious process can be extremely simplified; one simply considers that the host $\times$ parasite interaction is characterized by an 'intensity of infection', which is optimized by the parasite (within the limit imposed by the host) in order to maximize its transmission rate without harming too much the host. This problem, known as the 'evolution of virulence', has led to a large number of theoretical developments (e.g. Anderson and May 1979; May and Anderson 1979; Levin and Pimentel 1981; Bremermann and Pickering 1983; Frank 1992, 1996; Bull 1994; van Baalen and Sabelis 1995; Gandon et al. 2001). However, it may be often useful to describe more accurately the within-host infectious process, and to understand how such processes are shaped by natural selection exerted on parasites. In a simple situation, parasites replicate inside a host considered as a uniform environment, and are killed by replicating immune cells; the course of their infection is characterized by a peak dynamic. Natural selection exerted on the parasite optimizes the infection dynamic through the parasite's replication rate (several works have explored this case; Antia et al. 1994; Ganusov et al. 2002; Gilchrist and Sasaki 2002; André et al. 2003). In more realistic situations, parasites express habitat preferences within the heterogeneous within-host environment, that is, their affinity with various host tissues is under the pressure of natural selection. The course of infection is characterized by the dynamics of the parasite's density in each host tissue. Particular tissues may serve as entries inside the host's body; others as reservoirs of parasites, protected from the action of immunity (the cerebrospinal fluid for instance); some others may be the richest location of nutrients necessary for parasite growth and replication (for instance the digestive tube or the blood); and others may be best suited for dispersal from the host (for instance, the respiratory tract or the seminal fluid). The evolution of a parasite's strategy of host exploitation when several ecological niches are present inside the host (e.g. several tissues) has rarely been considered explicitly in models (but see Koella and Antia 1995). This question becomes even more intricate if several parasite strains can infect the host independently or if mutants appear during the infection and invade new host tissues. This phenomenon could be for instance involved in the rare appearance of very virulent meningitis owing to the invasion of the cerebrospinal fluid by Neisseria meningitidis usually carried asymptomatically in the nasopharynx (Levin and Bull 1994; Richardson et al. 2002), or in the development of AIDS owing to HIV (Tersmette et al. 1989; Schuitemaker et al. 1992). In all these cases, the infection must be considered as a whole ecosystem undergoing development through reproduction, mutation, migration inside the host, and niche specialization.

Parasite behaviour can also vary during the course of infection. The evolution of parasite strategies in this case can be investigated with 'optimal control theory'. Sasaki and Iwasa (1991) use this technique to model a situation where the parasite can modify its replication rate during the course of infection, while Koella and Antia (1995) consider also the evolution of the parasite's investment in dispersal. In both cases, it is shown that under a large number of circumstances the parasite could be selected to change its behaviour in the course of infection, switching from slow-replicating to faster-replicating phases. Besides, in biological terms, it is well known that the expression of bacterial virulence factors (permitting resource intake from the host and in turn replication) is often plastic and controlled by a system of communication among bacteria called quorum sensing (Williams et al. 2000). The 'assault' of bacteria on its host can hence be synchronized and conditional on a high density. Brown and Johnstone (2001) have built a kin selection model to study the evolution of such a conditional strategy (see discussion below concerning the kin selection aspect of this question, and Box 9.2).

Parasites also control the course of their infection through their rate of antigenic variation, permitting an escape from the adaptive immune system of their host. The mutation rate of parasites evolves to maximize the length of infection (targeted mutation can be on the whole genome in the case of viruses, and at particular antigenic loci in the case of bacteria and protozoa, see Moxon et al. 1994 for a review and Frank 2002: chs. 5 and 7). Various models have been built on this question in the specific case of

Box 9.2 Kin-selection models of the evolution of virulence and host manipulation

Kin selection is a fundamental evolutionary process (Frank 1998), and has increasingly been invoked in the development of models of parasite evolution. The classic tool for modelling kin selection is the technique of inclusive fitness (Hamilton 1964). Here, we briefly summarize the 'direct fitness' formulation of inclusive fitness (Taylor and Frank 1996; Frank 1998), an important and influential tool. First we express a general fitness function for a social parasite, then we apply the direct fitness technique to derive the ESS (evolutionary stable strategy) of exploitation.

Fitness functions in locally interacting groups can be usefully separated into two elements: within-group ('individual') and between-group ('group') fitness (Frank 1998; e.g. host manipulation, Brown 1999; virulence, Frank 1996; sibling competition, Godfray and Parker 1992; vigilance, McNamara and Houston 1992). The approach taken below is to construct a fitness function, w, consisting of an individual (I) and a group (G) component in a multiplicative form,

$$w(m, \bar{m}) = I(m)G(\bar{m}). \tag{1}$$

Here m is the trait of a focal individual, and $w(m, \bar{m})$ is the fitness of the focal individual (with strategy m), in an infrapopulation (group) in which the average strategy is $\bar{m}$. $I(m)$ equals the individual fitness function: if the trait is costly to the individual then it is a declining function of m (and conversely a rising function of m if the trait is directly beneficial to the individual). $G(\bar{m})$ equals the group fitness function, which again may be either a rising or falling function of $\bar{m}$.

At an evolutionary stable equilibrium, the unbeatable strategy m^* is found by solving $dw/dm|_{\bar{m} = m = m^*} = 0$. First find dw/dm with the chain rule, then replace the phenotypic derivative $\partial\bar{m}/\partial m$ with the corresponding relatedness coefficient (R), yielding

$$\left. \frac{\partial w}{\partial m} + R \frac{\partial w}{\partial \bar{m}} \right|_{m = \bar{m} = m^*} = 0. \tag{2}$$

Equation 2 implies that there is no marginal gain from a deviation from m^* at equilibrium. For more details, see Taylor and Frank (1996) and Frank (1998).

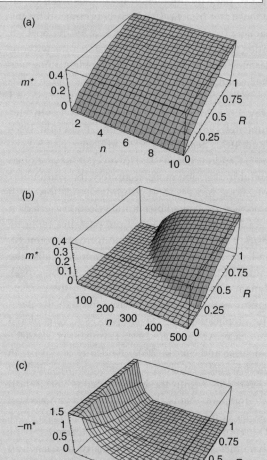

Figure 9.1 Evolutionarily stable individual 'collective action' strategies (m^*, Figs 9.1(a) and (b)) and 'tragedy of the commons' strategies ($-m^*$, Fig 9.1(c)) as a function of relatedness (R) and intensity of infection (n), following the model presented in Box 9.1. Fig. 9.1(a), obligate collective action, $p = 0$ (e.g. RNA replicase, Brown 2001). Fig. 9.1(b), threshold collective action, $p = 100$ (e.g. quorum-sensing traits, Brown and Johnstone 2001). Fig 9.1(c), tragedy of the commons, $p = 100$. In all figures, $c = 1$.

Source: Reworked from Brown *et al.* (2002).

Continued

By way of a simple example, consider a simple case of linear G and I functions, wherein the exploitation trait is costly to the individual, but beneficial to its group. This 'altruistic trait' framework has been used to consider the evolution of host manipulative traits that are costly to individuals to engineer (eg Immuno-manipulation, Brown 1999, Brown and Grenfell 2001; RNA replicase production, Brown 2001; siderophore production, West and Buckling 2003).

$$w(m, \bar{m}) = (1 - cm)(p + n\bar{m}), \qquad (3)$$

where c represents the cost of cooperation, p represents passive fitness (the fitness of a non-contributing individual in an infrapopulation of non-contributors) and n represents infrapopulation size or density. Substituting eq (3) into eq (2), we can solve to obtain the stable level of collective effort:

$$m^* = \frac{nR - cp}{cn(1 + R)}. \qquad (4)$$

So long as m is limited to positive values, we have a model of cooperative host-manipulation (cost to the individual, benefit to the group—i.e. 'collective action'). However, by considering the negative space of m, we recover a classic 'tragedy of the commons' model of virulence, where increasing the virulence trait (increasing the negative magnitude of m) has a positive impact on the within-host fitness, and a negative impact on the among-host fitness (Brown et al. 2002).

The positive (host-manipulation/collective action) and negative (classic virulence/tragedy of the commons) regions of m^* are plotted separately in Figure 9.1, as a function of infrapopulation size (n) and relatedness (R).

The model we present here is a very simple caricature, to which many refinements can be made. West and Buckling (2003) analyse a suite of related models (focusing on the 'collective action' dilemma of individual cost versus group benefit) using the more general tactic of non-specified individual and group functions. Furthermore, they consider the effect of additive versus multiplicative fitness functions, host mortality and the scale of competition on the trait ESS.

parasites (Nowak 1990; Kamp et al. 2002), but also more generally on the evolution of mutation rate in a changing environment (Leigh 1970; Taddei et al. 1997; Johnson 1999). Antigenic variation (through mutation) is a way to manipulate the host's immune system by sending it in the wrong direction. The parasite may also manipulate directly the immune system of its host by interfering specifically with some of its communication molecules. Numerous examples of such immune subversion are coming from the viral world (see reviews by Alcami and Koszinowski 2000; Tortorella et al. 2000; Murphy 2001), but some bacteria also interfere with immune regulation (Rottem and Naot 1998). Finally, the parasite may also manipulate its local environment with regard to competition with other parasites. Protecting its host against further parasite arrival, or eradicating all competitor parasites already present, may be a good strategy. This can be attained via an effect on the host's immune system. Brown and Grenfell (2001) modelled the potential for adaptive immune-manipulation by established macroparasite worms against larval challengers, illustrating

that this 'host vaccination' could be selected for under reasonable conditions of density-dependent fecundity, given an adequate dissociation between the antigenic profile of adult and larval stages.

To conclude this section, besides the traditional models on the evolution of virulence, we emphasize that the complexities of how parasites 'administrate' the body of their hosts, akin to 'military' strategies of invasion and occupation, should be more often considered explicitly in models.

9.2.2 Parasite manipulation of traits specific to the host

When the host is transformed into an efficient and durable parasite syringe, it is important for the parasite to control the behaviour and energy allocation of this syringe in order to maximize transmission events. To this end, the parasites engage in numerous tricks to distort or hijack dramatically their host's body, providing the most potent examples of Dawkins' notion of an extended phenotype (Dawkins 1982, 1990).

Over the past 30 years, a significant body of empirical work has focused on the changes in host behaviour following parasitism, and whether or not these changes represent parasite adaptations, host counter-adaptations, or non-adaptive side-effects (e.g. Rothschild 1962; Holmes and Bethel 1972; Dawkins 1990; Thomas *et al.* 2004). These changes in host phenotype vary greatly, from statistically discernable changes in size or habitat, to the emergence of completely novel behaviours (Poulin 1994*a*, 1995*b*; Thomas *et al.* 2004). If these effects are really parasite adaptations then they constitute the most striking examples of manipulation, as the parasites manage to unexpectedly influence traits that belong 'intimately' to their host, and not only some properties of the infectious process. Given the often-antagonistic interests of host and parasite, the host body can be viewed as an ecological and evolutionary battleground between host and parasite genotypes. Nonetheless, the simplest explanation for parasite-induced changes to the host phenotype is that they are simply non-adaptive side-effects of infection (Dawkins 1990; Poulin 1995*b*; but see Thomas *et al.* 2004). At this point it is worth considering a few case studies, to understand the attraction of the host-manipulation hypothesis. The great majority of empirical studies on host-manipulation have centred on the problem of jumping from one host to another, in species of parasites employing complex, multiple-host life cycles (see references in Poulin 1994*a*, 1995*b*, 2000; Thomas *et al.* 2004). The digenean *Dicrocoelium dendriticum* presents a well-known yet representative case (Carney 1969; Wickler 1976; Dawkins 1990). In order to complete its complex life cycle, this fluke faces the challenge of jumping from an ant to a sheep host. The behaviour of healthy, uninfected ants presents a formidable barrier to this transition, as having no selective interest in being consumed by grazing sheep, they are appropriately evasive. *Dicrocoelium dendriticum* apparently overcomes this barrier by lodging in the sub-oesophageal ganglion of the ant, leading to a significant behavioural alteration in its host. Infected ants climb to the top of grass blades, and lock their jaws to the very tip, in readiness for the next grazing sheep. Numerous other examples reveal a similar pattern of influence, with parasites creating impressive modifications in one host, to promote their ingestion by another. In some cases,

the mechanistic complexity of the modification is most impressive, as for instance when parasites secrete analogues of host hormones (e.g. production of a growth hormone analogue by the protozoan *Nosema* sp., leading to a doubling in size of their insect host (Fisher 1963). For other hormonal examples, see Beckage (1991, 1997), the apparent purposefulness of the modification lends the manipulation hypothesis its greatest support, as for instance when the onset of behavioural changes following infection coincides with the arrival of the parasite at a developmental stage suited for transmission. For example, the suppression of anti-predator responses in stickleback fish is only a property of infective plerocercoids (Tierney *et al.* 1993; see also Poulin *et al.* 1992).

As a result of these and many similar examples, changes in the behaviour of parasitized animals are commonly viewed as parasite adaptations to enhance transmission (see reviews by Dobson 1988*b*; Dawkins 1990; Moore and Gotelli 1990; Keymer and Read 1991; Poulin 1994 *a,b*, 1995*b*, 2000; Thomas *et al.* 2004). However, considerable confusion remains over how cases of adaptive host manipulation can be reliably recognized, given the contrasting possibility of host counter-adaptation, and the often more parsimonious explanation of non-adaptive side-effect (see Poulin 1995*b*; but see also Thomas *et al.* 2004). To date, models have only played a minor role in aiding the dissection of the adaptiveness of complex host $\times$ parasite interactions. Poulin (1994*b*) presented a graphical analysis of a number of key issues in host manipulation. More recently, interest has grown in understanding the social selective forces underpinning potentially manipulative behaviours (Brown 1999; Brown *et al.* 2001, 2002; West and Buckling 2003). In our view, there is still much potential for dynamically explicit models of parasite evolution in the setting of complex life cycles (see Choisy *et al.* 2003 for a starting point).

Note that most of the manipulative traits outlined above (in Sections 9.2.1 and 9.2.2) benefit not only the individual parasites that manipulate but also all their co-infecting neighbourhood, including potential 'cheats'; free-rider parasites that take the advantages of manipulation, but do not pay the costs. In the preceding discussions, the infection is considered as a unique individual in conflict with another (the host). However, the existence of competition between parasites within the host is also an important aspect of parasites' life which may have important consequences for manipulative traits (Box 9.2). Let us take the example of immune manipulation. Some viruses secrete proteins that alter the regulation of host defence; this time and energy consuming behaviour affects the entire host physiology and benefits all the viruses of the infection, whether or not they secrete the immune modulator. If the mutation rate (or immigration rate) within the virus population is too high, then the secretion behaviour may be lost because 'cheats', not secreting the immune modulators, are favoured by local competition (see Bonhoeffer and Nowak 1994). This question of multilevel selection is present concerning numerous traits of host exploitation and manipulation: virulence and within-host replication rate (e.g. Frank 1992; Nowak and May 1994), expression of virulence factors (Brown and Johnstone 2001; West and Buckling 2003), dispersal from the host, tissue preference within the host (see Levin and Bull 1994), host behavioural manipulation (Brown 1999). For instance Brown (1999) develops a simple evolutionary analysis of 'unbeatable strategies' of host manipulation, focusing in particular on the cooperative dilemmas of manipulating a single host as part of a potentially large group of parasite individuals. The relationship between the evolution of host manipulation and the evolution of virulence is discussed in Brown et al. (2002) and Box 9.2.

9.3 Parasite manipulation of global ecosystems

In the Section 9.2 we have explored how the host (seen as an ecosystem by itself) can be manipulated by the parasite in its own interest via several traits (replication rate, tissue affinity, mutation rate, secretion of virulence factors, interference with immunity, etc.). On a higher level (population of hosts in their external environment), the parasite also influences the ecosystem through an action on host behaviour, mortality, and reproduction. A major difference between these two levels of action is that, in the case of the within-host environment, the ensemble of the host's cells (the organism) is a tightly integrated unit of selection, whereas this is less likely to be the

case in aggregates of independently reproducing organisms. This difference has consequences both for the host and the parasite. First, regarding the host, immune cells work for the entire organism and not for themselves; as a consequence they do not function like independent predators: their replication rate is controlled by an upper level of integration (the organism); it does not simply rely on the quantity of parasites ('prey') they 'eat'. This decoupling permits the immune system to eradicate parasites entirely even when they get very rare, which is rarely possible in the case of predators with their prey. This major demographic difference between immune cells and predators has been studied and reviewed by Antia et al. (2003). Second, and more importantly in this chapter, regarding the parasites, the strong genetic relatedness between co-infecting parasites (and the emergence of the whole infection as a unit of selection), permits the evolution of cooperative strategies of host manipulation (as discussed in Section 9.2 and Box 9.2).

Regarding the ecosystem at the higher level, parasites can affect both the demography and evolution (genetic composition) of the population of hosts (see Chapter 3). In most cases this effect is a simple coincidental product of parasitism. However, if the population of hosts is structured (by distance isolation or environmental heterogeneity) then external environmental changes brought by a within-host community of parasites may in turn differentially affect genetically related parasites in other hosts. In such case, an adaptive manipulation of the upper-level ecosystem (i.e. of the hosts demography and evolution) may evolve. In a first simple analysis, this manipulation may affect the evolution of the parasite's strategy of host exploitation. Spatially explicit models show that parasites tend to be less virulent when the population of hosts is structured in space, the parasites being selected to maintain a high local density of available hosts that benefits their genetically related neighbours (Claessen and de Roos 1995; Rand et al. 1995; van Baalen 2002). This phenomenon might also be involved in the evolution of parasite strategies of host 'domestication', that is, parasite traits that influence the evolution of the local host population in a direction that benefits the local population of parasites. For instance, certain parasites may tend to be more virulent against hosts that resist their attack (the so-called mafia behaviour; Soler *et al.* 1998). Such behaviour is costly for a given parasite as it increases the host death rate; however it may be favourable for the local parasite population as it reduces the density of resistant hosts and therefore increases the density of sensible ones. This question requires a formal mathematical analysis (previous models, for example, Soler *et al.* 1998, only analyse the evolution of mafia behaviour in the case where the host has already evolved a plastic strategy of immune response). More generally, domestication in a structured environment might often be involved in the long-term coevolution of host/parasite systems, and especially in the emergence of symbiotic systems from initially parasitic ones.

In order to consider how parasite-induced changes in host phenotype can have wide-ranging impacts on host population and community structure, we will move to mobile genetic elements (MGEs) in bacteria, as bacterial systems permit rapid experimental investigation on a large populational scale. MGEs (eg bacteriophage, plasmids, transposons) can be propagated via a variety of mechanisms leading to both horizontal and vertical transmission (see Lawrence 1999). Host manipulation by MGEs is rife, as they can carry genes altering many aspects of the host phenotype. In contrast to the helminth-biased examples discussed in Section 9.2.2, many of these changes in host phenotype are unambiguously adaptations, displaying complex specialization and beneficial effects, at least to the MGE. The MGE genes underlying host-manipulative traits are referred to as 'conversion genes'.

Consider the case of colicinogenic plasmids, not least because of a rich and accelerating dialogue between theory and experiments. These are extrachromosomal DNA parasites of *Escherichia coli* that cause (at a low frequency) the explosive suicide of the host, and the release of antimicrobial toxins termed colicins (Riley and Wertz 2002; Riley *et al.* 2003). On an individual host level, this is a dangerous pathogen. What is more, on an individual plasmid level this behaviour is also extremely costly as the plasmid genes are exploded along with the host. To understand this system, a population-based perspective is essential. For a given lineage of colicinogenic plasmid carriers (carriers for short), only a small fraction release colicins via cell lysis. Furthermore, non-lysed carriers are immune to the colicin released by their lysed kin, thanks to the specific antidote coding genes carried by the plasmid. In a pure culture, this lysis represents a straightforward cost of parasitism to the host, and a more puzzling cost to the parasite itself. However, if the carrier lineage is in contact with a line of *E. coli* that is susceptible to the toxin (i.e. not carrying the same specific plasmid), then the release of colicin may act to reduce the density of susceptibles, and hence act as a spur to carrier growth, benefiting both the carrier bacteria and carried plasmid. The conditions determining the outcome of competition between carrier and susceptible lineages of *E. coli* where first sketched in a pioneering paper by Chao and Levin (1981). They illustrated that in a well-mixed environment (e.g. shaken liquid culture), carriers cannot invade a susceptible lineage if their initial density is below a critical threshold. However, they went on to show that the introduction of spatial structure can overcome this barrier to invasion. Carriers could invade a numerically dominant susceptible population if the bacteria were grown on static agar plates, even for ratios where invasion failed in the well-mixed environment.

Subsequent theoretical work illustrated that the localization of killing brought about by spatial structure ensured that the local density of toxin could be sufficiently high, even when carriers where extremely rare (Frank 1994; Durrett and Levin 1997). In contrast, in the well-mixed environment the density of toxin produced by rare carriers is rapidly diluted to ineffective levels. In sum, models of well-mixed competition illustrate that 'susceptible' is always a locally stable strategy (always resistant against vanishingly small invasions by carriers), but carriers can also be a locally stable strategy for certain parameters (Frank 1994; Durrett and Levin 1997, Brown, Le Chat, and Taddei submitted). Thus when both carriers and susceptibles are stable to invasion, the ecological end-point depends on the initial frequency of carriers to susceptibles in the local well-mixed environment. More recent work has focussed attention on how spatial structure can support diversity, once multiple competing strains of carriers are taken into account (Czárán *et al.* 2002; Kerr *et al.* 2002; Czárán and Hoekstra 2003). Thus we see that simple parasite-induced niche construction (e.g. production of colicins) can have wide-reaching impacts on the invasiveness of the host, and host community diversity (similar arguments can be advanced for phage-mediated competition; Brown, Le Chat, and Taddei submitted).

9.4 Transmission modes as ecological niches in the ecosystem of parasites

We shall now consider evolution in a community of parasites that compete for transmission among a population of hosts, as in the previous section. Within a host individual, parasitic strains compete to gain access to resources. This competition could yield a diversification of pathogenicity strategies (type of tissues preferentially attacked, etc.). The same sort of competition, and possibly the same sort of specialization, could in fact occur at the level of the population of hosts. At this level, parasites indeed compete to gain access to susceptible hosts. At the lower level of selection, life history traits that determine replication rate or resistance to the host's immune system will be selected. At the upper level, traits that determine what host or what trait of the host they use for their transmission could be under selection. Of course some parasite traits might be selected at both levels, and it is anyway generally assumed that tradeoffs exist between these two suites of traits. As a result, pathogens should evolve toward intermediate strategies where costs and benefits of virulence are balanced. In the following paragraphs we will see that under some form of competition evolution does not follow this simple evolutionary path. The average strategy might indeed deviate strongly from what would be expected if R_0 was maximized when selection yields extreme specialization on transmission modes.

As we have seen previously, pathogens can attack specifically some part, tissue, or organ of their host. Each host is, from the pathogens' point of view, a collection of resources that can be exploited, a set of different ecological niches that can be occupied. Not surprisingly, pathogenic strains that do not compete, because they do not exploit the same resource, can coexist within a single infected host. This is yet another manifestation of Gause's competitive exclusion principle. The same sort of specialization can occur among parasites that infect a population or a community of hosts. This point is clearly demonstrated by Lipsitch *et al.* (1996). These authors considered a mixed mode of transmission: vertically each time their host reproduces and horizontally by contact between infected and susceptible hosts. Lipsitch *et al.* (1996) found that two strains of pathogens cannot coexist within the same population of hosts, unless they have contrasted modes of transmission: parasites that are transmitted mostly vertically can coexist with others that transmit mostly horizontally (see Box 9.3). Again, as the authors point out in their paper, this result is a manifestation of the competitive exclusion principle: vertically transmitted pathogens exploit the capacity of their host to reproduce; horizontally transmitted pathogens exploit the capacity of their host to contact susceptible conspecifics. The competition between these two kinds of pathogens is therefore reduced and their coexistence is possible.

As the same reasoning applies at both levels of selection, we can generalize and propose that different modes of virulence and of transmission characterize distinct ecological niches. At this point the reader might wonder how the existence of several ecological niches might affect the evolutionary dynamics in pathogens. If pathogens all converge toward the niche where resources are the most abundant and where transmission is the most efficient, R_0 will be maximized as classically predicted. In some situations however, pathogens might not converge towards this 'ideal' niche, simply because it is overcrowded. Specialization might then appear, pathogens being selected to occupy niches where resources are maybe less abundant but where competition with other strains is reduced. We shall see in the following that selective pressures might be different among the niches pathogens can colonize. Evolution of pathogens will therefore depend first on what niche is

Box 9.3   The principle of adaptive dynamics—finding conditions necessary for diversification

We shall illustrate the principles of the adaptive dynamics technique using the example of a pathogen with a mixed mode of transmission, exerted from Lipsitch *et al.* (1996).

Defining the invasion function

Let u and v be the densities of infected and uninfected hosts respectively. Each pathogenic strain is characterized by a rate of horizontal transmission β and a virulence α: infected pathogens die at a rate $(1 + \alpha)\mu > \mu$. When infected hosts reproduce, pathogens are transmitted vertically in a proportion $1 - \delta$ of the offspring. This yields the following set of differential equations:

$$\frac{dv}{dt} = f(v + \delta u)(1 - u - v) - \mu v - \beta uv,$$

$$\frac{du}{dt} = f(1 - \delta) u (1 - u - v) - (1 + \alpha)\mu u + \beta uv.$$

We can use these equations to derive the equilibrium epidemiological state (u^r, v^r) of the host-pathogen community. We shall now consider a mutant pathogenic strain that appears in this equilibrium community. Assuming that mutants are initially rare ($u^m \approx 0$), we can write

$$\frac{du^m}{dt} = u^m \{f(1 - \delta)(1 - u^r - v^r) - (1 + \alpha^m)\mu + \beta^m v^r\}.$$

From this equation, we can define an invasion criterion. A mutant pathogen will invade in an equilibrium community if and only if

$$w = f(1 - \delta)(1 - u^r - v^r) - (1 + \alpha^m)\mu + \beta^m v^r > 0.$$

The possibility for a mutant to invade a resident population depends, of course, on its characteristics α^m and β^m. It also depends on that of the resident pathogen because u^r and v^r are functions of α^r and β^r.

Convergence toward singular points

In most epidemiological models, transmission comes at the cost of additional virulence. We will now assume that a increasing function relates α to β (e.g. $\alpha = a\beta^b$). As a result, the rate of pathogen transmission increases with β while the number of susceptible hosts, and therefore the

opportunities of transmission, decreases when β increases. Under such a trade-off model, there should be a value of β, a singular strategy, at which the benefits and the cost of virulence are balanced.

The signification of singular strategies is exemplified on Fig. 9.2. On this figure, each pathogenic strain is characterized by its β. Gray areas define the set of mutants (β^m on the y-axis) that can invade in a given resident population (β^r on the x-axis). The edge of this set is a curve on which the invasion function is zero (mutations are neutral). The singular strategy corresponds to the point where that edge crosses the line $\beta^m - \beta^r$. Note that in Figs. 9.2(a) and 9.2(b) the pathogen population always evolves toward the singular strategy. This convergence is not always guaranteed in adaptive dynamics models.

Branching points and the diversification of pathogens

Most generally, α is supposed to increase at an accelerating rate with β. Under this assumption, once the community of

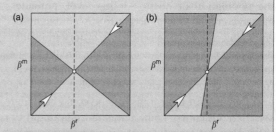

Figure 9.2 Two Pairwise invasibility plots.

Notes: The gray areas indicate set of mutant strategies (β^m) that can invade the resident strategy β^r, that is, those for which the invasion criterion w is positive. The circle and the dashed vertical line indicate the singular strategy, i.e. the level of transmissibility where costs and benefits of virulence compensate. Arrows indicate that in both cases A and B selection will bring pathogens at this point. In case A, once this point is reached, the resident population of pathogens is immune from invasion (all values of $\beta^m \neq \beta^r$ yield $w < 0$). The singular point is therefore an Evolutionary Stable Strategy (ESS). In case B, the resident population of pathogen corresponding to the singular point can be invaded by any mutant (all values of $\beta^m \neq \beta^r$ yield $w < 0$). In this later case the population of pathogen will become polymorphic for this level of transmissibility (see text and Fig. 9.3)

Continued

pathogen has reached the singular point, no mutant can invade: the pathogens have evolved toward a strategy that makes them immune from invasion. The singular point is then evolutionary stable (Fig. 9.2(a)).

We shall now assume that α increases at a decelerating rate with $\beta : \alpha = a\beta^b$ with $b < 1$. We will not discuss here what circumstances might justify this particular assumption. Our purpose is to show that, if valid, this particular form of trade-off produces a situation where the singular point, instead of being evolutionary stable, can be invaded by any mutant strategy (see Fig. 9.2(b)). Once the pathogen population has reached this point, it starts developing a polymorphism for transmissibility (and hence virulence). Eventually two strains with very contrasted transmission and virulence status differentiate and coexist within the same host population. This type of dynamics is illustrated in Fig. 9.3.

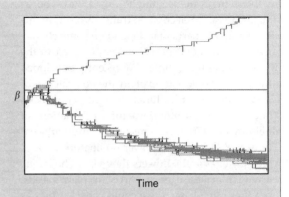

Time

Figure 9.3 A situation of evolutionary branching.

Notes: Each horizontal gray bar represents a pathogenic strain: its position along the y-axis gives its characteristic β; its thickness is proportional to the frequency of the strain in the pathogen population. Vertical lines indicate mutation events that give rise to a new pathogenic strain from a pre-existing one. The community of pathogens first evolves toward the singular strategy (central horizontal line). Once it has reached this point, a polymorphism develops and two strains emerge with low and high virulence level respectively.

available and second on the selective pressures that are characteristic of this niche.

We shall now go back to the example exerted from Lipsitch *et al.* (1996): pathogens can specialize on contrasted modes of transmission. This verbal prediction seems to match some experimental cases, for example, endophytes. Endophytes are pathogenic fungi of grass species. In some species one form of the fungus is vertically transmitted to offspring, colonizing seed tissues. A different form of the same species produces spores that are dispersed and can contaminate neighbouring plants. In this system horizontal transmission mechanically suppresses vertical transmission. Spores are indeed produced in the grass inflorescence, which effectively precludes production of seeds, and therefore vertical transmission of the fungus (Schardp 2001). Clearly, the two forms of the fungus do not exploit the same resource on the host. As we would predict, they are found to coexist in some populations of grass. Once specialization has occurred on transmission modes, traits that are not directly involved in transmission should experience different selective pressures in the two pathogenic strains. A clear

expectation in this situation is that, because vertical transmission aligns the pathogen's interest with that of its host, virulence should be counterselected in the vertically transmitted strain. This prediction is again supported by detailed studies of endophyte species. While the horizontally transmitted strains impose heavy costs on their hosts, the first of which being that they suppress their sexual reproduction, the vertically transmitted strain resemble mutualists. It has even been demonstrated that some of these strains induce greater host resistance to draught and herbivores. We have here a clear illustration of the evolutionary scenario we proposed earlier: the first traits to be selected are those that determine which niche a pathogen occupies; which traits are then selected, and in what direction, entirely depends on this niche.

The case of endophytes illustrates very clearly our point because the trait on which specialization occurs directly determines the selective pressures that act on virulence. The same situation probably arises in many other systems (e.g. bacterial plasmids). However, this correlation between selection on traits involved in specialization and traits involved in virulence might not always be as straightforward. Selection on traits involved in virulence might in particular depend on some characteristics of the host species, as emphasized in the first section of this chapter. We recently developed a model of this sort to explain the coexistence of several species of pollinators with contrasted strategies of oviposition in natural populations of globeflowers (*Trollius europaeus*, Ranunculacea) (Ferdy *et al.* 2002). Globeflower pollinators are flies which lay eggs in the flowers they visit. The larvae hatching from these eggs eat part of the plant's seeds. From these simple facts, we could consider flies as parasites as their reproduction is detrimental to their host. But among the species that can be found in natural populations of globeflowers, some visit young flowers and lay one egg at each visit while others visit wilted flowers in which they lay all their 15 eggs at once. The first species is often considered as a mutualist, as it contributes to pollination while imposing a minimum cost on its host. The second species, for symmetrical reasons, is considered as a cheater or a parasite. The trait that seems to be involved in specialization in this system is the age of flowers in which females lay their eggs. It has been shown that larvae that hatch from eggs laid at different times in a flower do not feed on the same part of the fruit. Therefore these larvae do not compete and the existence of flies that lay their eggs in flowers of different age can be interpreted as resulting from selection for competition avoidance. But then, how should we interpret the fact that the first of these flies lays one egg per flower while the second lays all their eggs in a single visit? We proposed that this results from a correlation between the age of the flower in which eggs are laid and their survival chances. Preliminary data indicate that eggs survive better in the first species that in the second, probably because the closed corolla of the globeflower protects them from parasitism and other external aggressions. Eggs that are laid in wilted flowers do not benefit from such protection. The result of this difference in survival probability is that, for a given number of eggs, the number of hatching larvae, and therefore the competition between them, is much higher when eggs are laid in young flowers than

when they are laid in old ones. Flies that lay their eggs in young flowers should therefore be selected to reduce competition among their offspring, which can be achieved by spreading eggs among many flowers. In this scenario flies first specialize on flowers of different age to escape from competition with other species. Flies visiting young flowers are then selected to spread their eggs among many flowers. This second step in the evolutionary dynamics of the system is mediated by the host whose corolla shape induces stronger competition among larvae in one of the two niches. It has recently been proposed that a similar scenario could apply to the much better known case of Yucca–Yucca moth interaction (Pellmyr and Huth 1994; Ferrière *et al.* 2002).

A further issue in this story concerns the evolution of corolla shape in globeflowers, which we propose to be a trait allowing the plant to manipulate competition among fly larvae. Of course, this trait cannot be selected because it induces the evolution of mutualism in flies. Rather, in our model, this trait is favoured simply as an anti-parasite strategy that allows the plant to kill fly larvae before they cause too much damage to its developing seeds. The fact that these traits select for mutualism in some fly species is fortunate to the host, but it is only accidental. Whatever traits are selected in the host, we see here that this selection will determine the ecosystem structure of pathogenic species.

The importance of the existence of distinct niches in host populations has recently been demonstrated in a model where specialization occurs on a trait that directly determines the degree of mutualism of one partner of an interacting pair of species (Ferrière *et al.* 2002). Clearly mutualism can here be interpreted as 'negative virulence' and all the arguments we developed earlier apply to both mutualists and pathogens. The work of Ferrière *et al.* (2002) shows that once pathogens split into several populations occupying distinct niches, the evolution of their virulence becomes in part determined by the structure of the community of mutualists/pathogens—namely the symmetry of competition. Mutualism here is not really stable because of some intrinsic characteristic. It is stable only when all other niches are occupied by pathogens. We could rephrase this by saying that in such systems, the question of the evolution of mutualism/benevolence becomes distinct from that of its stability.

9.5 Discussion and concluding remarks

9.5.1 Mutualism as negative virulence: host–parasite co-evolution

Relative to within-species interactions, asymmetries move centre-stage in the interspecific context of host–parasite interactions. The classical definition of a parasite as a symbiont that does harm (e.g. Poulin 1998a) freezes this asymmetry as a net flow of value from host to parasite. However, this simple categorization can hide a more interesting mix of antagonistic and overlapping interests (Dawkins 1990; Michalakis *et al.* 1992; Herre 1999; van Baalen and Jansen 2001). For instance, concomitant immunity to infection carries important benefits to both established parasites and their hosts (Brown and Grenfell 2001). Likewise, the production of antibacterial agents by *E. coli* may aid a vertebrate host. From a coevolutionary perspective, prudent exploitation offers benefits to both the host and to parasites, suggesting room for cooperation between host and parasite in fighting subsequent infections (van Baalen and Sabelis 1995). These examples highlight that parasite-induced host manipulation does not necessarily cause harm to the host (Brown *et al.* 2002). Viewing the host–parasite interaction as a complex mix of interests, or *desiderata* (Dawkins 1990) is an important step away from the medically influenced view of parasites being always virulent to some degree.

Not only is there a phenotypic continuum ranging from mutualism to parasitism, with commensalism as an intermediate stage, but also all these states are not just characteristics of host–parasite interactions. They characterize an interaction in a particular ecosystem. If ever some force was changing the composition of the community or the functioning of the ecosystem, a mutualistic association could start functioning as a parasitic one. In the example above, an *E. coli* strain that produces antibacterial agents is a mutualist if the host is infected by pathogenic bacteria; it becomes a parasite if these pathogens disappear. Fitness, parasitism, and virulence are not simple functions of the interacting genotypes: they also depend on the ecological community these genotypes live and reproduce in.

9.5.2 Host manipulation and community ecology

A final aim of this study is to provide a basis for future research investigating the community-level consequences of parasitism. Most of the existing evidence for parasitic impacts on community structure concern 'parasite arbitration', where a single parasite mediates apparent competition among multiple host species through differential host-susceptibility (Hudson and Greenman 1998; Tompkins *et al.* 2001; see Chapter 8). Thomas *et al.* (2000a) identify the potential for more indirect chains of influence between parasites and community structure, mediated by the host-phenotype. Life history traits of free-living species can be an important determinant of community structure, for instance changes in development time or dispersal can impact on competitive interactions and food web structure (Chase 1999; Tokeshi 1999). Thus via their effects on host phenotype (whether through parasite adaptations, host counter-adaptations, both, or neither), parasites have the capacity to indirectly influence the structure of both host and parasite communities. Some of the systems described in this chapter provide potential examples of such indirect effects on community structure. Bacterial collective action can lead to innumerable changes in their hosts, of significance to third-party species. For instance, human dental plaque is a product of complex interactions between the host and multiple bacterial species. A subset of these bacterial species actively creates a fibrous biofilm matrix, providing a modified environment within the host, enabling the establishment of a broad bacterial community (Costerton *et al.* 1995; Wood *et al.* 2000). As we have seen above, once the composition of the ecological community is modified, who is a parasite and who is not might also change. From an evolutionary point of view, once a complex community has established, many new ecological niches are open and selection can take many different directions. A simplistic view of the situation would be that each virulent strain opens a niche for a protective bacteria. The number of niches could in fact be much higher and the traits submitted to selection much more complex, because each member of the community can manipulate for its own interest a complex network of interactions between other members. Some experimental studies in macroparasites gives a flavour of this complexity: the host-manipulating trematode *Microphallus papillorobustus* offers increased transmission success to non-manipulative trematodes sharing the same

intermediate and definitive hosts (Thomas *et al.* 1997, 1998*b*; see Chapter 8). Models of the evolution of virulence started with pairs of interacting species. They then considered heterogeneity or polymorphism in these species. We are now at the point where we need to consider the community to understand the selective forces that act on parasites. The answer to many current questions on the evolution of parasites might come from this dawning convergence between evolutionary and community ecology.

CHAPTER 10

Parasitism in man-made ecosystems

François Renaud,[1] Thierry De Meeüs,[1] and Andrew F. Read[2]

Technological and cultural change in human populations is opening up new ecological niches for pathogens and parasites. The organisms that cause many of these "diseases of progress" have opportunities for global spread and access to host population densities unprecedented in human history. Understanding the natural history and evolutionary ecology of these pathogens needs to become a key part of public health planning.

10.1 Introduction

Like free-living organisms, parasites and pathogens can colonize and evolve in new environments. In this way, travel and technical developments (e.g. air conditioning, plane, boats, new economic links, etc.), medical and surgical developments (e.g. catheters, fibroscopy, prosthesis, organ transplants associated with anti-rejection medicine, immunosuppressive drugs, etc.) are generating new environments in hospital ecosystems which are colonized now by new parasite and pathogen flocks. Elsewhere, agricultural processes have widely disturbed ecological parameters in natural ecosystems for food development; they are responsible for the emergence and development of new parasite and pathogen species, and also for changes in host–parasite interactions. Through different examples, the aim of this chapter is to present and to discuss some phenomena and processes involved in the conquest by parasites and pathogens of man-made ecosystems. We could name diseases which are the result of pathogens colonizing man-made ecosystems as 'progress infectious diseases'.

To pass from 6 billion to 10 or 12 billion human inhabitants by the end of the twenty-first century

[1] Génétique et Evolution de Maladies Infectieuses GEMI/UMR CNRS-IRD 2724, Equipe: 'Evolution des Systèmes Symbiotiques', IRD, 911 Avenue Agropolis, B.P. 5045, 34032 Montpellier Cedex 1, France.

[2] Institutes of Evolution, Immunology and Infection Research, School of Biological Sciences, University of Edinburgh, EH9 3JT, Scotland.

represents one main subject of anxiety for scientists. Ten billion humans could not live on the earth with the lifestyle enjoyed by the 750 million people presently living in developed countries, because of lack of water, energy, quality, and quantity of space. Developed countries must contribute to the development of all countries to balance economy and life conditions. However (indeed, *if* ever) this is achieved, substantial environmental modification seems likely. Throughout history, mankind has severely modified the biosphere. Human impacts on ecosystems are as old as the human species. However, following industrialisation, the consequent increase in numbers of people and their ability to modify the biosphere, the extent and consequences of human impacts on ecosystems have accelerated. Impacts resulting from human activities occur in all parts of the biosphere, and at all kinds of temporal and spatial scales. (Dickinson and Murphy 1998). The ecological consequences of the unavoidable modifications of the future are hard to predict: we simply do not have a thorough understanding of the impact of global change on local environmental conditions and the evolution of biodiversity.

10.1.1 But what is an ecosystem?

Let us imagine a pond, for example. What animals might live here? Insects, worms, birds, fish, mice, muskrats, ducks, deer, wolves. What do these animals need to eat? Insects eat plants, fish eat worms and insects, birds eat fish, worms, and insects. Mice eat grain. Muskrats eat ducks, eggs, and chicks. Ducks eat insects and worms. Deer eat grass. Wolves eat mice, muskrats, and deer. All these animals rely on the pond for the water they need. The deer cut the grass, wolves remove the sick and weak deer from the herd. Muskrats regulate the duck population. All these animals rely on each other. Just like people in a human community. An ecosystem is a community too. Consider a pond community, with all its variety of plants, insects, birds, and other living things. What would happen to the community if the water vanished? What if the ducks all disappeared?

Other examples of ecosystems include forests, rivers, oceans, deserts, cold arctic tundra, high mountains, and rain forests. Different plants and animals grow in different ecosystems. Normally, the living things in the ecosystem balance in such a way that no living things take over the whole ecosystem and destroy it, at least not for a while. For example, the production of O_2 by the first photosynthetic algae had dramatic consequences on the anaerobic life that predominated at that time.

10.1.2 But why some deer are sick and weak within the herd?

May be they could be parasitized? We just want to underline here that the above description of an ecosystem, as we can read it in numerous books or websites, disregards systematically the fundamental role that parasites play in ecosystem functioning! Indeed, it is less poetic to speak about a tapeworm or a virus than a duck or a deer! Nevertheless, parasites are present in a large part of ecosystems, and the liver, kidneys, lungs, gills, gut, and pharyngeal sphere of a host constitute as many ecosystems for parasites and pathogens as do rivers, oceans, desert, high mountains, and jungles for free-living organisms.

10.1.3 But what is a parasite and/or a pathogen?

'An organism in or on another living organism obtaining from it part or all of its organic nutriments, commonly exhibiting some degree of adaptive

structural modifications, and causing some degree of real damage to its host' (Price 1980). So, the parasite/pathogen lives at the expense of the host, and this host's exploitation has automatic consequences on host biology and physiology, on host evolutionary biology and on evolutionary relationships between hosts and parasites (Renaud and De Meeüs 1991). Because the host represents the 'habitat/resource' system of the parasite, each modification on host ecosystem will have consequences on the parasite ecosystem, and because parasites affect host fitness, they act on host ecosystem too. Because living organisms are parasitized, we cannot consider ecosystem evolution without parasites (see Chapter 9).

It is undeniable that humans greatly disturb ecosystem equilibrium (deforestation, eutrophication, overgrazing, etc.), but the aim of this chapter is not to consider anthropogenic ecosystem disturbances and subsequent evolution of parasites and diseases which are presented in other chapters of this book (see Chapter 7). Instead, we will illustrate and discuss, through different examples, the impact of technical progress by humans on the evolutionary ecology of parasites and pathogens. How have parasites exploited new 'human-made' ecosystems, especially those concerning public health?

10.1.4 What is a 'human made' ecosystem?

It is an ecosystem artificially elaborated by humans in order to enhance their quality of life.

For example, let us imagine that the wheel was never invented! The wheel is everywhere on our cars, trains, planes, machines, wagons, and most factory and farm equipment. What could we do without wheels? But as important the wheel is, we do not know who exactly invented it. The oldest wheel found in archaeological excavations was discovered in Mesopotamia, and is believed to be more than 5500 years old. Eventually, wheels became covered with tyre in order to make the trip more easy and pleasant. But, it is almost impossible remove all the water a worn-out tyre contains. Consequently, old tyres become excellent habitat for the larvae of different mosquito species, especially *Aedes* spp. which are the vectors of the Dengue virus. Worn and waste tyres are being traded throughout the world, and are

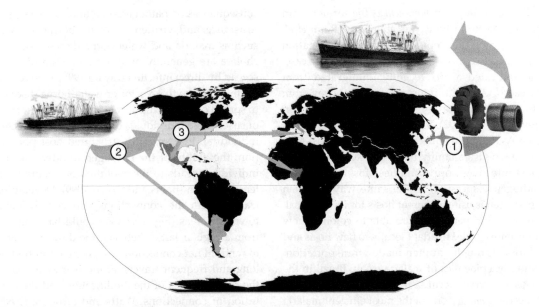

Figure 10.1 Example of hypothetical main routes of *Aedes albopictus*- infested tires.

Notes: 1: Japanese origin; 2: first colonization wave: US and South America in 1985/1986; 3: second colonization wave: Mexico, Africa, and Europe in 1990/1991.

Source: Data from Reiter (1998: 93, table 10).

responsible for the introduction of mosquitoes in different countries (Fig. 10.1). 'In short, it seems we must accept the establishment of exotic species as an inevitable consequence of modern transportation technology' (Reiter 1998). For example, inspections of containers arriving in US ports showed the presence of living *Aedes albopictus* and four other mosquito species in worn tyres from Japan (Craven *et al.* 1988). Japan is the biggest exporter of worn tyres in the world. *Ae. albopictus* is capable of vertical and horizontal transmission of the Dengue virus, and other important human arboviruses (Shroyer 1986). Thus pathogenic agents can take advantage of tyre trade. We can imagine that such trades could have also consequences on other vectors, such as the *Anopheles* mosquitoes which transmit malaria.

10.2 Economic and touristic human travels: enhancement of human contacts!

Boats not only carry the worn tyres which constitute new ecosystems for mosquito larvae, but they also have bilge. The classic bacterial disease, cholera,

entered both North and South America during the last century from the bilge-water of an Asian freighter. Indeed, molecular typing showed that the South American isolates were pandemic genotypes previously observed in Asia. Water bilge seems to constitute a very good ecosystem for the transported pathogens such as bacteria (Anderson 1991; Morse 1995). Cholera is not the only opportunistic pathogen which use such kinds of transport: an epidemic strain of *Neisseria meningitidis* seems to have disseminated rapidly along routes by travelling in ballast waters (Moore and Broome 1994).

Malaria parasites use mosquitoes to transmit between vertebrate hosts. For example, different mosquito species belonging to the genus *Anopheles* are the definitive hosts of the malaria agent *Plasmodium vivax* , one of the four species of human malaria. Human malaria is currently absent in western Europe, but an autochthonous case of *P. vivax* malaria occurred in Tuscany (Italy) in August 1997, decades after malaria eradication (Baldari *et al.* 1998)! The disease was diagnosed in a woman with no travel history who lived in a rural area where indigenous *Anopheles labranchiae*, the former main malaria vector in Italy, was abundant (Romi *et al.* 1997). A molecular epidemiological investigation concluded that this was an introduced malaria case, and indicated a girl recently immigrated from Punjab (India) and living about 500 m away from the patient, as the source of *P. vivax* infection (Severini *et al.* 2002). The parasite was able to pass from Asia to Europe because an infected host took a plane to visit a family member. Planes and boats constitute new opportunities for vector and pathogen dispersion, in the same way as when a pathogen is using different hosts for its dispersal. Parasites and pathogens are able to colonise new environments and to adapt locally to new hosts and vectors through human-made transportation. Similarly, epidemics of malaria in NE Brazil in the 1930s occurred because of the introduction of *Anopheles gambiae* , one of the most efficient malaria vectors, which probably arrived from a boat bringing mail from Africa (Killeen *et al.* 2002).

But what could be the consequences ever more frequent travel? What would be the possible consequences on pathogens evolution as regards to resistance and virulence? Vector-borne diseases such as malaria and water-borne diseases such as cholera are generally more virulent than diseases spread by direct infection (Ewald 1994). One reason for this may be that vector or water-borne diseases to spread over long distances, and causing infection of susceptible individuals distant from the infected individual. In a spatially structured host population, the ability of the pathogen to infect distant individuals leads to the evolution of a more virulent pathogen (Boots and Sasaki 1999). Developing travel alters the connections between different towns or areas (Fig. 10.2). We could have passed from a regular lattice between the different points to random net-connections which are permitted by long and frequent travels. From their analyses on the consequences of the modification and the evolution of connections, Watts and Strogatz (1998) suggested that infectious diseases spread more easily in small-word networks than in 'regular' lattices. We can predict that the increase in world travel would have strong consequences on

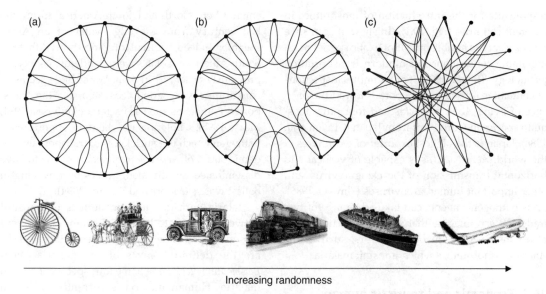

Increasing randomness

Figure 10.2 Connection topology from (a) regular ring lattice, to (b) small-world, to (c) a random network.

Notes: The intermediate connection is called 'small world' network, and infectious diseases spread more easily in small-world networks that in regular lattices. The different means of transport are presented in order to try to illustrate the development of 'travel man-made ecosystems' which permit to link different geographical points and modify the connections within and between populations. No company or factory could be incriminated here for disseminating pathogens.

Source: Modified from Watts, D. J. and Strogatz, S. H. (1998).

pathogens dispersal and consequently on the evolution of infectious diseases virulence and resistance, but this could be also applied for all categories of vectors which are responsible of pathogen carriages. It seems reasonable to assume that human societies in the past lived in larger with more isolated communities. But, modern social networks (Wasserman and Faust 1994) are known to be small words (Watts and Strogatz 1998), and it follows that infection networks may also show 'small world' connections in modern societies. When infections occur predominantly locally we predict a lower virulence than when transmissions occur predominantly randomly throughout within and between populations (Boots and Sasaki 1999).

10.3 Human comfort and industrialization

Throughout human evolution, people have always tried to enhance their comfort. Temperature and humidity have a significant effect on human comfort and health. The most comfortable humidity range is 40–60%, but air temperature and humidity are related in respect to comfort or perceived temperature. The combination of temperature and humidity where people report comfort is termed the 'comfort zone'. At the time we are writing this chapter, an epidemic legionnaire's disease occurs in North of France (Pas de Calais). Indeed, 85 persons where infected during January and February 2004, and among them 13 died (Fig. 10.3).

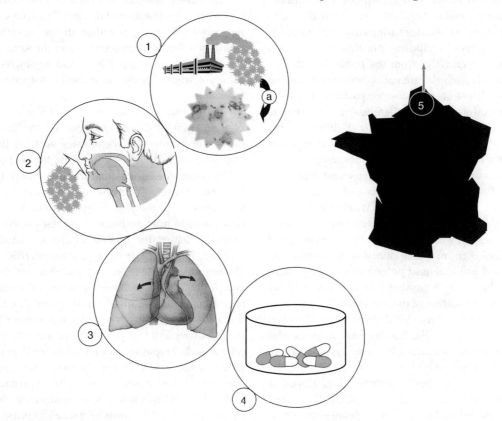

Figure 10.3 *Legionella* disease: infection routes.

Notes: 1: The pathogen responsible for the disease is a bacteria living in fresh water. The optimal growth temperature is between 35 and 40 °C. This pathogen is present in sanitary facilities (showers, taps, etc.), air conditioning, fountains, greenhouse, cooling towers, etc. For example, bacteria are presents in droplets coming from factory steam; 2: Bacteria contained in 'contaminated droplets' are inhaled by humans, and clinical symptoms arise after 2–10 days of incubation; 3: The serious form of the disease named 'Legionnaire's disease' generally arise in weakened patients (elderly, immunocompromised, etc.) which can evolve to a lethal pulmonary infection in about 15% of cases; 4: The treatment is based on antibiotics; 5: While writing this chapter, a *Legionella* epidemic was raging within the north of France (Department: Pas de Calais). More than 80 cases were registered during January 2004.

Legionnaires' disease is a lung infection pneumonia) caused by a *Legionella pneumophila*. The name of this bacterium was derived from the original outbreak at the 1976 American Legion Convention in Philadelphia (Hlady *et al.* 1993). These bacteria are readily found in natural aquatic environments and some species have been recovered from soil. *Legionella* parasitize Amoeba, and spread through cysts of these protozoans (Fliermans 1996). The pathogen can survive in a wide range of conditions, including temperatures of 0–$63\,°C$, pH of 5.0–8.5, and dissolved oxygen concentrations of 0.2–$15\,ppm$ in water. Temperature is a critical determinant for *Legionella* proliferation. *Legionella* and other micro-organisms become attached to surfaces in an aquatic environment forming a biofilm. *Legionella* has been shown to attach to and to colonize various materials found in water systems including plastics, rubber, and wood. But, crucially from the public health perspective, *Legionnella* are not only found in natural habitats. These bacteria develop particularly well in human infrastructures where water is present as saunas, for example (Den Boer *et al.* 1998). But the main human-made ecosystem they colonize seems to be the cooling systems found, for example, in factories, hotels, and hospitals (Alary and Joly 1992; Pedro-Botet *et al.* 2002; Sabria and Yu 2002). For instance, in January 2000, WorkSafe Western Australia reported a case of Legionnaires' disease from a teacher who had worked in a room supplied with cooled air from an evaporative air-conditioning unit, and had also used potting mix while gardening at home. Both potting mix and warm water allow multiplication of these bacteria.

Figure 10.3 describes the different steps of human infection by *Legionella*. But, not only workers who live in the contaminated building are concerned, indeed people working or living around the *Legionella* source can be infected (Fig. 10.4). For example, during the 2004 French epidemic discussed above, the infected patients were people living near a factory where the bacteria were identified in the cooling systems. Even though the link between the presence of the bacteria and human infections was not clearly established, the French government decided to stop immediately the factory activity.

Other microbes can contaminate air-conditioning units and cooling towers which can result in other health problems for workers and visitors such as respiratory sensitization and building related illness, or 'sick building syndrome'. It is thus essential to maintain a good indoor air quality at all times.

Water is essential for life! But drinking water networks are heterogeneous and constitute real biological reactors, that is to say an ecosystem between a mobile phase (i.e. water) and an appointed phase (i.e. biofilm). These networks are continually contaminated by microorganisms (i.e. bacteria, algae, protozoans, fungi, yeast, metazoans) and nutrients (i.e. organic dissolved matter) that passed through the treatment systems or gather from accidental procedures (i.e. breaks and repairs). Drinking water networks thus bring together all the favourable conditions for the maintenance and the spread of microbial systems diversified and organized in different trophic levels, and thus constitute real food webs.

Another example which could well illustrate the phenomenon of disease induced by the human desire for comfort is wastewater such as those encountered near houses in septic tanks. Septic tanks were designed to improve sanitation. Bacteria, viruses, protozoans, and worms are the types of pathogens in wastewater that are hazardous to human. Bacteria are responsible for several wastewater related diseases, including typhoid, bacillary dysentery, gastroenteritis, and cholera. Depending on the bacteria involved, symptoms can begin hours to several days after ingestion. Viruses cannot multiply outside their hosts, and wastewater is a hostile environment for them. But enough viruses survive in water to make people sick. Hepatitis A, polio, and viral gastroenteritis are a few of the diseases that can be contracted from viruses in wastewater. A protozoan is the cause of amoebiosis, also known as amoebic dysentery. Parasitic worms also dwell in untreated sewage. Tapeworms and pinworms are the most common parasites found in these wastewaters, from where their eggs can be ingested. Children and the elderly are the groups the most significantly affected by wastewater related diseases.

Figure 10.4 Persons exposed to Legionnaires' disease the case of factory cooling towers which could constitute a potential reproductive ground for *Legionella* bacteria, Pas de Calais—France 2004: >80 cases of *Legionella* disease.

Notes: 1: Technical personal intervening near the cooling towers; 2: persons working near the smoke; 3: persons living in buildings and houses near the factory: the contaminated air penetrating through windows and new air intakes.

10.4 Humans get sick, age, and die!

Modern humans try to live better and longer. These days, this desire has culminated in massive pharmaceutical and medical industries and associated science base, as well as on going interest in alternative medicine. Unsurprisingly, some pathogens take advantage of human illness and death, and of the new openings provided by medical science.

10.4.1 Sickness

'Despite a century of often successful prevention and control efforts, infectious diseases remain an important global problem in public health, causing over 13 million deaths each year. Changes in society, technology and the microorganisms themselves are

contributing to the emergence of new disease' Cohen (2000). During the twentieth century, many medical and public health officials were optimistic that most of infectious diseases could be eradicated. This has patently not occurred, and indeed that the ongoing emergence of new pathogens is a reality (Liautard 1997). Pathogens have plunged into the new ecological niches provided by new human behaviours and customs. It would be impossible and tedious to make an exhaustive review of these diseases, because they are so numerous. The development of medical technology in hospital ecosystems lead to the development of cohorts of opportunistic pathogens which exploit these new ecosystems. Diseases emerging in hospital ecosystems are known under the terminology 'Nosocomial infections' (hospital-acquired infections), derived from the

Greek 'nosokomeion', which means hospital. Hospitals are the source of many diseases because patients are often immunocompromised, and because infected people come to hospitals. The majority of nosocomial infections have an endemic origin (i.e. inside the hospital), where infection comes from a microorganism present in the ecosystem, and the surgical intervention renders it infectious. In an important article, Cohen (2000) reviewed the causes of nosocomial infections in modern medical environments. The classical case is represented by an inoffensive bacterium that is brought by the surgeon's lancet into the body of a patient and evolves to a septicaemia. Figure 10.5 illustrates different origins of nosocomial infections where medical tools can be incriminated. These infections can be due to microbes that have lived on or in (i.e. colonized) the patient for many years without harm before healthcare procedure provides a means of bypassing the patient's host defences (Farr 2003).

Nosocomial infections have been known for a long time in hospital ecosystems: Oliver Wendell

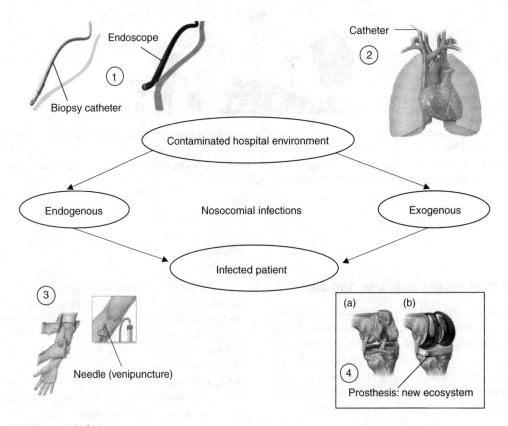

Figure 10.5 Nosocomial infections.

Notes: Following an invasive surgery (i.e. through the skin as illustrated in 1, 2, 3, for example), a patient can be infected by germs coming from an:

Endogenous origin: patient is infected by its own germs following a surgical act and/or because he displays a particular weakness. For example, a patient under artificial breathing can develop pneumonia from a germ of its own digestive tract, which can go up to the respiratory ways. The same phenomenon can be observed for an urinary infection from an urinary probe carrier.

Exogenous infection: Cross infections transmitted from a patient to another through hand contacts or medical tools. These infections can be originated from the germs inhabiting hospital workers, or linked to the contamination of the hospital environment (water, air, material, foods, etc.).

Prosthesis constitutes new ecosystems for pathogens. The figure shows a knee prosthesis (4): (a) before; and (b) after.

Holmes published a paper on this topic as early as 1843. Nosocomial infections can be directly linked to medical treatment, or can simply occur during hospitalization, independently of any medical act. They concern patients, but also workers present in this ecosystem. They can occur because the immune system is busy fighting some other chronic illness, or for people who are immunocompromised. People can be immunocompromised from certain diseases (e.g. AIDS), certain medications (e.g. immunosuppressants or chemotherapy), surgical recovery, or other serious medical complaints that limit the person's ability to fight against these infections (Berche *et al.* 2000).

These infections can spread by endogenous or exogenous ways (Fig. 10.5). Many types of pathogens are involved, including fungal infections (e.g. *Candida, Aspergillus, Fusarium*), bacterial and viral pneumonia (e.g. influenza, Staphylococci, *Pseudomonas*) which can be found in different organs, giving rise to urinary tract infections, surgical site infections, respiratory tract infections, blood stream infections, skin infections, gastrointestinal tract infections, central nervous system infections, and so on. Numerous surgical acts can initiate nosocomial infections, these fall into three main types: (i) urogenital probes lead to urinary infections, (ii) catheters are sources of systemic and local bacterial and viral infections, and (iii) artificial breathing systems are responsible of pulmonary infections. For a lead into the extensive literature on nosocomial infections see Arnow *et al.* (1993), Scheckler *et al.* (1998), Lucet (2000), Lemaitre and Jarlier (2000), Korinek (2000), and Joly and Astagneau (2000).

The frequency of nosocomial infections in France is typical of industrialized countries, with about 7% of hospitalized patients developing a nosocomial infection. In other countries it ranges from from 5% to 12%. In the United States, more than 2 million cases of nosocomial infections have been reported, leading to about 80,000 deaths and to 8000 additional days in hospital for 1000 infected patients, all at a cost of $US5 billion dollars in 1985 (Wenzel 1985; Haley 1991).

In this section on human sickness, we can illustrate the Machiavellianism of pathogens exploiting a human health problem: *drug addiction*. Many pathogens from viruses to worms use insect as vectors in order to infect new hosts. In the same way, we could do a comparison with medical tools that pathogens use as vectors. The best and saddest example could be the syringe which represents a wonderful vector for pathogens to pass from host to host. A non sterilized syringe represents a very efficient ecosystem exploited by many pathogens. One terrible recent example was provided by HIV and drug addiction. The HIV virus can be passed through different venous injections from different individuals sharing the same syringe. Unfortunately, this problem does not only occur within drug addicts; nurses in hospital have been contaminated by this parasite when taking a blood sample from infected patients. AIDS is one of the major diseass at the beginning of the twenty-first century, and we have to keep in mind the public scandal which occurred in France at the end of the twentieth century with contaminated HIV blood. Indeed, haemophiliacs need recurrent blood transfusions, and before the use of warmed blood elements, many of them were infected by HIV through the needle which served to transfuse them (Fig. 10.5). Most of them died, and this scourge continues to kill a lot of people in the world.

10.4.2 Ageing

Modern humans live longer, at least in industrialized countries, where the number of the elderly is rapidly increasing (Morris and Potter 1997). This leads to an increasingly large group of hosts ripe for exploitation. For example, ageing results in senescence of the gut-associated lymphoid tissue, and decreasing in gastric acid secretion (Feldman *et al.* 1996). The consequences of these physiological disturbances lead to an increase of susceptibility to pathogens. Indeed, as a low pH of the stomach represents an important barrier to entry of enteric pathogens, reduction in gastric acidity can increase the susceptibility to infection by these pathogens (Morris and Potter 1997). The communal living environment of some elderly, exacerbated by problems such as incontinence, further creates an habitat in which enteric and food-borne pathogens can spread rapidly (Benett 1993). In a study between 1968 and 1979 in the United States, Blaser and Feldman (1981) showed that the frequency of *Salmonella* bacteremia increased dramatically in the elderly compared with other age groups. *Salmonella* infections increase the risk of death, and elderly are often immunocompromised and they are assisted by a large cohort of medications. This treatment undeniably leads to the selection of drug resistance in many categories of pathogens. The grouping of patients in 'elderly care homes' may constitute production units of 'pathogen resistant ecosystems', which will represent a new and complex public health problem as the population ages. We expect that pathogens produced in these ecosystems, many of which may be drug resistant, will spill out to attack other age groups of the population (i.e. babies, toddlers, and children)? To our knowledge, policies have not yet considered this ecological problem. It is well known in evolutionary parasitology that an increase in the number of susceptible hosts can have dramatic consequences for the spread of pathogens in the whole population (Bird Influenza Virus below).

10.4.3 Death

Even if human have largely improved health care and have lengthened life expectancy, particularly in industrialized countries, death always occurs. Humans worship death and we can observe a lot of flower vases in some cemeteries. These vases can constitute ideal ecosystems for the development of different mosquito species larvae. In a very interesting paper, Lancaster *et al.* (1999) related the invasion of *Ae. albopictus* into an urban encephalitis focus, where flower vases ecosystems found in cemeteries could play an important role. Elsewhere, O'Meara *et al.* (1995) investigated the competition between *Aedes aegypti* and *Ae. albopictus* in a US cemetery named Rose Hill, and showed that until the summer of 1990, only *Ae. aegypti* inhabited vases at Rose Hill. *Ae. albopictus* was first collected that summer and by 1994, had become the most prevalent species (i.e. greatest percentage of vases occupied). Juliano (1998) sampled three cemeteries in Southern Florida where *Aedes* inhabit

water-filled stone cemetery vases. From manipulative field experiments, he analysed the mechanisms involved in the competition between *Ae. aegypti* and *Ae. albopictus*. He showed that, at least at the three sites tested, *Ae. albopictus* was more competitive than *Ae. aegypti*. This invader superiority was attributed to better resource acquisition in these ecosystems. Knowing that these mosquito species are vectors of different pathogenic viruses, these cemetery ecosystems can play an undeniable role in mosquito transmitted diseases, at least in the area concerned.

10.4.4 Surgical progress

Like different mechanical parts of our car, many organs of our body can be replaced by better functioning parts. Xenotransplantation is the graft (i.e. skin, tissue) or the transplant (i.e. organ) into humans of tissue or organs from animals (Wadman 1996). Xenotransplantation seems a very real possibility now that the generation of transgenic pigs as potential organ donors for humans has been achieved. Baboons are also likely to be used, especially so for bone marrow grafting. But, both baboons and pigs may silently harbour a great variety of viruses belonging mostly but nor exclusively to the *Herpesviridae*, *Retroviridae*, and *Papoviridae* families. All these viruses are potentially able to infect deeply immunocompromised patients. This may lead to the emergence of deadly viral infections, the so-called 'xenozoonosis', among the recipients and/or the general population (Chastel 1996). Indeed, the majority of viruses that emerged during these last 30 years displayed a zoonotic (mainly simians) origin (Morse and Schluederberg 1990). For example, the *Herpesvirus simiae* , specific to asian cerpothitecicidae of the genus Macaca and where it seems to be a largely benign virus, becomes very dangerous for human when injected by biting or by accidental injection by infected syringe or needle (Artensein *et al.* 1991). DNAs represent a well-established molecular species ecosystem where a large number of selfish DNAs (including number of viruses) are evolving and pass through generations. Each species harbours its own selfish DNAs, and they constitute, for example, a large part of the human genome (De Meeûs *et al.* 2003). Therefore, transmission of virus through grafts is confirmed, and allografts were reported to be at the origin of primary infections by, at least, the cytomegalovirus, Epstein–Barr virus, VIH, and hepatitis C (Chastel 1998). We could be thus confronted to a new virulent variant of these viruses or to a genetic matching between close human and simians viruses (i.e. Herpesvirus, Retrovirus). We cannot exclude the possibility of an outbreak of a genetic chimera between human and animal viruses, or of the 'complementation' of a defective virus. Xenotransplants can come from transgenic animal in order to avoid graft rejection by human recipient. 'I view xenograft tissues as essentially very complex vectors for shuttling new viruses in humans' (Allan 1995).

Prostheses are another example of surgical progress. The addition of exogenous material leads to the establishment of new ecosystems inside the body (Fig. 10.5). These new niches can be colonized by pathogens. For example, prosthetic joints, prosthetic implants, and vascular prosthetic materials are a 'nest' for many pathogens such as group C *Streptococcus, Staphylococcus epidermidis, Staphylococcus aureus, Mycobacterium tuberculosis, Histoplasma capsulatum* (Gillespie 1997; Kleshinski *et al.* 2000).

10.4.5 Hygienic progress

Polio was almost unknown until the dramatic epidemics that terrorised the developed countries in the twentieth century. This terror led to irrational responses, such as aggression to immigrants of shantytowns. Before hygiene was common, infants were safely immunized against polio by maternal milk. However, once hygiene standards were high in developed countries, individuals were first exposure to polio occured at older ages, when the clinical complications are likely. Thus, changing hygiene habits has allowed the poliovirus to the exploitation of new habitats, with dramatic consequences for the host (Schlein 1998; Seytre and Schaffer 2004).

Medical progress has led to women being attacked by pathogens due to menses (i.e. tampon use) and contraceptive methods (i.e. Intra-Uterine contraceptive Devices—IUDs or coil). The insertion of tampons or coils in female genital systems represents a new opportunity for pathogens. Urinary tract infections of women are common, and a source of considerable expense. The possibility that tampon usage is a risk factor for recurrent urinary tract infection has not been studied in detail, but it has been associated with bacterial vaginosis. The tampon may facilitate the spread of bacteria from the vagina to the urethra and bladder (Doran 1998). Elsewhere, it was thought that IUD infections spread through lymphatic canals to produce a perisalpingitis similar to that of postabortal or postpartum infections. Even if it was not demonstrated that IUDs are directly responsible of these infections, their role remains to be determined (Schwarz 1999).

10.5 Human need to eat!

Life, reduced to its simplest expression, could be caricatured through two main functions for each individual in each plant and animal species: survival and reproduction. Food is the fuel for this, but eating is not without risk. Our new food habitats have opened up new niches for many pathogens.

10.5.1 Tinned food

A major problem to which human populations were and are still confronted is to food preservation. Different methods of food conservation were developed in human societies, but two of them are the more used, at least in industrialized countries: tinned and cooled food.

Listeriosis, a serious infection caused by food contaminated by the bacterium *Listeria monocytes*, has recently been recognized as an important health problem in the United States and European countries (Lorber 1997; Silver 1998). Listeriosis is a disease that is enhanced by alimentary progress. This bacterium is frequently found in soil and water, and becomes pathogenic for human when ingested at high densities. Naturally contaminated food never presents dangerous densities of this bacterium, but this pathogen experiences an increase of its population demography at low temperatures. Refrigerators constitute a favourable ecosystem for these bacterial populations where they can reach the infectious quantity dose for human.

Human brucellosis or 'Bang's disease' which was discovered a century ago remains poorly known and difficult to treat. Pathogens responsible for the disease are bacteria belonging to the genus *Brucella*, a strictly aerobic coccobacillus. *Brucella* can enter the body via the skin, respiratory tract, or digestive tract. Once there, this intracellular organism can enter the blood and the lymphatic canals where it multiplies inside the phagocytes. The disease spreads through animal contacts or contaminated food, especially cheese! Cheese permits milk to be preserved, and several hundred of people are infected each year in France. The disease causes nausea, meningitis, hepatitis, and miscarriages (Straight and Martin 2002).

Botulism is a food-borne disease; the agent of the disease is an anaerobic bacteria *Clostridium botulinium* with a spore-forming rod that produces a neurotoxin. The spores are heat-resistant and can survive in foods that are incorrectly or minimally processed. The disease is caused by the neurotoxin produced by *C. botulinium* that is present in the food (Smith and Sugiyama 1988). The organism and its spores are widely distributed in nature (e.g. cultivated and forest soils, coastal waters, gut of fish and mammals, gills and viscera of crabs and other shellfishes), but the types of foods involved in botulism vary according to regional food preservation and eating habits. Almost any type of food that is not too acidic (pH above 4.6) can support the growth and toxin production of *C. botulinium*. Botulinal toxin has been evidenced in a considerable variety of foods, such as canned corn, soups, ham, sausage, smoked, and salted fish. The incidence of the disease is low, but the mortality is high if not immediately and properly treated.

10.5.2 Intensive farming

High intensity farming has been very good at producing large amounts of cheap food, but has also opened up new parts of the human–pathogen ecosystem. Salmonellosis is one of the most common food-borne illness causing enteric infections in developed countries. The pathogens are bacteria (*Salmonella*) which consist of a range of very related bacteria, many of which are pathogenic to humans and animals (Thorns 2000). The strains which are implicated in the diseases are generally different serovars of *Salmonella enterica* that caused diseases of the intestine, as suggested by their name. For example, *S. enterica* serovar typhi is the causative agent of typhoid fever. It is very common in developing countries, where it causes a serious and often fatal disease. *Salmonella* bacteria primarily invade the wall of the intestines causing inflammation and damage. Infection can spread in the body through the bloodstream to other organs such as liver, spleen, lung, joints, placenta, or foetus, and the membrane surrounding the brain. Toxic substances produced by bacteria can be released and affect the rest of the body. *Salmonella* has evolved mechanisms to escape our immune system (Olsen *et al.* 2001). In the liver, bacteria can grow again, and be released back into the intestine. *S. enterica* serovar enteritidis has become the single most common cause of poisoning in the United States in the last 20 years. *Salmonella* are found in contaminated food, with recent increases in the number *S. enteritidis* a consequence of mass production chicken farms. When tens or hundreds of thousands of chickens live together, die together and are processed together, a *Salmonella* infection can rapidly spread throughout the whole food chain, and hence *Salmonella* can be rapidly dispersed among million of people.

Changed farming practices have also led to new opportunities for non-food-borne human pathogens. In the Asillo zone, located at a very high altitude of 3910 m in the Peruvian Altiplano, high levels of human infection by *Fasciola hepatica* (i.e. the liver fluke) were linked to man-made irrigation zones. Man-made irrigation areas are built only recently to which both liver fluke and lymnaeid snails (i.e. the first intermediate host of the liver fluke) have quickly adapted. Such man-made water resources in high altitude of Andean countries dangerously inflate the parasitic risk, because the lack of drinking water and running water inside dwellings forces inhabitants to obtain water from irrigation canals and drainage channels (Esteban *et al.* 2002). Elsewhere, it is largely recognized that a significant amount of malaria transmission in Africa and Madagascar is due to human activities. Modifications of the environment resulting from land use often create or alter habitats for mosquito vectors, and may indirectly affect parasite development rates, and the lifespan of mosquito (Service 1991; Ault 1994; Coluzzi 1994).

Marine ecosystems are also on the danger list due to pollution and aquaculture which modify all parameters of natural equilibrium. The organic pollution exerts both oppressing and stimulating influences, with industrial waste depressing the formation and function of parasite systems. Some current aquaculture practices are environmentally benign, others, especially those in some of the fastest growing portions of the industry, can degrade water quality, transmit diseases to wild populations, disrupt marine ecosystems, and spread invasive parasites and pathogens species (Maender 2002; Young 2003, see also Chapter 7).

We think that it is now important to present a current very deep problem which links public and veterinary health. At the time we are writing this chapter at the beginning of 2004, public opinion and World Health Organization is strongly focused on two animal zoonotic diseases which have arisen recently, particularly in Asia. 'The terror of the unknown is seldom better displayed than by the response of a population to the appearance of an epidemic, particularly when the epidemic strikes without apparent cause'. This quote from Kass (1977) concerned the emergence of legionnaires' disease, and it well describes public response to the recent emergence of an atypical pneumonia named as Severe Acute Respiratory Syndrome (SARS). SARS was first recognized in the Guangdong Province of China, in November 2002. Subsequent to its introduction in Hong Kong in mid-February 2003, the virus spread to more than thirteen countries and caused disease across five continents. According to the World Health Organisation (WHO) in January 2004, a cumulative total of eight thousand SARS cases with more than

800 deaths had been reported. A novel coronavirus was identified as the human etiological agent of SARS, causing a similar disease in cynomolgous macaques (Peiris *et al.* 2003). Because cases where SARS was first diagnosed occurred in restaurant workers handling wild mammals and exotic food, scientists focused on wild animals recently captured and marketed for consumption. Their work provides evidence that SARS shifts from animals to humans, possibly frequently (Guan *et al.* 2003). Elsewhere, Stanhope *et al.* (in press) have confirmed this host shift, because they could identified that SARS-CoV has a recombinant history with lineages of types I and III virus, concomitant with the reassortment of bird and mammalian coronaviruses. Food marketing trades could thus provide the opportunity animal ScoV-like viruses to amplify and to be transmitted to new hosts, including humans. Even if the natural reservoir is not clearly identified, market animals (civets, raccoon dog, and ferret badgers for example) might be compatible hosts that increase the opportunity for transmission of the virus to humans. Markets constitute thus man-made ecosystems favourable to a new set of pathogenic agents. Thousand of these aforesaid animal species have been slaughtered as a preventive measure. But, is it the most efficient solution?

The Avian Influenza outbreak which is currently raging in south east Asia is related to SARS only in demonstrating how the so-called zoonotic diseases in animals can become a threat to human health. It clearly illustrates how man-made intensive farming ecosystems can represent spring boards for new pathogens. Highly pathogenic avian influenza A viruses of subtypes H5 and H7 belong to the Orthomyxoviridae family, and are the causative agent of fowl plague in poultry. Type A influenza viruses are those which affect humans, but also pigs, horses, and some marine mammals (whales and seals). There are three types (A, B, and C) of influenza viruses, depending on the antigens detected in the virus capsid. The antigens that are used to recognize the different viruses belong to two kinds of glycoproteins, hemagglutinin (HA), and neuramidase (NA). There are fifteen different HA antigens (H1 to H15) and nine different NA antigens (N1 to N9) for influenza A. Human disease historically has been caused by three subtypes of HA (H1, H2, and H3) and two subtypes of NA (N1 and N2), which were responsible for million of deaths around the world, through different pandemics. All known subtypes of influenza A can be found in birds, but only subtypes H5 and H7 have caused severe disease outbreaks in bird populations (Fouchier *et al.* 2004). Figure 10.6 illustrates the life cycle and possible routes taken by these pathogens; migratory birds and waterfowl are thought to serve as reservoir hosts for influenza A virus in nature (Murphy and Webster 1996), but waterfowls generally do not suffer from the disease when infected with avian influenza.

The viruses easily circulate among birds worldwide as they are very contagious for birds, and can be deadly, particularly for domesticated birds like chickens and turkeys. The disease spreads rapidly within poultry flocks and between farms. Direct contact of domestic flocks with wild migratory waterfowls has been implicated as a frequent cause of epidemics, and live bird markets are implicated in the spread of the disease. Avian influenza A virus may initiate new pandemics in humans because the human population is serologically naive toward most HA and NA subtypes (Fouchier *et al.* 2004). Until recently, it was considered that pigs were the obligate intermediate host for transmission of these virus types to humans (Yasuda *et al.* 1991; Webster 1997, Fig. 10.6). Past influenza pandemics have led to high levels of illness, death, social disruption and economic loss (Fig. 10.6), but in general, avian influenza viruses do not replicate efficiently or cause disease in humans (Bear and Webster 1991). However, the highly pathogenic influenza virus subtype H5N1, first documented in Hong Kong in 1997, was transmitted in 2003 from bird to humans, and was responsible for a very serious outbreak, the seriousness of which is still unclear at time of writing. Nevertheless, this epidemic is an illustration of how intensive poultry farming ecosystems (i.e. increasing host density to increase food productivity) are responsible for these new outbreaks, at least in the countries where this influenza H5N1 virus finds its origin (Fig. 10.6).

10.5.3 But how can viruses of high and low virulence coexist?

This is a case of a classical question in evolutionary biology: what conditions are required to maintain polymorphism (genetic and/or phenotypic variability), in space and time, within and among populations of all kind of species? Literature on the topic is rich (e.g. De Meeüs *et al.* 1993 and 1995; De Meeüs and Renaud 1996 for examples). We will not attempt here to make a review on this topic, but in their paper published in 2004, Boots, Hudson, and Sasaki developed a theoretical model to envisage the conditions for maintenance and spreading of low versus high virulent type of pathogens (e.g. viruses). The model clearly shows that large shifts in pathogen virulence are related to host population structure (i.e. demography and genetic). Poultry flock structures managed by humans represent new ecological and demographic configurations for the evolution and emergence of new virulent pathogen strains, which enter more and more in contact with human populations.

Given the increasing long-distance movement of people and domestic animals around the modern world, our results have important implications for emerging diseases in general. Recombination among avirulent (and therefore possibly previously undetected) strains of viruses and other pathogens may produce new virulent strains that may spread through vertebrate host populations because they have shifted to a new evolutionary stable state. (Boots *et al.* 2004)

What is happening with the current influenza outbreak has been clearly predicted by theoretical population biology models. This stresses the need there is that such approaches should be taken now into account in future public and veterinary health control.

Pathogens and parasites can use different man-made ecosystems to spread and threaten human's health. Indeed, as illustrated in the case of influenza virus, the pathogen can first exploit the intensive farming processes (i.e. food production) in order to emerge and rapidly infect humans (Fig. 10.6), and second, hitch-hike the man-made 'travel ecosystems' (i.e. plane, boat, or train) to spread among populations and become pandemic (Fig. 10.2).

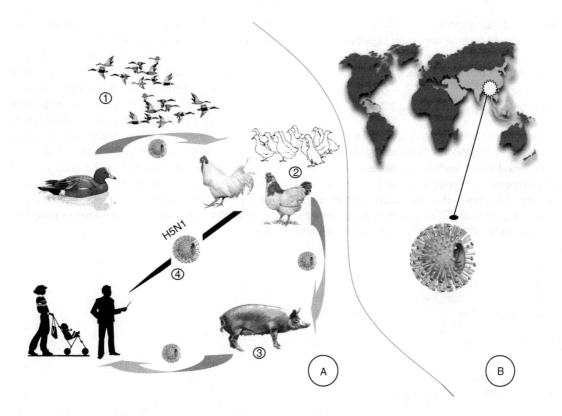

Figure 10.6 Avian influenza: (a) infecting routes and (b) geographical outbreaks in 2004 and the Avian influenza virus.

Notes: (a) Waterfowl are thought to be the natural reservoir of Avian influenza A viruses (1). Viruses replicate in the intestines as well as the respiratory tract of birds.

During migratory processes of birds, poultry flocks become infected when contacts between them and naturally infected birds are established. In the 2004 outbreak, very large quantities of virus were excreted in the faeces of infected farming birds, resulting in widespread contamination of the environment (2). This presence of numerous H5N1 subtype created one of the most important risks for human exposure and subsequent infection.

Some findings support the hypothesis that the pig was a 'mixing vessel', able to produce new virus subtypes by genetic reassortment that can infect humans. Until recently, it was supposed that pigs were obligatory intermediate hosts for human infections (3). Nevertheless, epizoonose of Avian Influenza in different Asian countries in 2004 confirmed the possibility of direct human infection from birds, via the H5N1 Influenza virus subtype (4). Indeed, if Avian Influenza viruses lack the receptors needed to infect mammals efficiently, the infection of humans observed during the 2004 and two previous H5N1 outbreaks demonstrates that transmission from birds to mammals, including humans, can occur despite the lack of receptors. (b) The influenza outbreak in 2004 affected more than ten Asian countries.

From the 20 February 2004, Thai authorities reported that 147 patients were admitted in the hospital since the beginning of the zoonotic outbreak, and eight died. The main problem for all countries around the world is that this Asian influenza outbreak became pandemic. There were three pandemics in the twentieth century. All of them spread worldwide within one year:

– 1918–19: 'Spanish flu' [H1N1]: 20–50 million people died worldwide. Nearly half of those who died were young, and healthy adults.
– 1957–58: 'Asian flu' [H2N2]: First identified in China, the Asian flu spread in the United States and caused about 70,000 deaths.
– 1968–69: 'Hong Kong flu' [H3N2]: First identified in Hong Kong, the virus spread in the United States and caused about 35,000 deaths.

More than seven billion poultry were slaughtered in infected Asian countries which suggest the farming of billions of bird in this geographic area. This demographic situation is favourable for (i) the emergence and the spread of highly virulent strains of pathogens within and between stock breeding, and (ii) the transfer between host species, including humans.

In front of such processes, we do not know which zoonoses will become important in public health in the future, and we must be vigilant over the emergence of new pathogens.

10.6 Concluding remarks

It was not our intention to produce an exhaustive review of all situations where pathogens and parasites found new infectious routes associated with human customs, development, and technology, and we are conscious of many gaps (for instance, Western beds provide an excellent ecosystem for ectoparasites such as lice, fleas, ticks, sarcoptes, bed bugs, etc.). But our goal was to present some aspects which we believe will become more and more topical given current trends. We believe future key questions will be how societies could and should manage (i) population ageing, (ii) need of food access, (iii) earth demographic growth, (iv) people density and urbanization, (v) new technical and medical tools. This is a challenge for which we must get prepared. Pathogens are everywhere and can adapt to a wide panel of environments (even computers). Indeed, we may be just at the beginning of the evolution of pathogens and parasites evolution in our biosphere ecosystem.

CONCLUSION

Parasites, communities, and ecosystems: conclusions and perspectives

Gary G. Mittelbach

What is a parasite?

Most ecologists harbour a classic, taxonomic view of parasites; that is, parasites are protozoans or metazoans that occupy and harm their free-living hosts. Visions of worms come to mind. Yet, as the chapters in this book forcefully argue, the definition of a parasite is much broader than this traditional view (e.g. Guégan et al., Chapter 2). In fact, Moore (2002) suggests that parasites are like pornography—they elude definition (but we may know them when we see them). Clearly, many bacteria, viruses, fungi, and other symbiotic organisms may be viewed as parasites. If we adopt this broad perspective, then we soon realize that most of the world's species are probably parasitic (Price 1980; Toft 1986). Further, as Brown et al. (Chapter 9) note, 'If an organism is not a parasite, then it harbors parasites, as do many parasites themselves'. Thus, we reach the inevitable conclusion that the majority of species interactions also involve parasitism.

Community ecologists have long focused on species interactions as major determinants of the distribution and abundance of organisms. More recently, ecosystem ecologists also have begun to recognize the importance of species and species interactions to the functioning of ecosystems (e.g. Loreau et al. 2001; Tilman et al. 2001; Loreau et al. Chapter 1). Yet, studies of parasites continue to

W. K. Kellogg Biological Station and Department of Zoology
Michigan State University, Hickory Corners, MI.

lie outside mainstream community and ecosystem ecology. As Loreau et al. (Chapter 1) discovered, '...the journal Ecosystems has not published a single paper containing the words parasite, parasitism or parasitoid in its title, keywords or even abstract' since it was founded in 1998. No doubt, the journal Ecosystems has covered topics that would be included in our broad definition of parasites (e.g. mycorrhizal fungal associations). But, the point remains that parasites are not included in the everyday thinking of most community and ecosystem ecologists. Why is this the case?

As the authors of this book point out, there are many possible reasons. For one, parasites are hard to see. They are often small and live inside other organisms, and ecologists do not notice them when we routinely sample communities and ecosystems. Second, if we do find parasites, they can be repulsive. When our field sampling turns up a sick or infected animal (or even plant), we often shy away. Third (and most important, I think), is that the effects of parasites may be subtle. For example, we may see predation in action, but we are unlikely to notice that the reason one prey was eaten and the other was not was because of a difference in parasite load or disease. Or, we may discover that interspecific competition determines the distribution of plant species along an environmental gradient, but we do not see that it was a fungal infection that tipped the competitive balance between the species. One of the important messages of this book is that parasites influence the interactions between species and that while the mechanisms may be subtle, the results can be profound.

How can ecologists better incorporate the effects of parasites into studies of communities and ecosystems? The contributors to this book have identified many potential avenues for progress. In this concluding chapter, I gather together some of the ideas developed in this book and (along with the authors) try to highlight future directions for study. Some of these 'future directions' are new, exciting areas of active research. Others are old, tough nuts to crack—challenging problems that will yield only to continued hard work and careful thought.

It was Mark Twain who quipped that America and England are 'Two nations separated by a common language'. Likewise, building stronger linkages between ecology and parasitology will require, in part, better communication. Ecologists and parasitologists/epidemiologists are traditionally trained in separate departments, with little overlap in coursework and little overlap in vocabulary. As a consequence, ecologists are often stymied by the wealth of specialized terms used by parasitologists to describe the complex life cycles of their organisms. Similarly, ecologists have their own specialized jargon to describe the myriad ways in which species may interact with each other and with their environment. In summarizing the research themes below, I try to note examples of where ecologists and parasitologists have been studying the same processes, but describing it with a different language.

How do parasites affect communities and ecosystems?

Trait-mediated effects

Parasites can strongly affect the dynamics of their host populations and a rich tradition of theory and empirical data underlies the study of host/parasite dynamics (Møller, Chapter 3). In the past, most studies have considered only the direct effects of parasites on host mortality or host fecundity. However, there is now a growing body of work documenting how parasites affect host behaviours

as well (e.g. Brown *et al.* Chapter 9; see also Poulin 1994a; Moore 2002; Thomas *et al.* 2004). In many cases, parasites alter host behaviours in ways that increase the host's probability of being eaten, thereby increasing the parasite's likelihood of transmission to the next host. Thus, parasites may affect host densities through direct impacts on host mortality or fecundity, or indirectly, through changes in host behaviour, morphology, or physiology. Community ecologists have classified the analogous impacts of predators on their prey as either density-mediated or trait-mediated effects (Abrams *et al.* 1996). Interestingly, the typical trait-mediated effects of the parasite on its host tend to increase host mortality, while the typical trait-mediated effects of a predator on its prey result in a decrease in mortality. There is currently great interest in understanding the relative importance of trait-mediated and density-mediated effects of predators on the structure of ecological communities (e.g. Werner and Peacor 2003 and other papers in this special feature in *Ecology*). Surprising, however, the impacts of parasites on host behaviours are rarely, if ever, mentioned as examples of trait-mediated indirect effects. This surely needs to be remedied, both by ecologists and parasitologists.

Cascading effects and food webs

How might the effects of parasites on host densities and host behaviours impact communities and ecosystems? In perhaps the simplest case, if parasites affect the abundance of a keystone species (sensu Paine 1966), the impact may cascade dramatically through the community. Loreau *et al.* (Chapter 1) suggest that these effects will be most dramatic when the affected host species occurs high in the food web. On the other hand, we would also expect to see strong, community-wide effects when the affected host species is an abundant foundation species (sensu Bruno *et al.* 2003). For example, the near elimination of the American chestnut (*Castanea dentate*) by a fungal pathogen in the early 1900s dramatically altered the structure of eastern US deciduous forests. How this changed the functioning of the ecosystem is unknown. However, given the ability of some pathogens to completely decimate their hosts, the cascading effects of parasites (defined in the broad sense) may be much larger than those commonly attributed to textbook examples of keystone predators.

Recent interest in the impacts of biodiversity on ecosystem functioning suggests further mechanisms by which parasite effects on host species abundance may impact ecosystems. As Loreau *et al.* (Chapter 1) note, 'An important limitation of virtually all recent theoretic and experimental studies on the effects of biodiversity on ecosystem function and stability is that they have concerned single trophic levels—primary producers for the most part'. However, we know that interactions between trophic levels may affect the biomass, productivity, and functional composition of different trophic levels, including primary producers. Therefore, incorporating multiple trophic levels into theoretical and experimental studies of biodiversity and ecosystem functioning is a critical next step (e.g. Downing and Leibold 2002; Duffy *et al.* 2003; Thébault and Loreau 2003; Petchey *et al.* 2004). However, incorporating parasites into ecology's well-developed framework of food chains and food webs is challenging to say the least (Sukhdeo and Hernandez, Chapter 4). In those cases where parasites kill their hosts, the problem is somewhat easier. Parasitoids act much like 'standard' ecological predators and a substantial body of theory and empirical studies describes this 'consumer–resource interaction' (Murdoch *et al.* 2003). Consumer–resource interactions constitute the building blocks of theoretical and real-world food chains and food webs. However, as noted earlier, the effects of many parasites are sublethal. Incorporating these sublethal effects of parasites may require approaches like those recently used by ecologists to incorporate the trait-mediated effects of 'standard' ecological predators into food webs (e.g. Abrams 1995, 2004). The complex life cycles of many parasites, with their multiple hosts and multiple life stages, further complicates the prospect of effectively incorporating this important class of organisms into ecological food webs. Sukhdeo and Hernandez (Chapter 4) present some ideas for how to attack the difficult problem of including parasites in food webs.

Shared parasites, apparent competition, and parasite-mediated coexistence

About 25 years ago, Holt (1977) first developed the mathematical theory showing that species sharing a predator may negatively affect each other's abundance and thereby 'appear' to compete. Since that time, the theory of 'apparent competition' has come to occupy a fundamental place in community ecology and numerous studies have documented its importance in nature. As Thomas *et al.* (Chapter 8) and others note, hosts that share a parasite species may interact via apparent competition just as prey species that share a predator. However, the impact of shared parasites on their host species can be much more complex than the impact of shared predators on their prey, due to the fact that parasites may require intermediate hosts, that hosts may recover from infection (and become immune), and that transmission rates may vary among host species (Holt *et al.* 1994). As Thomas *et al.* (Chapter 8) note, 'It is important to emphasize that changes in prey abundance though the actions of a non-specific parasite can lead to a range of indirect effects that are predicted to enhance or destroy host species diversity. Understanding these effects is the challenge of empirical community ecology.'

Predators may enhance species diversity as well as reduce it and ecologists have identified a number of way in which predators may promote the coexistence of competing species. These include: (1) selective predation on the competitive dominant (Paine 1966; Lubchenco 1978); (2) frequency-dependent predation where mortality falls selectively on the more abundant species (Roughgarden and Feldman 1975; Vance 1978); and (3) predation that opens up 'patches' for species that exhibit a tradeoff between colonizing ability and competitive ability (Slatkin 1974; Caswell 1978; Connell 1978). Each of these classic predator-mediated coexistence mechanisms may apply to parasites as well as predators.

While parasites are commonly viewed as harmful to the host (e.g. as 'shared enemies' above), Brown *et al.* (Chapter 9) note that ' . . . this simple categorization can hide a more interesting mix of antagonistic and overlapping interests'. For example, the boundary between parasite and mutualist may be easily blurred. In the case of mycorrhizal fungi that infect the roots of plants, the interaction between the plant and the fungus is mutualistic in relatively low-resource environments. The fungus benefits the plant in obtaining soil nutrients and the plant benefits the fungus by 'leaking' root exudates. However, in high-resource environments, the fungus becomes a parasite on the plant. Thus, we need to view the interaction between host and parasite as context dependent in the same way that ecologists see context dependency in the interactions between species within a community (i.e. interactions are affected by the nature of the abiotic environment, the presence/absence of other species, etc.). 'Viewing the host-parasite interaction as a complex mix of interests, or *desiderata* (Dawkins 1990) is an important step away from the medically influenced view of parasites being always virulent to some degree' (Brown *et al.* Chapter 9).

Species invasions and parasite escape

Recent evidence suggests that parasites may play a major role in determining the magnitude and impact of species invasions. Species invasions, along with habitat destruction, are perhaps the two greatest threats to biodiversity on a global scale. In the United States alone the economic cost of invasive species is estimated at >$100 billion dollars annually. The enemy release hypothesis proposes that one of the factors underlying the successful introduction and spread of exotic species is that they may leave many of their natural enemies behind. The enemy release hypothesis has a long history in ecology (Darwin 1859; Elton 1958) and initially ecologists focused on species escapes from herbivore and predator 'enemies' (Keane and Crawley 2002). However, recent studies suggest that the escape from parasite 'enemies' may be equally or more important.

Mitchell and Power (2003), Torchin *et al.* (2003), and Torchin and Mitchell (2004) show that on average, introduced plants and animals escape more than half of their native parasites and this escape is due to both a reduction in the number of parasite species infecting introduced species and a reduction in the percentage of individuals infected (parasite prevalence). Further, while introduced species may acquire new parasite species from the introduced range, on average they acquire <25% of the parasite species they escape (Mitchell and Power 2003; Torchin *et al.* 2003). How important is this release from parasites in contributing to the demographic success and expansion of exotics in their new habitat? The answer awaits experimental tests (e.g. Callaway *et al.* 2004), however, the correlative patterns are suggestive (Torchin and Mitchell 2004). Also, because introduced species have been shown to accumulate parasite species through time (Cornell and Hawkins 1993; Torchin *et al.* 2001), it is tempting to ascribe the commonly observed boom and bust of many introduced species to any early escape from natural enemies (parasites) and a subsequent increase in acquired enemies through time. The success and failure of purposeful species introductions also provides evidence of the role of parasites in species invasions, as shown by Møller's (Chapter 3) work on the immune defences of bird species introduced to New Zealand in the late 1800s. Although it is too early to say how important parasites are in determining the successful establishment and demographic response of introduced species, this is clearly an area where parasites may play an important role in communities and ecosystems.

The ecosystems of parasites

Ecosystem types

In the discussions above, I have been mostly concerned with the roles that parasites play in communities and ecosystems. However, what about the ecosystems of parasites themselves? This is a complex and fascinating topic, driven in part by the fact that parasites occupy two classes of ecosystems. As Brown *et al.* (Chapter 9) note, we can think of the host as the first ecosystem that the parasite occupies. Numerous interactions may occur between parasite species within the host and parasites may transform or 'engineer' their host ecosystem in complex and interesting ways (Thomas *et al.* Chapter 8; Tinsley, Chapter 6). The second ecosystem that parasites occupy is the larger 'global' ecosystem that contains the host population(s). This ecosystem includes all the complex spatial and temporal dynamics associated with each host population (Brown *et al.* Chapter 9, Holt and Boulinier, Chapter 5). The field of parasitology tends to focus on the first 'host' ecosystem, while studies of the 'global' ecosystem tend to fail within the domain of epidemiology and more recently the rapidly developing field of the ecology of infectious diseases (Grenfell and Dobson 1995). Understanding the ecology of parasites and their impacts necessarily includes studying both ecosystems.

Ecosystems and environmental change

The chapters by Lafferty and Kuris (Chapter 7), Renaud *et al.* (Chapter 10), and Tinsley (Chapter 6) highlight important ways in which environmental change and human-made environments impact the ecosystems in which parasites occur. In recent years, ecologists have documented how species ranges are shifting in response to climate change (Parmesan and Yohe 2003; Root *et al.* 2003), concluding that such shifts will have profound effects on species interactions and biodiversity (Thomas *et al.* 2004). The study by Thomas *et al.* (2004) predicts that 18–35% of species in their sample regions will be 'committed to extinction' by 2050 due to climatic effects on species's ranges. Likewise, climate change may alter host and parasite geographical ranges, with potentially dramatic consequences for disease outbreaks (Harvell *et al.* 2002). Many aspects of host/parasite ecology are climate dependent and forecasting the effects of climate variation on infectious diseases is an area of active research (e.g. Patz 2002; Rodo *et al.* 2002).

In addition to altering the natural ecosystems of parasites, human have created entirely new and novel ecosystems for parasites, to which they have rapidly adapted. The chapter on 'Parasitism in man-made ecosystems' by Renaud *et al.* (Chapter 10) presents a series of fascinating (and frightening) examples of new ecosystems and new ecological niches (ranging from hospitals, to cemeteries, to air conditioning units) that have been colonized by opportunistic pathogens. Understanding the ecology of parasites in these new ecosystems represents one of the many emerging links between ecology and public health.

Spatial scale, meta-populations and meta-communities, and the effects of local and regional processes

As Guégan *et al.* (Chapter 2) and Holt and Boulinier (Chapter 5) note, the fields of ecology, parasitology, and epidemiology are each profoundly influenced by spatially dependent processes. Both free-living and parasitic organisms may function as meta-populations (Grenfell and Harwood 1997; Hanski 1999). Moreover, just as the dispersal of individuals may link the dynamics of populations in a meta-population, the dispersal of species may link the dynamics of communities across a landscape. The 'meta-community' model is an area of active research in ecology (e.g. Mouquet and Loreau 2003; Leibold *et al.* 2004), with theory predicting that the composition of any local community is the result of within-community interactions and the dispersal of species between communities. In addition, the response of communities and ecosystems to environmental change (e.g., global warming) may depend critically on the types of species present and their traits. When new species are introduced from the regional species pool, communities and ecosystems have the potential to function as complex adaptive systems responding to these environmental changes (Norberg *et al.* 2001; Norberg 2004). To date, meta-community theory has only considered free-living organisms, but of course parasites might exhibit meta-community properties as well. The assemblage of parasite species within a host may be viewed as a community of parasites that interact with their environment (host) and with each other. The outcome of these interactions will depend on the traits of the species present and on the potential for new species to invade the 'community'.

Many of the spatial biodiversity patterns observed in free-living organism (e.g. species–area relationships, species richness–isolation relationships, local versus regional richness relationships) are found in parasitic organisms as well (Guégan *et al.* Chapter 2). The best-known biodiversity pattern, the increase in species richness observed when travelling from the poles to the tropics (i.e. the latitudinal gradient), was described more than

200 years ago (Forster 1778; von Humboldt 1808) and has been documented in a wide variety of free-living taxa (see Willig *et al.* 2003). Broad-scale studies of parasite species richness are far less common and there is disagreement over how well parasites fit the classic pattern of increasing diversity with decreasing latitude (e.g. Rohde 1999; Willig *et al.* 2003). In Chapter 2, Guégan *et al.* discuss a new data set by Guernier *et al.* (2004) that examines the geographical distribution of >200 species of human pathogens (bacteria, viruses, fungi, protozoa, and helminthes). This extensive data set shows that pathogen species richness is strongly correlated with latitude in the classical pattern and that the effect of latitude remains after correcting for cofactors such as area and socio-demographic variables. Understanding the mechanisms driving broad-scale diversity gradients is an area of intense ecological research. As Guégan *et al.* note, parasite diversity may simply follow host species diversity, or instead, the same general processes may drive broad-scale diversity gradients in both free-living and parasitic groups. Further studies on broad-scale diversity patterns of parasites and pathogens are needed, as are theoretical and empirical studies that examine the effects of interacting spatial scales.

Final comments

While writing this chapter I thought many times of Hutchinson's little paper on 'Copepodology for the ornithologist' (Hutchinson 1951). Hutchinson wrote this paper (remembered more for its title than its content) because he wanted to share what he had recently learned from studying copepods with a broader audience of evolutionary biologists. In his introductory paragraph, Hutchinson notes that 'ornothologists and other students of terrestrial ecology' were unlikely to read an account of copepods in his forthcoming *Treatise on Limnology*. It is important that we remember Hutchinson's example. We all tend to become specialists in our own fields and the daily deluge of scientific papers overwhelms our abilities to keep up even in our own area. But, breakthroughs often come about when we recognize the generality of pattern or mechanism. 'Parasitology for the ecologist?' 'Ecology for the parasitologist?'—readers of this book will have discovered some of both.

References

Abrams, P. A. (1995). Implications of dynamically variable traits for identifying, classifying, and measuring direct and indirect effects in ecological communities. *American Naturalist* **146**: 112–134.

Abrams, P. A. (2004). Trait-initiated indirect effects due to changes in consumption rates in simple food webs. *Ecology* **85**: 1029–1038.

Abrams, P. A., Holt, R. D., and Roth, J. D. (1998). Apparent competition or apparent mutualism? Shared predation when populations cycle. *Ecology* **79**: 201–212.

Abrams, P. A., Menge, B. A., Mittelbach, G. G., Spiller, D. A., and Yodzis, P. (1996). The role of indirect effects in food webs. In: *Food webs: integration of patterns and dynamics* (ed. G. A. Polis and K. O. Winemiller). New York, Chapman and Hall, pp. 371–395.

Adamo, S. A. (1999). Evidence for adaptive changes in egg laying in crickets exposed to bacteria and parasites. *Animal Behaviour* **57**: 117–124.

Adamson, M. L. and Caira, J.N. (1995). Evolutionary factors influencing the nature of parasite specificity. *Parasitology* **109**: S85–S95.

Aeby, G. S. (2002). Trade-offs for the butterflyfish, *Chaetodon multicinctus*, when feeding on coral prey infected with trematode metacercariae. *Behavioral Ecology and Sociobiology* **52**: 158–165.

Agnew, P., Bedhomme, S. , Haussy, C., and Michalakis, Y. (1999). Age and size at maturity of the mosquito *Culex pipiens* infected by the microsporidian parasite *Vavraia culicis*. *Proceedings of Royal Society of London Series B— Biological Sciences* **266**: 947–952.

Ågren, G. I. and Bosatta, E. (1996). *Theoretical ecosystem ecology*. Cambridge, Cambridge University Press.

Aguirre, A. A., Ostfeld, R. S., Tabor, G. M., House, C. and Pearl, M. C. (2002). *Conservation Medicine: ecological health in practice*. New York, Oxford University Press.

Aho, J. M. (1990). Helminth communities of amphibians and reptiles: comparative approaches to understanding patterns and processes. In: *Parasite communities: patterns and processes* (ed. G. Esch, A. O. Bush, and J. M. Aho). London, Chapman and Hall, 157–195.

Aho, J. M. and Bush, A.O. (1993). Community richness in parasites of some freshwater fishes from North America species. In: *Diversity in ecological communities. Historical and geographical perspectives*. (ed. R. E. Ricklefs and D. Schluter). Chicago, The University of Chicago Press, pp. 185–193.

Alary, M. and Joly, J. R. (1992). Factors contributing to the contamination of hospital water distribution systems by legionellae. *Journal of Infectious Diseases* **165**: 565–569.

Alcami, A. and Koszinowski, U. H. (2000). Viral mechanisms of immune evasion. *Trends in Microbiology* **8**: 410–418.

Alker, A. P., Smith, G. W. and Kim, K. (2001). Characterization of *Aspergillus sydowii* (Thom et Church), a fungal pathogen of Caribbean sea fan corals. *Hydrobiologia* **460**: 105–111.

Allan, J. S. (1995). Xenograft transplantation and the infectious disease conundrum. *ILAR Jounal* **37**: 37–48.

Allen, M. F. (1991). *The Ecology of Mycorrhizae*. Cambridge, Cambridge University Press.

Allen, T. F. H. and Hoekstra, T.W. (1992). *Toward a unified ecology*. Columbia, NY, Columbia University Press.

Allen, T. F. H., King, A.W., Milne, B., Johnson, A and Turner, S. (1993). The problem of scaling. *Evolutionary Trends Plants* **7**: 3–8.

Altizer, S. (2001). Migratory behaviour and host–parasite coevolution in natural populations of monarch butterflies infected with a protozoan parasite. *Evolutionary Ecology Research* **3**: 611–632.

Amundsen, P. A. and Kristoffersen, R. (1990). Infection of whitefish (*Coregonus lavaretus* L. s.l.) by *Triaenophorus crassus* Forel (Cestoda: Pseudophyllidea): a case study in parasite control. *Canadian Journal of Zoology* **68**: 1187–1192.

Anderson, C. (1991). Cholera epidemic traced to risk miscalculation. *Nature* **354**: 255.

Anderson, R. C. (1972). The ecological relationships of meningeal worm and native cervids in North America. *Journal of Wildlife Diseases* **8**: 304–310.

Anderson, R. C. (1988). Nematode transmission patterns. *Journal of Parasitology* **74**: 30–45.

Anderson, R. M. (1989). Discussion: ecology of pests and pathogens. In: *Perspectives in ecological theory*. (ed. J. Roughgarden, R.M. May, and S.A. Levin). Princeton, Princeton University Press, pp. 348–362.

Anderson, R. M. and Gordon, D.M. (1982). Processes influencing the distribution of parasite numbers within host populations with special emphasis on parasite-induced host mortalities. *Parasitology* **85**: 373–398.

Anderson, R. M. and May, R. M. (1978). Regulation and stability of host–parasite population interactions. I. Regulatory mechanisms. *Journal of Animal Ecology* **47**: 219–247.

Anderson, R. M. and May, R. M. (1979). Population biology of infectious diseases: part I. *Nature* **280**: 361–367.

Anderson, R. M. and May, R.M. (1981). The population dynamics of microparasites and their invertebrates hosts. *Philosophical Transactions of the Royal Society of London Series B—Biological Sciences* **291**: 451–524.

Anderson, R. M. and May, R. M. (1986). The invasion, persistence and spread of infectious diseases within animal and plant communities. *Philosophical Transactions of the Royal Society of London Series B—Biological Sciences* **314**: 533–570.

Anderson, R. M. and May, R. M. (1991). *Infectious diseases of humans: dynamics and control*. Oxford, UK, Oxford University Press.

André, J. B., Ferdy, J. B., and Godelle, B. (2003). Within-host parasite dynamics, emerging trade-off, and evolution of virulence with immune system. *Evolution* **57**: 1489–1497.

Antia, R., Levin B. R. and May, R. M. (1994). Within-host population dynamics and the Evolution and Maintenance of Microparasite Virulence. *American Naturalist* **144**: 457–472.

Antia, R., Bergstrom, C. T., Pilyugin, S. S., Kaech, S. M., and Ahmed, R. (2003). Models of Cd8+Responses: 1. What Is the Antigen-Independent Proliferation Program. *Journal of Theoretical Biology* **221**: 585–598.

Antonovics, J., Iwasa, Y., and Hassell, M. M. (1995). A generalized model of parasitoid, venereal, and vector-based transmission processes. *American Naturalist* **145**: 661–665.

Arneberg, P., Folstad, I., and Kartyer, A.J. (1996). Gastrointestinal nematodes depress food intake in naturally infected reindeer. *Parasitology* **112**: 213–219.

Arneberg, P., Skorping, A., Grenfell, B., and Read, A. F. (1998a). Host densities as determinants of abundance in parasite communities. *Proceedings of the Royal Society of London Series B- Biological Sciences* **265**: 1283–1289.

Arneberg, P., Skorping, A. and Read, A. F. (1998b). Parasite abundance, body size, life histories, and the energetic equivalence rule. *American Naturalist* **151**: 497–513.

Arnow, P. M., Quimosing, E. M. and Beach, M. (1993). Consequences of intravascular catheter sepsis. *Clinical Infectious Diseases* **16**: 778–784.

Aron, J. L. and Patz, A. (2001). *Ecosystem change and public health. A Global Perspective*. Baltimore, MD, The Johns Hopkins University Press.

Artensein, A. W., Hicks, C. B., Goodwin, B. S., and Hilliard, J. K. (1991). Human infection with B virus following a needlestick injury. *Review of Infectious Diseases* **13**: 288–291.

Arthur, J. R. (1997). Recent advances in the use of parasites as biological tags for marine fish. In: *Diseases in Asian Aguaculture III* (ed. T. W. Flegel TW and I. H. MacRae). Manila, Fish Health Section, Asian Fisheries Society, pp. 141–154.

Atmar, W. and Patterson, B. D. (1995). The nestedness temperature calculator a visual basic program, including 294 presence-absence matrices. AICS Research Inc., University Park, NM and The Field Museum, Chicago, IL, USA. Available: http://aics-research.com/nestedness/tempcalc.html via the Internet.

Augspurger, C. K. (1983). Seed dispersal by the tropical tree, *Platypodium elegans*, and the escape of its seedlings from fungal pathogens. *Journal of Ecology* **71**: 759–771.

Augspurger, C. K. (1984a). Seedling survival of tropical tree species: interactions of dispersal distance, light gaps, and pathogens. *Ecology* **65**: 1705–1712.

Augspurger, C. K. (1984b). Pathogen mortality of tropical tree seedlings: experimental studies of the effects of dispersal distance, seedling density, and light conditions. *Oecologia* **61**: 211–217.

Ault, S. K. (1994). Environmental management: A re-emerging vector control strategy. *American Journal of Tropical Medicine and Hygiene* **50**: 35–49.

Awachie, J. B. E. (1966). Observations on *Cyathocephalus truncatus* Pallas, 1781 (Cestoda: Sathebothriidea) in its intermediate and definitive hosts, in a trout stream, North Wales. *Journal of Helminthology* **40**: 1–10.

Ayensu, E., Claasen, D. V., Collins, M., Dearing, A., Fresco, L., Gadgil, M., Gitay, H., Glaser, G., Lohn, C. L., Krebs, J., Lenton, R., Lubchenco, L., McNeely, J. A., Mooney, H. A., Pinstrup-Andersen, P., Ramos, M., Raven, P., Reid, W. V., Samper, C., Sarukhan, J., Schei, P., Tundisi, J. G., Watson, R. T., Xu, G. H., and Zakri, A. H. (1999). International ecosystem assessment. *Science* **286**: 685–686.

Bakke, T. A., Harris, P. D., and Cable, J. (2002). Host specificity dynamics: observations on gyrodactylid monogeneans. *International Journal for Parasitology* **32**: 281–308.

Baldari, A., Tamburro, G., Sabatinelli, R., Romi, R., Severini, C., Cuccagna, G., Fiorilli, M.P. Allegri, C., Buriani, C., and Toti, M. (1998). Malaria in Maremma, Italy. *Lancet* **351**: 1246–1247.

Balmford, A., Green, R. E., and Jenkins, M. (2003). Measuring the changing state of nature. *Trends in Ecology and Evolution* **18**: 326–333.

Barbehenn, K. R. (1969). Host–parasite relationships and species diversity in mammals: an hypothesis. *Biotropica* **1**: 29–35.

Barker, D. E., Marcogliese D. J., and Cone D. K. (1996). On the distribution and abundance of eel parasites in Nova Scotia: local versus regional patterns. *Journal of Parasitology* **82**: 697–701.

Bartlett, M. S. (1957). Measles periodicity and community size. *Journal of Royal Statistical Society A* **120**: 48–70.

Bartley, D., Jackson, F., Coop, R. L., Jackson, E., Johnston, K. and Mitchell, G. B. (2001). Anthelmintic-resistant nematodes in sheep in Scotland. *Veterinary Record* **149**: 94–95.

Baron, J. and Galvin, K. A. (1990). Future directions of ecosystem science. *Bioscience* **40**: 640–642.

Barone, J. A. (1998). Host specificity of folivorous insects in a moist tropical forest. *Journal of Animal Ecology* **67**: 400–409.

Bartoli, P. and Boudouresque, C. F. (1997). Transmission failure of parasites (Digenea) in sites colonized by the recently introduced invasive alga *Caulerpa taxifolia*. *Marine Ecology Progress Series* **154**: 253–260.

Bear, A. S. and Webster, R. G. (1991). Replication of avian Influenza viruses in Humans. *Archives of Virology* **119**: 37–42.

Beckage, N. E. (1991). Host–parasite hormonal relationships: a common theme? *Experimental Parasitology* **72**: 332–338.

Beckage, N. E. (1997). *Parasites and pathogens: effects on host hormones and behaviour.* New York, Chapman and Hall.

Beck, M. A. and Levander, O. A. (2000). Host nutritional status and its effect on a viral pathogen. *Journal of Infectious Diseases* **182**: S93–S96.

Beer, S. A. and German, S. M. (1993). Ecological prerequisites of worsening of the cercariosis situation in cities of Russia (Moscow Region as an example). *Parazitologiya (St Petersburg)* **27**: 441–449.

Begon, M., Harper, J. L., and Towsend, C. R. (1990). *Ecology: Individuals, populations and communities.* 2ndedn.Boston, MA, Blackwell Scientific Publications.

Begon, M., Harper, J. L., and Townsend, C. R. (1996). *Ecology.* Oxford, Blackwell.

Bergersen, E. P. and Anderson, D. E. (1997). The distribution and spread of *Myxobolus cerebralis* in the United States. *Fisheries* **22**: 6–7.

Bell, R. G. (1998). The generation and expression of immunity to *Trichinella spiralis* in laboratory rodents. *Advances in Parasitology* **41**: 149–217.

Bennett, R. C. (1993). Diarrhea among residents of long-term care facilities. *Infectious Control of Hospital Epidemiology* **14**: 397–404.

Bentz, S., Leroy, S., du Preez, L., Mariaux, J., Vaucher, C., and Verneau, O. (2001). Origin and evolution of African *Polystoma* (Monogenea: Polystomatidae) assessed by molecular methods. *International Journal of Parasitology* **31**: 697–705.

Berche, P., Brun-Buisson, C., and Jarlier, V. (2000). Les infections nosocomiales. *Médecine thérapie* **1**: 1–120.

Bergey, L., Weis, J. S., and Weis, P. (2002). Mercury uptake by the estuarine species *Palaemonetes pugio* and *Fundulus heteroclitus* compared with their parasites, *Probopyrus pandalicola* and *Eustrongylides* sp. *Marine Pollution Bulletin* **44**: 1046–1050.

Beyers, R. J. and Odum, H. T. (1993). *Ecological microcosms.* New York, Springer.

Black, F. (1966). Measles endemicity in insular populations: critical community size and its evolutionary implication. *Journal of Theoretical Biology* **11**: 207–211.

Black, G. A. (1983). Taxonomy of a swimbladder nematode, *Cystidicola stigmatura* (Leidy), and evidence of its decline in the Great Lakes. *Canadian Journal of Fisheries and Aquatic Sciences* **40**: 643–647.

Blanford, S., Thomas, M. B., Pugh, C., and Pell, J. K. (2003). Temperature checks the Red Queen? Resistance and virulence in a fluctuating environment. *Ecology Letters* **6**: 2–5.

Blaser, M. J. and Feldman, R. A. (1981). *Salmonella bacteremia*: reports to the Centers for Disease Control. *Journal of Infectious Diseases* **143**: 743–746.

Blouin, M.S., Yowell, C. A., Courtney, C. H., and Dame, J.B. (1995). Host movement and the genetic structure of populations of parasitic nematodes. *Genetics* **141**: 1007–1014.

Blumberg, D. (1991). Seasonal variations in the encapsulation of eggs of the encyrtid parasitoid *Metaphycus stanleyi* by the pyriform scale, *Protopulvinaria pyriformis*. *Entomologia Experimentalis Et Applicata* **58**: 231–237.

Boettner, G. H., Elkinton, J. S., and Boettner, C. J. (2000). Effects of a biological control introduction on three non-target native species of saturniid moths. *Conservation Biology* **14**: 1798–1806.

Bolker, B. and Grenfell, B. (1995). Space, persistence and dynamics of measles epidemics. *Philosophical Transactions of the Royal Society of London Series B-Biological Sciences* **348**: 309–320.

Bonhoeffer, S. and Nowak, M. A. (1994). Intra-host versus inter-host selection: viral strategies of immune function impairment. *Proceedings of the National Academy of Science of the USA* **91**: 8062–8066.

Bonsall, M. B. (2003). The role of variability and risk on the persistence of shared-enemy, predator–prey assemblages. *Journal of Theoretical Biology* **221**: 193–204.

Bonsall, M. B. and Hassell, M. P. (1997). Apparent competition structures ecological assemblages. *Nature* **388**: 371–373.

Bonsall, M. B. and Hassell, M. P. (1998). Population dynamics of apparent competition in a host–parasitoid assemblage. *Journal of Animal Ecology* **67**: 918–929.

Boots, M. and Sasaki, A. (1999). 'Small worlds' and the evolution of virulence: infection occurs locally and and at a distance. *Proceedings of the Royal Society of London B* **266**: 1933–1938.

Boots, M., Hudson, P. J., and Sasaki, A. (2004). Large shifts in pathogen virulence relate to host population structure. *Science* **303**: 842–844.

Boulétreau, M., Fouillet, P., and Allemand, R. (1991). Parasitoids affect competitive interactions between the sibling species, *Drosophila melanogaster* and *D. simulans*. *Redia* **84**: 171–177.

Boulinier, T., Ives, A. R., and Danchin, E. (1996). Measuring aggregation of parasites at different host population levels. *Parasitology* **112**: 581–587.

Boulinier, T., McCoy, K. D., and Sorci, G. (2001). Dispersal and parasitism. In: *Dispersal* (ed. J. Clobert, E. Danchin, A. Dhondt, and J. D. Nichols). Oxford, Oxford University Press, pp. 169–179.

Boulinier, T., Sorci, G., Monnat, J.-Y., and Danchin, E. (1997). Parent–offspring regression suggests heritable susceptibility to ectoparasites in a natural population of kittiwake *Rissa tridactyla*. *Journal of Evolutionary Biology* **10**: 77–85.

Bouma, M. J. and Dye, C. (1997). Cycles of malaria associated with El Niño in Venezuela. *Journal of the American Medical Association* **278**: 1772–1774.

Branagan, D. and Hammond, J. A. (1965). Rinderpest in Tanganyika: A review. *Bulletin of Epizootic Diseases of Africa* **13**: 225–246

Brändle, M. and Brandl, R. (2001). Species richness of insects and mites on trees: expanding Southwood. *Journal of Animal Ecology* **70**: 491–504.

Brändle, M. and Brandl, R. (2003). Species richness on trees: a comparison of parasitic fungi and insects. *Evolutionary Ecology Research* **5**: 941–952.

Bray, R. A., Littlewood, D. T. J., Herniou, E. A., Williams, B., and Henderson, R. E. (1999). Digenean parasites of deep-sea teleosts: a review and case studies of intrageneric phylogenies. *Parasitology* **119**: S125–S144.

Bremermann, H. and Pickering, J. (1983). A Game-Theoretical Model of Parasite Virulence. *Journal of Theoretical Biology* **100**: 411–426.

Brooks, D. R. and McLennan D. A. (1991). *Phylogeny, ecology, and behavior: a research program in comparative biology*. Chicago, University of Chicago Press.

Broutin, H., Simondon, F., and Guégan, J.-F. (2004). Whooping cough metapopulation dynamics in tropical conditions: disease persistence and impact of vaccination. *Proceedings of the Royal Society of London Series B— Biological Science., (Suppl.) Biology Letters* **271**: 5302–5305.

Brown, A. F. and Pascoe, D. (1989). Parasitism and host sensitivity to cadmium: an acanthocephalan infection of the freshwater amphipod *Gammarus pulex*. *Journal of Applied Ecology* **26**: 473–487.

Brown, C. R. and Brown, M. B. (1992). Ectoparasitism as a cause of natal dispersal in cliff swallows. *Ecology* **73**: 1718–1723.

Brown, J. H. (1995). *Macroecology*. Chicago, University of Chicago Press.

Brown, J. K. M. and Hovmeller, M. S. (2002). Aerial dispersal of pathogens on the global and continental scales and its impact on plant disease. *Science* **297**: 537–541.

Brown J.H. and Kodric-Brown, A. (1977). Turnover rates in insular biogeography: effect of immigration on extinction. *Ecology* **58**: 445–449.

Brown, S. P. (1999). Cooperation and conflict in host-manipulating parasites. *Proceedings of the Royal Society of London Series B- Biological Sciences* **266**: 1899–1904.

Brown, S. P. (2001). Collective action in an RNA virus. *Journal of Evolutionary Biology* **14**: 821–828.

Brown, S. P. and Grenfell, B. T. (2001). An unlikely partnership: An unlikely partnership: Parasitism, concomitant immunity and host defence. *Proceedings of the Royal Society of London Series B—Biological Sciences* **268**: 2543–2549.

Brown, S. P. and Johnstone, R. A. (2001). Cooperation in the dark: signalling and collective action in quorum-sensing bacteria. *Proceedings of the Royal Society of London Series B—Biological Sciences* **268**: 961–967.

Brown, S.P., Hochberg, M.E., and Grenfell, B. T. (2002). Does multiple infection select for increased virulence? *Trends in Microbiology* **10**: 401–405.

Brown, S. P., Loot, G., Grenfell, B. T., and Guegan, J. F. (2001). Host manipulation by *Ligula intestinalis*—accident or adaptation? *Parasitology* **123**: 519–529.

Brownlee, D. J. A., Fairweather, I., Johnston, C.F., Thorndyke, M.C., and Skuce, P. J. (1995). Immunocytochemical demonstration of a SALMFamide-like neuropeptide in the nervous system of adult and larval stages of the human blood fluke, *Schistosoma mansoni*. *Parasitology* **110**: 143–153.

Bruno, J. F., Stachowicz, J. J., and Bertness, M. D. (2003). Inclusion of facilitation into ecological theory. *Trends in Ecology and Evolution* **18**: 119–125.

Buchmann, K. and Lindenstrøm, T. (2002). Interactions between monogenean parasites and their fish hosts. *International Journal of Parasitology* **32**: 309–319.

Bull, J. J. (1994). Perspective: virulence. *Evolution* **48**: 1423–1437.

Bundy, D. A. P. (1981). Swimming behaviour of the cercaria of *Transversotrema patialense*. *Parasitology* **82**: 319–334.

Burdon, J. J. and Chilvers, G. A. (1982). Host density as a factor in plant disease ecology. *Annual Review of Phytopathology* **20**: 143–166.

Burdon, J. J. and Thrall, P. H. (2001). The demography and genetics of host–pathogen interations. In: *Integrating Ecology and Evolution in a Spatial Context* (ed. J. Silvertown and J. Antonovics). Oxford, UK., Blackwell, pp. 197–218.

Buron, I. de, and Morand, S. (2004). Deep-sea hydro-thermal vent parasites: why do we not find more? *Parasitology* **128**: 1–6.

Burrows, R., Hofer, H. and East, M. L. (1994). Demography, extinction and intervention in a small population: the case of Serengeti wild dogs. *Proceedings of the Royal Society of London, Series B–Biological Sciences* **256**: 281–292.

Bush, A. O. (1990). Helminth communities in avian hosts: determinants of pattern. In: *Parasite communities: patterns and processes*. (ed. G. W. Esch, A. O. Bush, and J. M. Aho). London, Chapman & Hall, pp. 198–232.

Bush, A. O. and Holmes, J. C. (1986a). Intestinal parasites of lesser scaup ducks: patterns of association. *Canadian Journal of Zoology* **64**: 132–141.

Bush, A. O. and Holmes, J. C. (1986b). Intestinal parasites of lesser scaup ducks: an interactive community. *Canadian Journal of Zoology* **64**: 142–152.

Bush, A. O., Heard, R. W. Jr., and Overstreet, R. M. (1993). Intermediate hosts as source communities. *Canadian Journal of Zoology* **71**: 1358–1363.

Bush, A. O., Fernandez, J. C., Esch, G. W., and Seed, J. R. (2001). *Parasitism. The diversity and ecology of animal parasites*. Cambridge, Cambridge University Press.

Bustnes, J. O., Galaktionov, K. V., and Irwin, S. W. B. (2000). Potential threats to littoral biodiversity: is increased parasitism a consequence of human activity? *Oikos* **90**: 189–190.

Cable, J. and Harris, P. D. (2002). Gyrodactylid developmental biology: historical review, current status and future trends. *International Journal of Parasitology* **32**: 255–280.

Cable, J. and Tinsley, R. C. (1991). Intra-uterine larval development in the polystomatid monogeneans *Pseudodiplorchis americanus* and *Neodiplorchis scaphiopodis*. *Parasitology* **103**: 253–266.

Cable, J. and Tinsley, R. C. (1992). Unique ultrastructural adaptations of *Pseudodiplorchis americanus* (Polystomatidae : Monogenea) to a sequence of hostile conditions following host infection. *Parasitology* **105**: 229–241.

Cable, J., Harris, P. D., and Bakke, T. A. (2000). Population growth of *Gyrodactylus salaris* (Monogenea) on Norwegian and Baltic Atlantic salmon (*Salmo salar*) stocks. *Parasitology* **121**: 621–629.

Callaway, R. M., Thelen, G. C., Rodriguez, A., and Holben, W. E. (2004). Soil biota and exotic plant invasion. *Nature* **427**: 731–733.

Calow, P. (1979). Costs of reproduction—a physiological approach. *Biological Review* **54**: 23–40.

Calvete, C., Estrada, R., Lucientes, J., Estrada, A., and Telletxea, I. (2003). Correlates of helminth community in the red-legged partridge (*Alectoris rufa* L.) in Spain. *Journal of Parasitology* **89**: 445–451.

Calvete, C., Blanco-Aguiar, J. A., Virgos, E., Cabezas-Diaz, S., and Villafuerte, R. (2004). Spatial variation in helminth community structure in the red-legged partridge (*Alectoris rufa* L.): effects of definitive host density. Parasitology, **129**: 101–113.

Campbell, R. A. (1983). Parasitism in the deep sea. In: *The Sea, Vol 8. Deep sea Ecology*. (ed. G.T. Rowe). New-York, Wiley-interscience, pp. 473–552.

Campbell, R. A., Haedrich, R. L. and Munroe, T. A. (1980). Parasitism and ecological relationships among deep-sea benthic fishes. *Marine Biology* **57**: 301–313.

Canadell, J. G., Mooney, H. A., Baldocchi, D. D., Berry, J. A., Ehleringer, J. R., Field, C. B., Gower, S. T., Hollinger, D. Y., Hunt, J. E., Jackson, R. B., Running, S. W., Shaver, G. R., Steffen, W., Trumbore, S. E., Valentini, R., and Bond, B. Y. (2000). Carbon metabolism of the terrestrial biosphere: a multitechnique approach for improved understanding. *Ecosystems* **3**: 115–130.

Carney, W. P. (1969). Behavioural and morphological changes in carpenter ants harbouring dicrocoelid metacercariae. *American Midland Naturalist* **82**: 605–611.

Carpenter, S. R. and Turner, M. G. (1998). At last: a journal devoted to ecosystem science. *Ecosystems* **1**: 1–5.

Carpenter, S. R. and Turner, M. G. (2000a). Hares and tortoises: interactions of fast and slow variables in ecosystems. *Ecosystems* **3**: 495–497.

Carpenter, S. R. and Turner, M. G. (2000b). Opening the black boxes: ecosystem science and economic valuation. *Ecosystems* **3**: 1–3.

Cassey, P. (2002). Life history and ecology influences establishment success of introduced land birds. *Biological Journal of the Linnean Society* **76**: 465–480.

Castle, M. D. and Christensen, B. M. (1990). Hematozoa of wild turkeys from the Midwestern United States: translocation of wild turkeys and its potential role in the introduction of *Plasmodium kempi*. *Journal of Wildlife Diseases* **26**: 180–185.

Caswell, H. (1978). Predator-mediated coexistence: a non-equilibrium model. *American Naturalist* **112**: 127–154.

Cates, R. G. (1980). Feeding patterns of monophagous, oligophagous, and polyphagous insect herbivores: the effect of resource abundance and plant chemistry. *Oceologia* **46**: 22–31.

Chandra, C. V. and Khan, R. A. (1988). Nematode infestation of fillets from Atlantic cod, *Gadus morhua*, of Eastern Canada. *Journal of Parasitology* **74**: 1038–1040.

Chao, L. and Levin, B. R. (1981). Structured habitats and the evolution of anticompetitor toxins in bacteria. *Proceedings of the National Academy of Sciences of the USA* **78**: 6324–6328.

Chapin, F. S., Matson, P. A., and Mooney, H. A. (2002). *Principles of terrestrial ecosystem ecology*. New York, Springer.

Chapin, F. S., Torn, M. S., and Tateno, M. (1996). Principles of ecosystem sustainability. *American Naturalist* **148**: 1016–1037.

Chapman, H. D. and Wilson, R. A. (1973). The propulsion of the cercariae of *Himasthla secunda* (Nicoll) and *Cryptocotyle lingua*. *Parasitology* **67**: 1–15.

Chase, J. M. (1999). Food web effects of prey size refugia: variable interactions and alternate stable equilibria. *American Naturalist* **154**: 559–570.

Chastel, C. (1996). The Dilemma of Xenotransplantation. *Emerging Infectious Diseases* 2.

Chastel, C. (1998). Xénotransplantation et risque viral. *Virologie* 2: 385–392.

Cheng, T. (1986). *General Parasitology.*, 2nd edn. Orlando, Academic Press.

Choisy, M., Brown, S. P., Lafferty, K. D., and Thomas, F. (2003). Evolution of trophic transmission in parasites: why add intermediate hosts? *American Naturalist* **162**: 172–181.

Choudhury, A. and Dick, T. A. (2000). Richness and diversity of helminth communities in tropical freshwater fishes: empirical evidence. *Journal of Biogeography* **27**: 935–956.

Chubb, J. C. (1979). Seasonal occurrences of helminths in freshwater fishes. Part II. Trematoda. *Advances in Parasitology* **17**: 141–313.

Chubb, J. C. (1980). Seasonal occurrence of helminths in freshwater fishes. Part III. Larval Cestoda and Nematoda. *Advances in Parasitology* **18**: 1–120.

Claessen, D. and de Roos, A. M. (1995). Evolution of virulence in a host–pathogen system with local pathogen transmission. *Oikos* **74**: 401–413.

Clayton, D. H., Bush, S. E., Goates, B. M., and Johnson, K. P. (2003). Host defense reinforces host-parasite cospeciation. *Proceedings of the National Academy of Sciences of the USA* **100**: 15694–15699.

Cleaveland, S., Hess, G.R., Dobson, A. P., Laurenson, M. K., McCallum, H.I., Roberts, M.G., and Wooddroffe, R. (2002). The role of pathogens in biological conservation. In: *The ecology of wildlife diseases* (ed. P.J. Hudson, A. Rizzoli, B. T. Grenfell, H. Heesterbeek, and A. P. Dobson). Oxford, Oxford University Press, pp. 139–150.

Cliff, A., Haggett, P., Ord, J., and Versey, G. (1981). *Spatial diffusion: an historical geography of epidemics in an island community*. Cambridge, Cambridge University Press.

Closs, G. (1991). Multiple definitions of food web statistics: an unnecessary problem for food web research. *Australian Journal of Ecology* **16**: 413–415.

Cohen, J. E. (1977). Ratio of prey to predators in community food webs. *Nature* **270**: 165–167.

Cohen, J. E. (1978). *Food webs and Nice Space*. Princeton, Princeton University Press.

Cohen, J. E. (1989). Food webs and community structure. In: *Perspectives in ecological theory.* (ed. J. Roughgarden, R. M. May and S. A. Levin). Princeton, Princeton University Press, pp. 181–202.

Cohen, A. N. and Carlton, J. T. (1998). Accelerating invasion rate in a highly invaded estuary. *Science* **279**: 555–558.

Cohen, J. E., Beaver, R. A., Cousins, S. H., DeAngelis, D. L., Goldwasser, L., Heong, K. L., Holt, R.D., Kohn, A. J., Lawton, J. H., Martinez, N., O'Malley, R., Page, L. M., Patten, B. C., Pimm, S. L., Polis, G. A., Rejmanek, M., Schoener, T. W., Schoenly, K., Sprules, W. G., Teal, J. M., Ulanowicz, R. E., Warren, P. H., Wilbur, H. M., Yodzis, P. (1993). Improving food webs. *Ecological Monographs* **74**: 252–258.

Cohen, M. L. (2000). Changing patterns of infectious disease. *Nature* **406**: 762–767.

Collares-Pereira, M., Korver, H., Terpstra, W. J., Santos-Reis, M., Ramalhinho, M. G., Mathias, M. L., Oom, M. M., Fons, R., Libois, R., and Petrucci-Fonseca, F. (1997). First epidemiological data on pathogenic leptospires isolated on the Azorean islands. *European Journal of Epidemiology* **13**: 435–441.

Coluzzi, M. (1994). Malaria and the Afrotropical ecosystems: Impact of man-made environmental changes. *Parassitologia* **36**: 223–227.

Combes, C. (1980). Les mécanismes de recrutements chez les métazoaires parasites et leur interprétation en termes de stratégies démographiques. *Vie et Milieu* **30**: 55–63.

Combes, C. (1991). Ethological aspects of parasite transmission. *American Naturalist* **138**: 866–880.

Combes, C. (1995). *Interactions durables. Ecologie et Evolution du parasitisme*. Paris, Masson.

Combes, C. (1996). Parasites, biodiversity and ecosystem stability. *Biodiversity and Conservation* **5**: 953–962.

Combes, C. (2001). *Parasitism: the ecology and evolution of intimate interactions*. Chicago, University of Chicago Press.

Combes C. and Morand, S. (1999). Do parasites live in extreme environments? Constructing hostile niches and living in them. *Parasitology* **119**: S107–S110.

Combes, C., Fournier, A., Moné, H., and Théron, A. (1994). Behaviours in trematode cercariae that enhance parasite transmission: patterns and processes. *Parasitology* **109**: S3–S13.

Combes, C., Bartoli, P., and Theron, A. (2002). Trematode transmission strategies. In: *The behavioural ecology of parasites* (ed. E. E. Lewis, J. F. Campbell, and M. V. K. Sukhdeo). New York, CAB International, pp. 1–12.

Conder, G. A. (2002). Chemical control of animal-parasitic nematodes. In: *The biology of nematodes* (ed. D. L. Lee). London, Taylor and Francis, pp. 521–529.

Connell, J. H. (1971). On the role of natural enemies in preventing competitive exclusion in some marine animals and in rain forest trees. In: *Dynamics of populations* (ed. P.J. den Boer and G. R. Gradwell). Wageningen, The Netherlands, Centre for Agricultural Publishing and Documentation. pp. 298–313.

Connell, J. H. (1978). Diversity in tropical rain forests and coral reefs. *Science* **199**: 1302–1310.

Cook, R. R. and Quinn, J. F. (1998). An evaluation of randomization models for nested species subsets analysis. *Oecologia* **113**: 584–592.

Cook, T., Folli, M., Klinck, J., Ford, S., and Miller, J. (1998). The relationship between increasing sea-surface temperature and the northward spread of *Perkinsus marinus* (Dermo) disease epizootics in oysters. *Estuarine Coastal and Shelf Science* **46**: 587–597.

Cornell, H. V. (1985). Species assemblages of cynipid gall wasps are not saturated. *American Naturalist* **126**: 565–569.

Cornell, H. V. (1993). Unstaurated patterns in species assemblages: the role of regional processes in setting local species richness. In: *Diversity in ecological communities. Historical and geographical perspectives* (ed. R. E. Ricklefs and D. Schluter). Chicago, The University of Chicago Press, pp. 243–252.

Cornell, H. V. and Hawkins, B. A. (1993). Accumulation of native parasitoid species on introduced herbivores: a comparison of host as natives and hosts as invaders. *American Naturalist* **141**: 847–865.

Cornell, H. V. and Karlson, R. H. (1997). Local and regional processes as controls of species richness. In: *Spatial ecology . The role of space in population dynamics and interspecifc interactions* (ed. D. Tilman and P. Kareiva). Princeton, Princeton University Press, pp. 250–268.

Cort, W. W., McMullen, D. B., and Brackett, S. (1937). Ecological studies on the cercariae in *Stagnicola emarginata angulata* (Sowerby) in the Douglas Lake region. *Journal of Parasitology* **23**: 504–552.

Cosgrove, C. L. and Southgate, V. R. (2002). Mating interactions between *Schistosoma mansoni* and *S. margrebowiei*. *Parasitology* **125**: 233–243.

Costanza, R. (2000). Social goals and the valuation of ecosystem services. *Ecosystems* **3**: 4–10.

Costerton, J. W., Lewandowski, Z., Caldwell, D. E., Korber, D. R., and Lappinscott, H. M. (1995). Microbial biofilms. *Annual Review of Microbiology* **49**: 711–745.

Cox, F. E. G. (1993). *Modern Parasitology*. Oxford, Blackwell Science.

Craven, R. B., Eliason, D. A., Francy, D. B., Reiter, P., Campos, E. G., Jakob, W. L., Smith, G. C., Bozzi, C. J., Moore, C. G., Maupin, G. O., and Monath, T. P. (1998). Importation of *Aedes albopictus* and other exotic mosquito species into the United States in used tires from Asia. *Journal of the American Mosquito Control Association* **4**: 138–142.

Creel, S. and Creel, N. (1996). Limitation of African wild dogs by competition with larger carnivores. *Conservation Biology* **10**: 526–538.

Creswell, J. E., Vidal-Martinez, V. M., and Crichton, N. J. (1995). The investigation of saturation in the species richness of communities: some comments on methodology. *Oikos* **72**: 301–304.

Criscione, C. D. and Blouin, M. S. (2004). Life cycles shape parasite evolution: comparative population genetics of salmon trematodes. *Evolution* **58**: 198–202.

Cummings, D. A.T., Irizarry, R. A., Huagn, N. E., Endy, T. P., Nisalak, A., Ungchusak, K., and Burke, D. S. (2004). Traveling wave in the occurrence of dengue haemorrhagic fever in Thailand. *Nature* **427**: 344–347.

Cunningham, A. A. (1996). Disease risks of wildlife translocations. *Conservation Biology* **10**: 349–353.

Cunningham, A. A. and Daszak, P. (1998). Extinction of a species of land snail due to infection with a microsporidian Parasite. *Conservation Biology* **12**: 1139–1144.

Curriero, F. C., Patz, J. A., Rose, J. B., and Lele, S. (2001). The association between extreme precipitation and waterborne disease outbreaks in the United States, 1948–1994. *American Journal of Public Health* **91**: 1194–1199.

Curtis, L. A. and Hurd, K. M. (1983). Age, sex and parasites: spatial heterogeneity in a sandflat population of *Hyanassa obsolete*. *Ecology* **64**: 819–828.

Curtis, T. P., Sloan, W. T., and Scannell, J. W. (2002). Estimating prokaryotic diversity and its limits. *Proceeding of the National Academy of Science of the USA* **99**: 10494–10499.

Czárán, T. L. and Hoekstra, R. F. (2003). Killer-sensitive coexistence in metapopulations of micro-organisms. *Proceedings of the Royal Society of London Series B— Biological Sciences* **270**: 1373–1378.

Czárán, T. L., Hoekstra, R. F., and Pagie, L. (2002). Chemical warfare between microbes promotes biodiversity. *Proceeding of the National Academy of Science of the USA* **99**: 786–790.

Darwin, C. (1859). *On the origin of species by means of natural selection*. John Murray, London.

Daszak, P. and Cunningham, A. A. (2002). Emerging infectious diseases. In: *Conservation medicine. Ecological health in practice* (ed. A. A. Aguirre, R. S. Ostfeld, G. M. Tabor, C. House, and M. C. Pearl). Oxford, Oxford University Press, pp. 40–61.

Daszak, P., Cunningham, A. A., and Hyatt, A. D. (2000). Emerging infectious diseases of wildlife: threats to biodiversity and human health. *Science* **287**: 443–449.

Daszak, P., Cunningham, A. A., and Hyatt, A. D. (2001). Anthropogenic environmental change and the emergence of infectious diseases in wildlife. *Acta Tropica* **78**: 103–116.

Dawah, H. A., Hawkins, B. A., and Claridge, M. F. (1995). Structure of the parasitoid communities of grass-feeding chalcid wasps. *Journal of Animal Ecology* **64**: 708–720.

Dawkins, R. (1982). *The Extended Phenotype*. Oxford, W. H. Freeman.

Dawkins, R. (1990). Parasites, desiderata lists and the paradox of the organism. *Parasitology* **100**: S63–S73.

DeAngelis, D. (1992). *Dynamics of nutrient cycling and food webs*. New York, Chapman and Hall.

de Mazancourt, C., Loreau, M., and Abbadie, L. (1998). Grazing optimization and nutrient cycling: when do herbivores enhance plant production? *Ecology* **79**: 2242–2252.

De Meeûs, T. and Renaud, F. (1996). Evolution of adaptative polymorphism in spatially heterogeneous environments. In: *Aspects of the genesis and maintenance of biological diversity* (ed. Hochberg, M., Clobert, J., and Barbault R.). Oxford University Press.

De Meeûs, T., Michalakis, I., Renaud, F., and Olivieri, I. (1993). Polymorphism in heterogeneous environments. Habitat selection and sympatric speciation. *Evolutionary Ecology* **7**: 175–198.

De Meeûs, T., Hochberg, M. E., and Renaud, F. (1995) Maintenance of two genetic entities by habitat selection. *Evolutionary Ecology* **9**: 131–138.

De Meeûs, T., Michalakis, Y., and Renaud, F. (1998). Santa Rosalia revisited: or why are there so many kinds of parasites in 'The Garden of Earthly Delights?' *Parasitology Today* **14**: 10–13.

De Meeûs, T., Durand, P. and Renaud, F. (2003). Species concept: what for? *Trends in Parasitology* **19**: 425–427.

Den Boer, J. W., Yzerman E., Van Belkum A., Vlaspolder F. and Van Breukelen F. J. (1998). Legionnaire's disease and saunas. *The Lancet* **351**: 114.

DeRuiter, G. A., Neutel, A. M., Moore, J. C. (1995). Energetics, patterns of interaction strengths, and stability in real ecosystems. *Science* **269**: 1257–1260.

Des Clers, S. and Wootten, R. (1990). Modelling the population dynamics of the sealworm *Pseudoterranova decipiens*. *Netherlands Journal of Sea Research* **25**: 291–299.

Desowitz, R. S. (1991). *The malaria capers*. New York, Norton.

Despommier, D. D. (1998). How does *Trichinella spiralis* make itself at home? *Parasitology Today* **14**: 318–323.

DeStewart, R. L., Ross, P. S., Voss, J. G., and Osterhaus, A. D. M. E. (1996). Impaired immunity in harbour seals (*Phoca vitulina*) fed environmentally contaminated herring. *Veterinary Quarterly* **18**: S127–S128.

Di Castri, F. (2000). Ecology in a context of economic globalization. *Bioscience* **50**: 321–332.

Dickinson, G. and Murphy, K. (1998). *Ecosystems*. New York, Routledge Introductions to Environment Series— Environmental Science.

Diekmann, O. and Hesterbeek, J. A. P. (2000). *Mathematical epidemiology of infectious diseases. Model building, analysis and interpretation*. Chichester, UK,Wiley & Son Ltd.

Diekmann, O., Hesterbeck, J. A. P., and Metz, J. A. J. (1990). On the definition and the computation of the basic reproductive rate ratio R_0 in heterogeneous populations. *Journal of Mathematical Biology* **28**: 365–382.

Diggle, P., Moyeed, R., Rowlingson, B., and Thomson, M. (2002). Childhood malaria in the Gambia: a case-study in model-based geostatistics. *Applied Statistics* **51**: 493–506.

Doak, D. F., Bigger, D., Harding, E. K., Marvier, M. A., O'Malley, R. E., and Thomson, D. (1998). The statistical inevitability of stability–diversity relationships in community ecology. *American Naturalist* **151**: 264–276.

Dobson, A. P. (1988*a*). Restoring island ecosystems: the potential of parasites to control introduced animals. *Conservation Biology* **2**: 31–39.

Dobson, A. P. (1988*b*). The population biology of parasite-induced changes in host behaviour. *Quarterly Review of Biology* **63**: 139–165.

Dobson, A. P. (1995*a*). The ecology and epidemiology of rinderpest virus in Serengeti and Ngorongoro crater conservation area. In: *Serengeti II: research, management and conservation of an ecosystem.* (ed. A. R. E. Sinclair and P. Arcese). Chicago, The University of Chicago Press, pp. 485–505.

Dobson, A. P. (1995*b*). Rinderpest in the Serengeti ecosystem: the ecology and control of a keystone virus. In: *Proceedings of a Joint Conference American Association of Zoo Veterinarians, Wildlife Disease Association, and American Association of Wildlife Veterinarians* (ed. R. E. Junge), pp. 518–519.

Dobson, A. P. (2003). Metalife! *Science* **301**: 1488–1490.

Dobson, A. P. and Carper, E. R. (1992). Global warming and potential changes in host-parasite and disease-vector relationships. In: *Global warming and biological diversity.* (ed. R. L. Peters and T. E. Lovejoy). New Haven, CT, Yale University Press, pp. 201–217.

Dobson, A. P. and Carper, R. (1993). Climate change and human health: biodiversity. *Lancet* **342**: 1096–1031.

Dobson, A. P. and Crawley, M. J. (1994). Pathogens and the structure of plant communities. *Trends in Ecology and Evolution* **9**: 393–398.

Dobson, A. P. and Hudson, P. J. (1986). Parasites, disease and the structure of ecological communities. *Trends in Ecology and Evolution* **1**: 11–15.

Dobson, A. P. and Hudson, P. J. (1992). Regulation and stability of a free-living host–parasite system—*Trichostrongylus tenuis* in red grouse. II. Population models. *Journal of Animal Ecology* **61**: 487–498.

Dobson, A. P. and May, R. M. (1987). The effects of parasites on fish populations—theoretical aspects. *International Journal of Parasitology* **17**: 363–370.

Dobson, A. P., Pacala, S. V., Roughgarden J. D., Carper, E.R., and Harris, E.A. (1992*a*). The parasites of *Anolis* lizards in the northern Lesser Antilles. *Oecologia* **91**: 110–117.

Dobson, A. P., Hudson, P. J., and Lyles, A. M. (1992*b*). Macroparasites: worms and others. In: *Natural enemies. The population biology of predators, parasites diseases.*

(ed. M. J. Crawley). Oxford, Blackwell Scientific Publications, pp. 329–348.

Dogiel, V. A. and Lutta, A. (1937). Mortality among spiny sturgeon of the Aral Sea in 1936. *Rybn Khoz* **12**: 26–27.

Doran, J. (1998). Recurrent cystitis associated with tampon use. *Irish Medical Journal* **91**: 3.

Downing, A. L. and Leibold, M. A. (2002). Ecosystem consequences of species richness and composition in pond food webs. *Nature* **416**: 837–841.

Drake, J. M. (2003). The paradox of the parasites: implications for biological invasion. *Proceedings of the Royal Society of London Series B-Biological Sciences (Suppl.). Biology Letters* **270**: S133–S135.

Dublin, H. T., Sinclair, A. R. E., Boutin, S., Anderson, E., Jago, M., and Arcese, P. (1990). Does competition regulate ungulate populations? Further evidence from Serengeti, Tanzania. *Oecologia* **82**: 283–288.

Duffy, J. E., Richardson, J. P., and Canuel, E. A. (2003). Grazer diversity effects on ecosystem functioning in seagrass beds. *Ecology Letters* **6**: 637–645.

Duncan, R. P. (1997). The role of competition and introduction effort in the success of passeriform birds introduced to New Zealand. *American Naturalist* **149**: 903–915.

Durrett, R. and Levin, S. (1997). Allelopathy in spatially distributed populations. *Journal of Theoretical Biology* **185**: 165–171.

Dury, P. (1999). Etude comparative et diachronique des concepts ecosystem et écosystème. *Meta* **44**: 484–500.

Dye, C. (1992). The analysis of parasite transmission by bloodsucking insects. *Annual Review of Entomology* **37**: 1–19.

Dye, C. (1995). Microparasite group report. In: *Ecology of infectious diseases in natural populations* (ed. B. Grenfell and A. Dobson). Cambridge, Cambridge University Press, pp. 123–144.

Dye, C. and Reiter, P. (2000). Climate chance and malaria: temperatures without fevers? *Science* **289**: 1697–1698.

Dybdahl, M. F. and Lively, C. M. (1998). Host-parasite coevolution: evidence for rare advantage and time-lagged selection in a natural population. *Evolution* **52**: 1057–1066.

Dybdahl, M. F. and Storfer, A. (2003). Parasite local adaptation: Red Queen versus Suicide King. *Trends in Ecology and Evolution* **18**: 523–530.

Ehrlich, P. R. and Raven, P. H. (1964). Butterflies and plants: a study in coevolution. *Evolution* **18**: 586–608.

Elliot, S. L., Blanford, S., and Thomas, M. B. (2002). Host–pathogen interactions in a varying environment: temperature, behavioural fever and fitness. *Proceedings of the Royal Society of London Series B—Biological Sciences* **269**: 1599–1607.

Ellison, A. M., Gotelli, N. J., Brewer, J. S., Liane, D., Cochran-Stafira, J. M., Kneitel, T. E., Miller, T. E., Worley, A. C., and Zamora, R. (2002). Carnivorous plants as model in ecological systems. *Advances in Ecological Research* **33**: 1–74.

Elston, D. A., Moss, R., Boulinier, T., Arrowsmith, C., and Lambin, X. (2001). Analysis of aggregation, a worked example: numbers of ticks on red grouse chicks. *Parasitology* **122**: 563–569.

Elton, C. (1927). *Animal Ecology.* London, Sidgwick and Jackson.

Elton, C. S. (1958). *The ecology of invasions by animals and plants.* London, Methuen.

Elton, C. S. and Miller, R. S. (1954). The ecological survey of animal communities: with a practical system of classifying habitats by structural characters. *Journal of Ecology* **42**: 460–496.

Elton, C. S., Ford, E. B., and Baker, J. R. (1931). The health and parasites of a wild mouse population. *Proceedings of the Zoological Society of London* 657–721.

Engelberg, J. and Boyarsky, L. L. (1979). The non-cybernetic nature of ecosystems. *American Naturalist* **114**: 317–324.

Epstein, P. R. (1999). Climate and health. *Science* **285**: 347–348.

Epstein, P. R., Sherman, B., Spanger-Siegfried, E., Langston, A., Prasad, S. and McKay, B. (1998). Marine ecosystems: emerging diseases as indicators of change. Boston, MA, p.85. The Center for Health and the Global Environment, Harvard Medical School.

Erlich, P. R. and Raven, P. H. (1964). Butterflies and plants: a study in coevolution. *Evolution* **18**: 586–608.

Esch, G. W. (1971). Impact of ecological succession on the parasite fauna in centrarchids from oligotrophic and eutrophic ecosystems. *American Midland Naturalist* **86**: 160–168.

Esch, G. W. (1977). *Regulation of Parasite Populations.* New York, Academic Press.

Esch, G. W. and Fernandez J. C. (1993). *A Functional Biology of Parasitism. Ecological and evolutionary implications.* London, Chapman and Hall.

Esch, G. W., Gibbons, J. W., and Bourque, J. E. (1979). The distribution and abundance of enteric helminths in *Chrysemys s. scripta* from various habitats on the Savannah River plant in South Carolina. *Journal of Parasitology* **65**: 624–632.

Esch, G. W., Bush, A. O., and Aho, J. M. (1990). *Parasite communities: patterns and processes.* London, Chapman and Hall.

Esch, G. W., Curtis, L. A., and Barger, M. A. (2001). A perspective on the ecology of trematode communities in snails. *Parasitology* **123**: S57–S75.

Esteban, G. J., Gonzalez, C., Bargues, M. D., Angles, R., Sanchez, C., Naquira, C., and Mas-Coma, S. (2002). High Fasciolasis infection in children linked to man-made irrigation zone in Peru. *Tropical Medicine and International Health* **7**: 339–348.

Estes, J. A. and Palmisano, J. F. (1974). Sea otters: their role in structuring nearshore communities. *Science* **185**: 1058–1060.

Evans, B. J., Kelly, D. B., Tinsley, R. C., Melnick, D. J., and Cannatella, D. C. (2004). A mitochondrial DNA phylogeny of clawed frogs: phylogeography on sub-Saharan Africa and implications for polyploid evolution. *Molecular phylogenetics and evolution,* **33**: 197–213.

Evans, F. C. (1956). Ecosystem as the basic unit of ecology. *Science* **123**: 1127–1128.

Evans, N. A. (1982). Effects of copper and zinc on the life cycle of *Notocotylus attenuatus* (Digenea, Notocotylidae). *International Journal for Parasitology* **12**: 363–369.

Ewald, P. W. (1994). *Evolution of infectious disease.* Oxford University Press.

Farr, B. M. (2003). Prevention and control of nosocomial infections. *Business briefing: Global healthcare* **3**: 37–41.

Feldman, M., Cryer, B., McArthur, K. E., Huet, B. A., and Lee, E. (1996). Effects on aging gastritis on gastric acid and pepsin secretion in humans: a prospectice study. *Gastroenterology* **110**: 1043–1052.

Fenner, F. and Ratcliffe, F. N. (1965). *Myxomatosis.* Cambridge University Press, London.

Ferdy, J. B., Déprés, L., and Godelle, B. (2002). Evolution of mutualism between globe flowers and their pollinating flies. *Journal of Theoretical Biology* **217**: 219–234.

Ferguson, N. M., May R. M., and Anderson R. M. (1997). Measles: persistence and synchronicity in disease dynamics. In: *Spatial ecology. The role of space in population dynamics and interspecific interactions.* (ed. D. Tilman and P. Kareiva). Princeton, Princeton University Press, pp. 137–157.

Ferguson, N. M., Keeling, M. J., Edmunds, W. J., Gani, R., Grenfell, B. T., Anderson, R. M., and Leach, S. (2003). Planning for smallpox outbreaks. *Nature* **425**: 681–685.

Ferrière, R., Bronstein, J., Rinaldi, S., Law, R., and Gauduchon, M. (2002). Cheating and the evolutionary stability of mutualisms. *Proceedings of the Royal Society of London Series B—Biological Sciences* **269**: 773–780.

Fisher, F. M. (1963). Production of host endocrine substances by parasites. *Annals of the New York Academy of Science* **113**: 63–73.

Fliermans, C. B. (1996). Ecology of *Legionella*: from data to knowledge with a little wisdom. *Microbiology Ecology* **32**: 203–228.

Flor, H. H. (1942). Inheritance of pathogenicity in *Melampsora lini*. *Phytopathology* **32**: 653–659.

Flor, H. H. (1955). Host–parasite interaction in flax rust: its genetics and other implications. *Phytopathology* **45**: 680–685.

Forbes, M. R. L. (1993). Parasitism and host reproductive effort ? *Oikos* **67**: 444–450.

Forster, J. R. (1778). *Observations made during a voyage round the world, on physical geography, natural history, and ethic philosophy*. G. Robinson, London.

Fosbrooke, H. (1972). *Ngorongoro: the Eighth Wonder*. London, Deutsch.

Fouchier, R. A. M., Schneeberger, P. M., Rozendaal, F. W., Broekman, J. M., Kemink, S. A. G., Munster, V., Kuiken, T., Rimmelzwaan, G. F., Schutten, M., van Doornum, G. J. J., Koch, G., Bosman, A., Koopmans, M., Osterhaus, A. D. M. E. (2004). Avian influenza A virus (H7N7) associated with human conjunctivis and a fatal case of acute respiratory distress syndrom. *Proceeding of the National Academy of Sciences of the USA* **101**: 1356–1361.

Fox, A. and Hudson, P. J. (2001). Parasites reduce territorial behaviour in red grouse (*Lagopus lagopus scoticus*). *Ecology letters* **4**: 139–143

Fox, L. R. and Morrow, P. A. (1981). Specialization: species property or local phenomenon? *Science* **211**: 887–893.

Frank, S. A. (1992). A kin selection model for the evolution of virulence. *Proceedings of the Royal Society of London Series B—Biological Sciences* **250**: 195–197.

Frank, S. A. (1994). Kin selection and virulence in the evolution of protocells and parasites. *Proceedings of the Royal. Society of London, Series B—Biological Science* **258**: 153–161.

Frank, S. A. (1996). Models of Parasite Virulence. *Quarterly Review of Biology* **71**: 37–78.

Frank, S. A. (1997). Spatial processes in host–parasite genetics. In: *Metapopulation biology: ecology, genetics, and evolution*. (ed. I.A. Hanski and M.E. Gilpin). San Diego, CA, Academic Press, pp. 325–352.

Frank, S. A. (1998). *Foundations of social evolution*. Princeton University Press.

Frank, S. A. (2002). *Immunology and evolution of infectious diseases*. Princeton, Princeton University Press and Oxford.

Freeland, W. J. (1979). Primate social groups as biological islands. *Ecology* **60**: 719–728.

Freeland, W. J. (1983). Parasites and the coexistence of animal species. *American Naturalist* **121**: 223–236.

Freeland, W. J. and Boulton, W. J. (1992). Coevolution of food webs: parasites, predators and plant secondary compounds. *Biotropica* **24**: 309–327.

Frenzel, M. and Brandl, R. (2000). Phytophagous insect assemblages and the regional species pool: patterns and asymmetries. *Global Ecology and Biogeography* **9**: 293–303.

Galaktionov, K. (1993). Life cycles and distribution of seabird helminths in Arctic and subarctic regions. *Bulletin of the Scandinavian Society for Parasitology* **6**: 31–49.

Galatkionov, K. V. and Bustnes, J. O. (1999). Distribution patterns of marine bird digenean larvae in periwinkles along the southern coast of the Barents Sea. *Diseases of Aquatic Organisms* **37**: 221–230.

Gandon, S. (2002). Local adaptation and the geometry of host–parasite coevolution. *Ecology Letters* **5**: 246–256.

Gandon, S. and Michalakis, Y. (2002). Local adaptation, evolutionary potential, and host-parasite coevolution: interactions between migration, mutation, population size, and generation time. *Journal of Evolutionary Biology* **15**: 451–462.

Gandon, S., Capowiez, Y., Dubois, Y., Michalakis Y., and Olivieri I. (1996). Local adaptation and gene-for-gene coevolution in a metapopulation model. *Proceedings of the Royal Society of London Series B—Biological Sciences* **263**: 1003–1009.

Gandon, S., Mackinnon, M. J., Nee S., and Read, A. F. (2001). Imperfect vaccines and the evolution of pathogen virulence. *Nature* **414**: 751–756.

Gannicott, A. M. and Tinsley, R. C. (1998). Environmental effects on transmission of *Discocotyle sagittata* (Monogenea): egg production and development. *Parasitology* **117**: 499–504.

Ganusov, V. V., Bergstrom C. T., and Antia, R. (2002). Within-host population dynamics and the evolution of microparasites in a heterogeneous host population. *Evolution* **56**: 213–223.

Gardner, S. L. and Campbell, M. L. (1992). Parasites as probes for biodiversity. *Journal of Parasitology* **82**: 389–399.

Garzón-Ferreira, J. and Zea, S. (1992). A mass mortality of *Gorgonia ventalina* (Cnidaria: Gorgonidae) in the Santa Marta area, Caribbean coast of Colombia. *Bulletin of Marine Science* **50**: 522–526.

Gasparini, J., McCoy, K. D., Haussy, C., Tveraa, T., and Boulinier, T. (2001). Induced maternal response to the Lyme disease spirochaete *Borrelia burdgorferi* sensu lato in a colonial seabird, the *Kittiwake Rissa tridactyla*. Proceedings of the Royal Society of London. Series B—Biological Sciences **1467**: 647–650.

Gaston, K. J. (1994). *Rarity*. London, Chapman and Hall.

Gaston, K. J. (1996). The multiple forms of the interspecific abundance–distribution relationships. *Oikos* **86**: 195–207.

Gaston, K. J. and Blackburn, T.M. (2000). *Pattern and process in macroecology*. London Blackwell Science.

Gaston, K. J. and Blackburn, T. M. (2003). Dispersal and the interspecific abundance–occupancy relationship in British birds. *Global Ecology and Biogeography* **12**: 373–379.

Gause, G. F. (1934). *The struggle for existence*. Baltimore, Williams and Wilkins.

George-Nascimiento, M. A. (1987). Ecological helminthology of wildlife animal hosts from South America: a literature review and a search for patterns in marine food webs. *Revista Chilena de Historia Natural* **60**: 181–202.

Gilbert, L. E. (1975). Ecological consequences of a coevolved mutualism between butterflies and plants. In: *Coevolution of animals and plants* (ed. L. E. Gilbert and P. H. Raven). Austin, University of Texas Press, pp. 210–240.

Gilbert, L. E. (1977). The role of insect-plant coevolution in the organization of ecosystems. In: *Comportement des Insectes et Milieu Trophique.* (ed. V. Labyrie). Paris, CNRS: 399–413.

Gilbert, L. E. (1979). Development of theory in the analysis of insect–plant interactions. In: *Analysis of ecological systems.* (ed. D. J. Horn, R. D. Mitchell, and G. R. Stairs). Columbus, Ohio State University, pp. 117–154.

Gilbert, M. A. and Granath, W. (2003). Whirling disease of salmonid fish: life cycle, biology and disease. *Journal of Parasitology* **89**: 658–667.

Gilbert, L. E. and Singer, M. C. (1975). Butterfly ecology. *Annual Review of Ecology and Systematics* **6**: 365–397.

Gilbert, G. S., Hubbell, S. P. and Foster, R. B. (1994). Density and distance to adult effects of a canker disease of trees in a moist tropical forest. *Oecologia* **98**: 100–108.

Gilbert, L., Norman, R., Laurenson, K. M., Reid, H. W., and Hudson, P. J. (2001). Disease persistence and dynamics of a three-host community: an empirical and analytical study of large-scale wild populations. *Journal of Animal Ecology* **70**: 1053–1061.

Gilchrist, M. and Sasaki, A. (2002). Modeling host–parasite coevolution: a nested approach based on mechanistic models. *Journal of Theoretical Biology* **218**: 289–308.

Gillespie, W. J. (1997). Prevention and management of infection after total joint replacement. *Clinical Infectious Diseases* **25**: 1310–1317.

Giglioli, G. (1963). Ecological change as a factor in renewed malaria transmission in an eradicated area: a localized outbreak of *A. aquasalis* transmitted on the Demerara River estuary, British Guinea, in the fifteenth year of *A. darlingi* and malaria eradication. *Bulletin of the World Health Organization* **29**: 131–145.

Gilmartin, W. G., DeLong, R. L., Smith, A. W., Sweeney, J. C., Lappe, B. W. D., Riseborough, R. W., Griner, L. A., Dailey, M. D., and Peakall, D. B. (1976). Premature parturition of the California sea lion. *Journal of Wildlife Diseases* **12**: 104–115.

Godfray, H. C. and Parker, G. A. (1992). Sib competition, parent–offspring conflict and clutch size. *Animal Behaviour* **43**: 473–490.

Gog, J., Woodroffe, R., and Swinton, J. (2002). Disease in endangered metapopulations: the importance of alternative hosts. *Proceedings of the Royal Society of London Series B—Biological Sciences* **269**: 671–676.

Goldwasser, L. and Roughgarden, J. (1993). Construction and analysis of a large Caribbean food web. *Ecology* **74**: 1216–1233.

Golley, F. B. (1993). *A history of the ecosystem concept in ecology: more than the sum of the parts.* New Haven, Yale University Press.

Gomulkiewicz, R., Holt, R. D., and Barfield, M. (1999). The effects of density dependence and immigration on adaptation and niche evolution in a black-hole sink environment. *Theoretical Population Biology* **55**: 283–296.

Gomulkiewicz, R., Thompson, J. N., Holt, R. D., Nuismer, S. L., and Hochberg, M. E. (2000). Hot spots, cold spots, and the geographic mosaic theory of coevolution. *The American Naturalist* **156**: 156–174.

González, G., Sorci, G., Møller, A. P., Ninni, P., Haussy, C., and de Lope, F. (1999). Immunocompetence and condition-dependent sexual advertisement in male house sparrows (*Passer domesticus*). *Journal of Animal Ecology* **68**: 1225–1234.

Goodson, N. J. (1982). Effects of domestic sheep grazing on bighorn sheep populations: a review. *Proceedings Biennial Symposium of Northern Wild Sheep and Goat Council* **3**: 287–313.

Gosz, J. R. (1996). International long-term ecological research: priorities and opportunities. *Trends in Ecology and Evolution* **11**: 444.

Gosz, J. R. (1999). Ecology challenged? Who? Why? Where is this headed? *Ecosystems* **2**: 475–481.

Goüy de Bellocq, J., Morand, S., and Feliu, C. (2002). Patterns of parasite species richness of western Palearcticmicromammals: island effects. *Ecography* **25**: 173–183.

Goüy de Bellocq, J., Sara, M., Casanova, J. C., Feliu, C., and Morand, S. (2003). A comparison of the structure of helminth communities in the woodmouse, *Apodemus sylvaticus*, on islands of the western Mediterranean and continental Europe. *Parasitology Research* **90**: 64–70.

Graefe, G. W., Hohorst, W. and Dräger, H. (1967). Forked tail of the cercaria of *Schistosoma mansoni*—a rowing device. *Nature* **215**: 207–208.

Gregory, R. D. (1990). Parasites and host geographic range as illustrated by waterfowl. *Functional Ecology* **4**: 645–654.

Grenfell, B. T. and Dobson, A. P. (1995). *Ecology of infectious diseases in natural populations.* Oxford, Oxford University Press.

Grenfell, B. T. and Harwood, J. (1997). (Meta)population dynamics of infectious diseases. *Trends in Ecology and Evolution* **12**: 395–399.

Grenfell, B.T., Kleczkowski, A., Gilligan C. A., and Bolker B.M. (1995). Spatial heterogeneity, nonlinear dynamics and chaos in infectious diseases. *Statistical Methods in Medical Research* **4**: 160–183.

Grenfell, B. T., Bjornstad, O. N., and Kappey, J. (2001). Travelling waves and spatial hierarchies in measles epidemics. *Nature* **414**: 716–723.

Grenfell, B. T., Pybus, O. G., Gog, J. R., Wood, J. L. N., Daly, J. M., Mumford, J. A., and Holmes, E. C. (2004). Unifying the epidemiological and evolutionary dynamics of pathogens. *Science* **303**: 327–331.

Griffiths, D. (1999). On investigating local–regional species richness relationships. *Journal of Animal Ecology* **68**: 1051–1055.

Gryseels, B., Stelma, F., Talla, I., Polman, K., Van Dam, G., Polman, S., Sow, M., Diaw, R., Sturrock, E., Decam, C., Niang, M., Doehring-Schwerdtfeger, E., and Kardorff, R. (1994). Epidemiology, immunology and chemotherapy of *Schistosoma mansoni* infections in a recently exposed community in Senegal. *Tropical and Geographical Medicine* **46**: 209–219.

Guan, Y., Zheng, B. J., He, Y. Q., Liu, X. L., Zhuang, Z. X., Cheung, C. L., Luo, S. W., Li, P. H., Zhang, L. J., Guan, Y. J., Butt, K. M., Wong, K. L., Chan, K. W., Lim, W., Shortridge, K. F., Yuen, K. Y., Peiris, J. S., and Poon, L. L. (2003). Isolation and characterization of viruses related to the SARS coronavirus from animals in southern China. *Science* **302**: 276–278.

Guarda, J. A., Asayag, C. R., and Witzig, R. (1999). Malaria reemergence in the Peruvian Amazon region. *Emerging Infectious Diseases* **5**: 209–215.

Gubbins, S. and Gilligan, C. A. (1997). A test of heterogeneous mixing as a mechanism for ecological persistence in a disturbed environment. *Proceedings of Royal Society of London Series B—Biological Sciences* **264**: 227–232.

Guégan, J.-F., Lambert, A., Lévêque, C., Combes, C., and Euzet, L. (1992). Can host body size explain the parasite species richness in tropical freshwater fishes? *Oecologia* **90**: 197–204.

Guégan, J.-F. and Kennedy, C. R. (1993). Maximum local helminth parasite community richness in British freshwater fish: a test of the colonization time hypothesis. *Parasitology* **106**: 91–100.

Guégan, J.-F. and Kennedy, C. R. (1996). Parasite richness/sampling effort/host range: the fancy three-piece jigsaw puzzle. *Parasitology Today* **12**: 367–370.

Guégan, J.-F. and Hugueny, B. (1994). A nested parasite species subset pattern in tropical fish: host as major determinant of parasite infracommunity structure. *Oecologia* **100**: 184–189.

Guégan, J.-F., Thomas, F., de Meeûs, T., Hochberg, M.E., and Renaud, F. (2001). Human fertility and disease diversity. *Evolution* **55**: 1308–1314.

Guernier, V., Hochberg, M. E., and Guégan, J.-F. (2004). Ecology drives the worldwide distribution of human infectious diseases. *PloS Biology* **2**: 740–746.

Guth, D. J., Blankespoor, H. D., and Cairns, J. (1977). Potentiation of zinc stress caused by parasitic infection of snails. *Hydrobiologia* **55**: 225–229.

Haas, W. (1974). Analyse der Invasionsmechanismen der Cercarie von *Diplostomum spathaceum*—I. Fixation unde Penetration. *International Journal of Parasitology* **4**: 311–319.

Haas, W. (1976). Die Anheftung (Fixation) der Cercarie von *Schistosoma mansoni*. Einfluss natürlicher Substrate und der Temperatur. *Zeitschrift fur Parasitenkunde* **49**: 63–72.

Haas, W. (1992). Physiological analysis of cercarial behavior. *Journal of Parasitology* **78**: 243–255.

Hagen, J. B. (1992). *An entangled bank: the origins of ecosystem ecology*. New Brunswick, Rutgers University Press.

Hahn, B. H., Shaw, G. M., De Cock, K. M., and Sharp, P.M. (2000). AIDS as a zoonosis: scientific and public health implications. *Science* **287**: 607–614.

Hairston, N. G. and Hairston, N. G. Sr. (1993). Cause–effect relationships in energy flow, trophic structure, and interspecific interactions. *American Naturalist* **142**: 379–411.

Haldane, J. B. S. (1949). Disease and evolution. *La Ricerca Scientifica Supplemento* **19**: 68–76.

Haley, R. W. (1991). Measuring the cost of nosocomial infections: methods for estimating economic burden on the hospital. *American Journal of Medicine* **91**: 32–38.

Hall, S. J. and Raffaelli, D. G. (1991). Food-web patterns: lessons from a species-rich web. *Journal of Animal Ecology* **60**: 823–842.

Hall, S. J. and Raffaelli, D. G. (1993). Food webs: theory and reality. In: *Advances in ecological research* (ed. M. Begon and A.H. Fitter). London, Academic Press.

Halvorsen, O. and Bye, K. (1999). Parasites, biodiversity, and population dynamics in an ecosystem in the high arctic. *Veterinary Parasitology* **84**: 205–227.

Halvorsen, O., Stien, A., Irvine, J. Langvatn, R., and Albon, S. (1999). Evidence for continued transmission of parasitic nematodes in reindeer during the Arctic winter. *International Journal for Parasitology* **29**: 567–579.

Hamilton, W. D. (1964). The genetical evolution of social behaviour, I. *Journal of Theoretical Biology* **7**: 1–16.

Handman, E. and Bullen, D. V. R. (2002). Interaction of *Leishmania* with the host macrophage. *Trends in Parasitology* **18**: 332–334.

Hanski, I. A. (1999). *Metapopulation ecology*. Oxford, Oxford University Press.

Hanski, I. A. and Gilpin, M. E. (1997). *Metapopulation biology. Ecology, genetics, and evolution*. London, Academic Press.

Hanski I. A. and Gaggiotti, O. E. (2004). *Ecology, genetics and evolution of metapopulations*. New York, NY. Academic Press Ltd.

Haraguchi, Y. and Sasaki A. (2000). Evolution of parasite virulence and transmission rate in a spatially structure population. *Journal of Theoretical Biology* 203: 85–96.

Harder, T. C., Willhus, T., Leibold, W., and Liess, B. (1992). Investigations on the course and outcome of phocine distemper virus infection in harbor seals (*Phoca vitulina*) exposed to polychlorinated biphenyls. *Journal of Veterinary Medicine B* 39: 19–31.

Hart, B.L. (1994). Behavioural defense against parasites: interaction with parasite invasiness. *Parasitology* 109: 139–151.

Hartvigsen, R. and Kennedy, C. R. (1993). Patterns in the composition and richness of helminth communities in brown trout, *Salmo trutta*, in a group of reservoirs. *Journal of Fish Biology* 43: 603–615.

Harvell, C. D., Kim, K., Burkholder, J. M., Colwell, R. R., Epstein, P. R., Grimes, D. J., Hofmann, E. E., Lipp, E. K., Osterhaus, A., Overstreet, R. M., Porter, J. W., Smith, G. W. and Vasta, G. R. (1999). Emerging marine diseases—Climate links and anthropogenic factors. *Science* 285: 1505–1510.

Harvell, C. D., Mitchell, C. E., Ward, J. R., Altizer, S., Dobson, A. P., Ostfeld, R. S., and Samuel, M. D. (2002). Climate warming and disease risks for terrestrial and marine biota. *Science* 296: 2158–2162.

Hassell, M. P. and Wilson, H. B. (1997). The dynamics of spatially distributed host–parasitoid systems. In: *Spatial ecology. The role of space in population dynamics and interspecific interactions*. (ed. D. Tilman and P. Kareiva). Princeton, Princeton University Press, pp. 75–110.

Hauer, F. R. and Lamberti, G. A. (1996). *Methods in stream ecology*. San Diego, CA, Academic Press.

Haukisalmi, V. and Henttonen H. (1999). Determinants of helminth aggregation in natural host populations: individual differences or spatial heterogeneity? *Ecography* 22: 629–636.

Hawkins, B. A. (2004). Are we making progress toward understanding the global diversity gradient? *Basic and applied ecology* 5: 1–3.

Hawkins, B. A. and Compton, S. G. (1992). African fig wasp communities: vacant niches and latitudinal gradients in species richness. *Journal of Animal Ecology* 61: 361–372.

Hawkins, B. A., Field, R., Cornell, H. V., Currie, D. J., Guégan, J-F, Kaufman, D. M., Kerr, J.T., Mittlebach, G. G.,

Oberdorff, T., O'Brien, E. M., Porter, E. E. and Turner, J. R. G. (2003). Energy, water, and broad-scale georgraphic patterns of species richness. *Ecology* 84: 3105–3117.

Hay, S. I., Tucker, C. J., Rogers, D. J., and Picker, M. J. (1996). Remotely sensed surrogates of meteorological data for the study of the distribution and abundance of arthropod vectors of disease. *Annales of Tropical Medecine and Parasitology* 90: 1–19.

Hay, S. I., Rogers, D. J., Randolph, S. E., Stern, D. I., Cox, J., Shanks, G. D., and Snow, R. W. (2002). Hot topic or hot air? Climate change and malaria resurgence in East African highland. *Trends in Parasitology* 18: 530–534.

Hay, S., Renshaw, M., Ochola, S. A., Noor, A. M., and Snow, R. W. (2003). Performance of forecasting, warning and detection of malaria epidemics in the highlands of western Kenya. *Trends in Parasitology* 19: 394–399.

Hayes, M. L., Bonaventura, J., Mitchell, T. P., Prospero, J. M., Shinn, E. A., Van Dolah, F., and Barber, R. T. (2001). How are climate and marine biological outbreaks functionally linked ? *Hydrobiologia* 460: 213–220.

Hector, A., Schmid, B., Beierkuhnlein, C., Caldeira, M. C., Diemer, M., Dimitrakopoulos, P. G., Finn, J. A., Freitas, H., Giller, P. S., Good, J., Harris, R., Högberg, P., Huss-Danell, K., Joshi, J., Jumpponen, A., Körner, C., Leadley, P. W., Loreau, M., Minns, A., Mulder, C. P. H., O'Donovan, G., Otway, S. J., Pereira, J. S., Prinz, A., Read, D. J., Scherer-Lorenzen, M., Schulze, E.-D., Siamantziouras, A.-S. D., Spehn, E. M., Terry, A. C., Troumbis, A. Y., Woodward, F. I., Yachi, S., and Lawton, J. H. (1999). Plant diversity and productivity experiments in European grasslands. *Science* 286: 1123–1127.

Hector, A., Schmid, B., Beierkuhnlein, C., Caldeira, M. C., Diemer, M., Dimitrakopoulos, P. G., Finn, J. A., Freitas, H., Giller, P. S., Good, J., Harris, R., Högberg, P., Huss-Danell, K., Joshi, J., Jumpponen, A., Körner, C., Leadley, P. W., Loreau, M., Minns, A., Mulder, C. P. H., O'Donovan, G., Otway, S. J., Pereira, J. S., Prinz, A., Read, D. J., Scherer-Lorenzen, M., Schulze, E.-D., Siamantziouras, A.-S. D., Spehn, E. M., Terry, A. C., Troumbis, A. Y., Woodward, F. I., Yachi, S., and Lawton, J. H. (2000). No consistent effect of plant diversity on productivity: Response. *Science* 289: 1255a, www.sciencemag.org/cgi/content/full/289/5483/1255a.

Heeb, P., Werner, I., Mateman A. C., Kolliker, M., Brinkhof, M. W. G., Lessels, C. M., and Richner, H. (1999). Ectoparasites infestation and sex-biased local recruitment of hosts. *Nature* 400: 63–65.

Heide-Jorgensen, M. P., Harkonen, T., Dietz, R., and Thompson, P. M. (1992). Retrospective of the 1988 European seal epizootic. *Diseases of Aquatic Organisms* 13: 37–62.

Helluy, S. (1981). Parasitisme et Comportement. Etude de la métacercaire de *Microphallus papillorobustus* (Rankin 1940) et de son influence sur les gammares. PhD thesis, USTL Montpellier.

Hemmingsen, W. and MacKenzie, K. (2001). The parasite fauna of the Atlantic cod, *Gadus morhua* L. *Advances in Marine Biology* **40**: 1–80.

Henttonen, H., Fuglei, E., Gower, C. N., Haukisalmi, V., Ims, R. A., Niemimaa, J., and Yoccoz, N. G. (2001). *Echinococcus multilocularis* on Svalbard: introduction of an intermediate host has enabled the local life-cycle. *Parasitology* **123**: 547–552.

Herre, E. A. (1999). Laws governing species interactions? Encouragement and caution from figs and their associates. In: *Levels of selection in evolution* (ed. L. Keller). Princeton, New Jersey, Princeton University Press.

Herre, E. A., Knowlton, N., Mueller, U. G., and Rehner, S. A. (1999). The evolution of mutualism: exploring the paths between conflict and cooperation. *Trends in Ecology and Evolution* **14**: 49–53.

Hess, G. R. (1994). Conservation corridors and infectious disease: a cautionary note. *Conservation Biology* **8**: 256–262.

Hess, G. R. (1996*a*). Disease in metapopulation models: implications for conservation. *Ecology* **77**: 1617–1632.

Hess, G. R. (1996*b*). Linking extinction to connectivity and habitat destruction in metapopulation models. *American Naturalist* **148**: 226–236.

Hess, G. R., Randolph, S. E., Arneberg, P., Chemini, C., Furlanello, C., Harwood, J., Roberts, M. G., and Swinton, J. (2002). Spatial aspects of disease dynamics. In: *The ecology of wildlife diseases* (ed. P. J. Hudson, A. Rizzoli, B. T. Grenfell, H. Heeterbeek, and A. P. Dobson). Oxford, Oxford University Press, pp. 6–44.

Hethcote, H. W. and van Ark, J. W. (1987). Epidemiological models for heterogeneous populations: proportionate mixing, parameter estimation, and immunization programs. *Mathematical Bioscience* **84**: 85–118.

Heussler, V. T., Küenzi, P., and Rottenberg, S. (2001). Inhibition of apoptosis by intracellular protozoan parasites. *International Journal for Parasitology* **31**: 1166–1176.

Heyneman, D. (1979). Dams and disease. *Human Nature* **2**: 50–57.

Higashi, M. and Burns, T. P. (1991). Enrichment of ecosystem theory. In *Theoretical studies of ecosystems: the network perspective* (ed. M. Higashi and T. P. Burns). Cambridge, Cambridge University Press, pp. 1–38.

Hill, J. K., Thomas, C. D., and Huntley, B. (1999). Climate and habitat availability determine 20th century changes in a butterfly's range margin. *Proceedings of the Royal Society of London B—Biological Science* **266**: 1197–1206.

Hillebrand, H. and Blenckner, T. (2002). Regional and local impact on species diversity—from pattern to processes. *Oecologia* **132**: 479–491.

Hillebrand, H., Watermann, F., Karez, R., and Berninger, U. G. (2001). Differences in species richness patterns between unicellular and multicellular organisms. *Oecologia* **126**: 114–124.

Hlady, W. G., Mullen, R. C., Mintz, C. S., Shelton, B. G., Hopkins, R. S., and Daikos, G. L. (1993). Outbreak of Legionnaire's disease linked to a decorative fountain by molecular epidemiology. *American Journal of Epidemiology* **138**: 555–562.

Hobbie, J. E., Carpenter, S. R., Grimm, N. B., Gosz, J. R., and Seastedt, T. R. (2003). The US Long Term Ecological Research Program. *Bioscience* **53**: 21–32.

Hochachka, W. M. and Dhondt, A. (2000). Density-dependent decline of host abundance resulting from a new infectious disease. *Proceedings of the National Academy of Sciences of the USA* **97**: 5303–5306.

Hochberg, M. E. and Holt, R. D. (2002). Biogeographical perspectives on arms races. In: *Adaptive Dynamics of Infectious Disease* (ed. U. Dieckmann, J. A. J. Metz, M. W. Sabelis, and K. Sigmund). Cambridge, Cambridge University Press, pp. 197–209.

Hochberg, M. E. and Ives, A. R. (1999). Can natural enemies enforce geographical range limits? *Ecography* **22**: 268–276.

Hochberg, M. E. and Møller, A. P. (2001). Insularity and adaptation in coupled victim-enemy associations. *Journal of Evolutionary Biology* **14**: 539–551.

Hochberg, M. E., Michalakis, Y., and de Meeüs, T. (1992). Parasitism as a constraint on the rate of life-history evolution. *Journal of Evolutionary of Biology* **5**: 491–504.

Hochberg, M. E. and van Baalen, M. (1998). Antagonistic coevolution over productivity gradients. *American Naturalist* **152**: 620–634.

Hochberg, M. E., Gomulkiewicz, R., Holt, R. D., and Thompson, J. N. (2000). Weak sinks could cradle mutualistic symbioses—strong sources shuld harbour parasitic symbioses. *Journal of Evolutionary Biology* **13**: 213–222.

Hoegh-Guldberg, O. (1999). Climate change, coral bleaching and the future of the world's coral reefs. *Marine and Freshwater Research* **50**: 839–866.

Holmes, J. C. (1961). Effects of concurrent infections on *Hymenolepis diminuta* (Cestoda) and *Moniliformis dubius* (Acanthocephala). I. General effects and comparison with crowding. *Journal of Parasitology* **47**: 209–216.

Holmes, J. C. (1982). Impact of infectious disease agents on the population growth and geographical distribution of animals. In: *Population biology of infectious diseases* (ed. R. M. Anderson and R. M. May). Berlin, Springer-Verlag, pp. 37–51.

Holmes, J. C. (1990). Helminth communities in marine fishes. In: *Parasite Communities: Patterns and Processes* (ed. G. W. Esch, A. O. Bush, and J. M. Aho). London, Chapman and Hall.

Holmes, J. C. (1996). Parasites as threats to biodiversity in shrinking ecosystems. *Biodiversity and Conservation* 5: 975–983.

Holmes, J. C. and Bethel, W. M. (1972). Modification of intermediate host behaviour by parasites. In: *Behavioural aspects of parasite transmission* (ed. E. U. Canning and C. A. Wright). London, Academic Press.

Holmes, O. W. (1843). The contagiousness of puerperal fever. *The New England Quaterly Journal of Medicine and Surgery* 1: 503–530.

Holt, R. D. (1977). Predation, apparent competition and the structure of prey communities. *Theoretical Population Biology* 12: 197–229.

Holt, R. D. (1984). Spatial heterogeneity, indirect interactions, and the coexistence of prey species. *American Naturalist* 124: 377–406.

Holt, R. D. (1985). Population dynamics in two–patch environments: Some anomalous consequences of an optimal habitat distribution. *Theoretical Population Biology* 28: 181–208.

Holt, R. D. (1993). Ecology at the mesoscale: the influence of regional processes on local communities. In: *Species Diversity in Ecological Communities. Historical and Geographical Perspectives* (ed. R. E. Ricklefs and D. Schluter). Chigago, The University of Chicago Press, 77–88.

Holt, R. D. (1997). From metapopulation dynamics to community structure: some consequences of spatial heterogeneity. In: *Metapopulation biology* (ed. I. Hanski and M. Gilpin). New York, Academic Press, pp. 149–164.

Holt, R. D. (1999). A biogeographical and landscape perspective on within-host infection dynamics. *Microbial Biosystems: New Frontiers*. Proceedings of the 8th International Symposium on Microbial Ecology. Canada Society for Microbial Ecology, Halifax (ed. C. R. Bell, M. Brylinsky, and P. Johnson-Green).

Holt, R. D. (2002). Food webs in space: on the interplay of dynamic instability and spatial processes. *Ecological Research* 17: 261–273.

Holt, R. D. and Hochberg, M. E. (2002). Virulence on the edge: a source–sink perspective. In: *Adaptive Dynamics of Infectious Disease* (ed. U. Dieckmann, J. A. J. Metz, M. W. Sabelis, and K. Sigmund). Cambridge, Cambridge University Press, pp. 104–119.

Holt, R. D. and Hoopes, M. F. In review. Food web dynamics in a metacommunity context: modules and beyond. In: *Metacommunities* (ed. M. Holyoak, M. Leibold, and R. D. Holt). University of Chicago Press, Chicago, in press.

Holt, R. D. and Lawton, J. H. (1993). Apparent competition and enemy-free space in insect host–parasitoid communities. *American Naturalist* 142: 623–645.

Holt, R. D. and Lawton, J. H. (1994). The ecological consequences of shared natural enemies. *Annual Review of Ecology and Systematics* 25: 495–520.

Holt, R. D. and Loreau, M. (2002). Biodiversity and ecosystem functioning: the role of trophic interactions and the importance of system openness. In: *The functional consequences of biodiversity: empirical progress and theoretical extensions* (ed. Kinzig, A., Tilman, D., and S. W. Pacala). Princeton, Princeton University Press, pp. 246–262.

Holt, R. D. and Pickering, J. (1985). Infectious disease and species coexistence: a model of Lotka-Volterra form. *American Naturalist* 126: 196–211.

Holt, R. D. and Polis, G. A. (1997). A theoretical framework for intraguild predation. *American Naturalist* 149: 745–764.

Holt, R. D., Grover, J., and Tilman, D. (1994). Simple rules for interspecific dominance in systems with exploitative and apparent competition. *American Naturalist* 144: 741–771.

Holt, R. D., Lawton, J.H., Polis, G. A., and Martinez, N. (1999). Trophic rank and the species–area relation. *Ecology* 80: 1495–1504.

Holt, R. D., Gomulkiewicz, R. and Barfield, M. (2003). The phenomenology of niche evolution via quantitative traits in a 'black-hole' sink. *Proceedings of the Royal Society of London Series B—Biological Sciences* 270: 215–224.

Hoopes, M. F., Holt R. D. and Holyoak, M. In review. The effects of spatial processes on two-species interactions. In: *Metacommunities* (ed. M. Holyoak, M. Leibold, and R. D. Holt). University of Chicago Press, Chicago.

Houghton, J. T., Filho, L. G. M., Callandar, B. A., Harris, N., Kattenberg, A. and Maskell, K. (1996). *Climate models—projections of future climate*. New York, Cambridge University Press.

Houlahan, J. E., Findlay, C. S., Meyer, A. H., Kuzmin, S. L., and Schmidt, B. R. (2001). Global amphibian population declines—Reply. *Nature* 412: 500.

Hubbell, S. P. (2001). *The unified neutral theory of biodiversity and biogeography*. Princeton, Princeton University Press.

Hudson, P. J. (1986a). The effect of a parasitic nematode on the breeding production of red grouse. *Journal of Animal Ecology* 55: 85–92.

Hudson, P. J. (1986*b*). Bracken and ticks on grouse moors in the north of England. In: *Bracken: ecology, land use and control technology* (ed. R. T. Smith). Parthenon Press, Camforth, 161–170.

Hudson, P. J. (1992). *Grouse in space and time.* Fordingbridge, The Game Conservancy Trust.

Hudson, P. J. (2004). Conservation medicine: synthesis or crisis discipline? *Trends in Ecology and Evolution* **18**: 616.

Hudson, P. J. and Bjørnstad, O. N. (2003). Vole stranglers and lemming cycles. *Science* **302**: 797–798.

Hudson, P. J. and Dobson, A. P. (1995). Macroparasites: observed patterns in naturally fluctuating animal populations. In: *Ecology of infectious diseases in natural populations* (ed. B.T. Grenfell and A. P. Dobson). Cambridge, Cambridge University Press, pp. 144–176.

Hudson, P. J. and Dobson, A. P. (2001). Harvesting unstable populations: Red grouse *Lagopus lagopus scoticus* (Lath.) in the United Kingdom. *Wildlife Biology* **7**: 189–196.

Hudson, P. and Greenman, J. (1998). Competition mediated by parasites: biological and theoretical progress. *Trends in Ecology and Evolution* **13**: 387–390.

Hudson, P. J., Dobson, A. P., and Newborn, D. (1992*a*). Do parasites make prey vulnerable to predation? Red grouse and parasites. *Journal of Animal Ecology* **61**: 681–692.

Hudson, P. J., Newbon, D., and Dobson, A. P. (1992*b*). Regulation and stability of a free-living host–parasite system—*Trichostrongylus tenuis* in red grouse. I. Monitoring and parasite reduction experiments. *Journal of Animal Ecology* **61**: 477–486.

Hudson, P. J., Dobson, A. P., and Newborn, D. (1998). Prevention of population cycles by parasite removal. *Science* **282**: 2256–2258.

Hudson, P. J., Dobson, A. P., and Newborn, D. (1999). Population cycles and parasitism. *Science* **286**: 2425.

Hudson, P., Rizzoli, A., Grenfell, B.T., Heesterbeek, H. and Dobson, A. P. (2001). *Ecology of Wildlife diseases.* Oxford, Oxford University Press.

Hudson, P. J., Dobson, A. P., Cattadori, I. M., A. P., Newborn, D., Haydon, D., Shaw, D., Benton, T.G., and Grenfell, B. T. (2002). Trophic interactions and population growth rates: describing patterns and identifying mechanisms. *Philosophical Transactions of the Royal Society* **37**: 1259–1272.

Hughes, R. N. and Answer, P. (1982). Growth, spawning and trematode infection of *Littorina littorea* (L.) from an exposed shore in North Wales. *Journal of Molluscan Studies* **48**: 321–330.

Hugueny, B. and Guégan, J. F. (1997). Community nestedness and the proper way to assess statistical significance by Monte-Carlo tests: some comments on Worthen and Rohde' (1996) paper. *Oikos* **80**: 572–574.

Huspeni, T. (2000). A molecular genetic analysis of host specificity, continental geography, and recruitment dynamics of a larval trematode in a salt marsh snail, Santa Barbara, University of California.

Huspeni, T. C. and Lafferty, K. D. (2004). Using larval trematodes that parasitize snails to evaluate a salt-marsh restoration project. *Ecological Applications*, **14**: 795–804.

Huston, M. A. (1997). Hidden treatments in ecological experiments: re-evaluating the ecosystem function of biodiversity. *Oecologia* **110**: 449–460.

Huston, M. A., Aarsen, L. W., Austin, M. P., Cade, B. S., Fridley, J. D., Garnier, E., Grime, J. P., Hodgson, J., Lauenroth, W. K., Thompson, K., Vandermeer, J. H., and Wardle, D. A. (2000). No consistent effect of plant diversity on productivity. *Science* **289**: 1255a, (www.sciencemag.org/cgi/content/full/289/5483/1255a).

Hutchings, M. R., Kyriazakis, I., Papachristou, T. F, Gordon, I. J., and Jackson, F. (2000). The herbivores'dilemma: trade-offs between nutrition and parasitism in foraging decisions. *Oecologia* **124**: 242–251.

Hutchinson, G. E. (1951). Copepodology for the ornithologist. *Ecology* **32**: 571–577.

Huxham, M., Raffaelli, D., and Pike, A. (1995). Parasites and food web patterns. *Journal of Animal Ecology* **64**: 168–176.

Huxham, M., Beany, S., and Raffaelli, D. (1996). Do parasites reduce the chances of triangulation in a real food web? *Oikos* **76**: 284–300.

Ives, A. R. and May, R. M. (1985). Competition between species in a patchy environment: relations between microscopic and macroscopic models. *Journal of Theoretical Biology* **115**: 65–92.

Ives, A. R., Klug, J. L., and Gross, K. (2000). Stability and species richness in complex communities. *Ecology Letters* **3**: 399–411.

Jaarsma, N. G., deBoer, S. M., Townsend, C. R., Thompson, R. M., and Edwards, E. D. (1998). Characterization of the food web in two New Zealand streams. *New Zealand Journal of Marine and Freshwater Research* **32**: 271–286.

Jackson, J. A. and Tinsley, R. C. (1998*a*). Reproductive interference in concurrent infections of two *Protopolystoma* species (Monogenea: Polystomatidae). *International Journal for Parasitology* **28**: 1201–1204.

Jackson, J. A. and Tinsley, R. C. (1998*b*). Incompatibility of *Protopolystoma xenopodis* (Monogenea: Polystomatidae) with an octoploid *Xenopus* species (Anura) from southern Rwanda. *International Journal for Parasitology* **28**: 1195–1199.

Jackson, J. A. and Tinsley, R. C. (2001). *Protopolystoma xenopodis* (Monogenea) primary and secondary infections in *Xenopus laevis*. *Parasitology* **123**: 455–463.

Jackson, J. A. and Tinsley, R. C. (2002). Effects of environmental temperature on the susceptibility of *Xenopus laevis* and *X. wittei* (Anura) to *Protopolystoma xenopodis* (Monogenea). *Parasitology Research* **88**: 632–638.

Jackson, J. A. and Tinsley, R. C. (2003a). Postlarval *Protopolystoma* spp. kidney infections in incompatible *Xenopus* spp. induce weak resistance to heterospecifics. *Parasitology Research* **90**: 429–434.

Jackson, J. A. and Tinsley, R. C. (2003b). Parasite infectivity to hybridising host species: a link between hybrid resistance and allopolyploid speciation? *International Journal for Parasitology* **33**: 137–144.

Jackson, J. A. and Tinsley, R. C. (2003c). Density-dependence of postlarval survivorship in primary infections of *Protopolystoma xenopodis*. *Journal of Parasitology* **89**: 958–960.

Jackson, J. A., Tinsley, R. C., and Hinkel, H. H. (1998). Mutual exclusion of congeneric monogenean species in a space-limited habitat. *Parasitology* **117**: 563–569.

Jackson, J. B. C., Kirby, M. X., Berger, W. H., Bjorndal, K. A., Botsford, L. W., Bourque, B. J., Bradbury, R. H., Cooke, R., Erlandson, J., Estes, J. A., Hughes, T. P., Kidwell, S., Lange, C. B., Lenihan, H. S., Pandolfi, J. M., Peterson, C. H., Steneck, R. S., Tegner, M. J. and Warner, R. R. (2001). Historical overfishing and the recent collapse of coastal ecosystems. *Science* **293**: 629–638.

Jaenike, J. (1994). Aggregations of nematode parasites among *Drosophila*: proximate causes. *Parasitology* **108**: 569–577.

Jaenike, J. (1995). Interactions between mycophagous *Drosophila* and their nematode parasites: from physiological to community ecology. *Oikos* **72**: 235–244.

Jaenike, J. (1996). Population level consequences of parasite aggregation. *Oikos* **76**: 155–160.

Jansen, V. A. A. and de Roos, A. M. (2000). The role of space in reducing predator–prey cycles. In: *The geometry of ecological interactions*. (ed. U. Dieckmann, R. Law, and J. A. J. Metz). Cambridge, Cambridge University Press, pp. 183–201.

Janssen, H., Scheepmaker, M., Couwelaar, M. V., and Pinkster, S. (1979). Biology and distribution of *Gammarus aequicauda* and *G. insensibilis* (Crustacea, Amphipoda) in the lagoon system of Bages-Sigean (France). *Bijdragen tot de Dierkunde* **49**: 42–70.

Janzen, D. H. (1970). Herbivores and the number of tree species in tropical forests. *American Naturalist* **104**: 501–528.

Janzen, D. H. (1980). Specificity of seed-attacking beetles in a Costa Rican deciduous forest. *Journal of Ecology* **68**: 929–952.

Jensen, T., Jensen, K. T., and Mouritsen, K. N. (1998). The influence of the trematode *Microphallus claviformis* on two congeneric intermediate host species (*Corophium*): infection characteristics and host survival. *Journal of Experimental Marine Biology and Ecology* **260**: 349–352.

Johnson, L. L. (2001). An analysis in support of sediment quality thresholds for polycyclic aromatic hydrocarbons to protect estuarine fish, U.S. Dept. Commer., NOAA: 30.

Johnson, N. C., Graham, J. H., and Smith, F. A. (1997). Functioning and mycorrhizal associations along the mutualism–parasitism continuum. *New Phytologist* **135**: 575–586.

Johnson, P. T. J., Lunde, K. B., Thurman, E. M., Ritchie, E. G., Wray, S. N., Sutherland, D. R., Kapfer, J. M., Frest, T. J., Bowerman, J., and Blaustein, A. R. (2002). Parasite (*Ribeiroia ondatrae*) infection linked to amphibian malformations in the western United States. *Ecological Monographs* **72**: 151–168.

Johnson, T. (1999). Beneficial Mutations, hitchhiking and the evolution of mutation rates in sexual populations. *Genetics* **151**: 1621–1631.

Jolles, A. E., Sullivan, P., Alker, A. P., and Harvell, C. D. (2002). Disease transmission of *aspergillosis* in sea fans: inferring process from spatial pattern. *Ecology* **83**: 2373–2378.

Joly, C. and Astagneau, P. (2000). Les infections sur cathéter de longue durée. *Médecine thérapie* **1**: 68–74.

Jones, C. G. and Lawton, J. H. (1995). *Linking species and ecosystems*. New York, Chapman and Hall.

Jones, C.G., Lawton, J.H., and Shachak, M. (1994). Organisms as ecosystems engineers. *Oikos* **69**: 373–386.

Jones, C.G., Lawton, J. H., and Shachak, M. (1997). Positive and negative effects of organisms as physical ecosystem engineers. *Ecology* **78**: 1946–1957.

Jones, J. B., Hyatt, A. D., Hine, P. M., Whittington, R. J., Griffin, D. A., and Bax, N. J. (1997). Special topic review: Australasian pilchard mortalities. *World Journal of Microbiology & Biotechnology* **13**: 383–392.

Jones, L. D., Gaunt M, Hails, R. S., Laurenson, K., Hudson, P. J., Reid, H., Henbest, P., and Gould, E. A. (1997). Efficient transfer of louping ill virus between infected and uninfected ticks cofeeding on mountain hares (*Lepus timidus*). *Medical & Veterinary Entomology* **11**: 172–176.

Juliano, S. A. (1998). Species introduction and replacement among mosquitoes: interspecific resource competition or apparent competition? *Ecology* **79**: 55–268.

Kalavati, C. and Anuradha, I. (1992). Species association patterns among myxosporeans parasitic to mullet, *Mugil cephalus*, found in the backwaters of Visakhapatnam, east coast of India. *Proceedings of the Zoological Society of Calcutta* **45**: 107–113.

Kamp, C., Wilke, C. O., Adami, C., and Bornholdt, S. (2002). Viral evolution under the pressure of an adaptive immune system—optimal mutation rates for viral escape. *Complexity* **8**: 28–33.

Kapel, C. M. O., Pozio, E., Sacchi, L., and Prestrud, P. (1999). Freeze tolerance, morphology, and RAPD-PCR identification of *Trichinella nativa* in naturally infected arctic foxes. *Journal of Parasitology* **85**: 144–147.

Kass, E. H. (1977). Legionnaires' disease. *New England Journal of Medicine* **297**: 1229–1230.

Kattenberg, A., Giorgi, F., Grassl, H., Meehl, G. A., Mitchell, J. F. B., Stouffer, R. J., Tokioka, T., Weaver, A. J. and Wigley, T. M. L. (1996). Climate Models—Projections of Future Climat. In: *Climate Change 1995: The science of climate change. Contribution of Working Group 1 to the Second Assessment Report of the Intergovernmental Panel on climate change* (ed. J. T. Houghton, L. G. M. Filho, B. A. Callandar, N. Harris, A. Kattenberg, and K. Maskell). New York, Cambridge University Press, pp. 285–357.

Kawecki, T. J. and Holt, R.D. (2002). Evolutionary consequences of asymmetric dispersal rates. *American Naturalist* **160**: 395–407.

Keane, R. M. and Crawley, M. J. (2002). Exotic plant invasions and the enemy release hypothesis. *Trends in Ecology and Evolution* **17**: 164–170.

Keas, B. E. and Blankespoor, H. D. (1997). The prevalence of cercariae from *Stagnicola emarginata* (Lymnaeidae) over 50 years in northern Michigan. *Journal of Parasitology* **83**: 536–540.

Kearn, G. C. (1998). *Parasitism and the platyhelminths*. London, Chapman and Hall.

Kearn, G. C. (1999). The survival of monogenean (platyhelminth) parasites on fish skin. *Parasitology* **119**: S57–S88.

Keeling, M. J. (1997). Modelling the persistence of measles. *Trends in Microbiology* **5**: 513–518.

Keeling, M. J. (1999). Spatial models of interacting populations. In: *Advanced ecological theory* (ed. J. McGlade). Oxford, Blackwell, pp. 64–99.

Keeling, M. J. (2000). Evolutionary dynamics in spatial host-parasite systems. In: *The Geometry of Ecological Interactions* (ed. U. Dieckmann, R. Law, and J. A. J. Metz). Cambridge University Press, pp. 271–291.

Keeling, M. J. and Grenfell, B. T. (1997). Disease extinction and community size: modelling the persistence of measles. *Science* **275**: 65–67.

Keeling, M. J. and Grenfell, B. T. (2002). Understanding the persistence of measles: reconciling theory, simulation and observation. *Proceedings of the Royal Society of London B-Biological Sciences* **269**: 335–343.

Keeling, M. J., Wilson, H. B., and Pacala, S. W. (2000). Reinterpreting space, time lags, and functional responses in ecological models. *Science* **290**: 1758–1761.

Keller, L. (1999). *Levels of selection in evolution*. Princeton, NJ: Princeton.

Kennedy, C. E. J. and Southwood, T. R. E. (1984). The number of species of insects associated with British trees: a reanalysis. *Journal of Animal Ecology* **53**: 455–478.

Kennedy, C. R. (1975). *Ecological animal parasitology*. Oxford, UK. Blackwell Scientific Publications.

Kennedy, C. R. (1978). The parasite fauna of resident char *Salvelinus alpinus* from Arctic islands, with special reference to Bear Island. *Journal of Fish Biology* **13**: 255–263.

Kennedy, C. R. (1995). Richness and diversity of macroparasite communities in tropical eels *Anguilla reinhardtii* in Queensland, Australia. *Parasitology* **111**: 233–245.

Kennedy, C. R. and Bush, A. O. (1994). The relationship between pattern and scale in parasite communities: a stranger in a strange land. *Parasitology* **109**: 187–196.

Kennedy, C. R. and Guégan, J.-F. (1994). Regional versus local helminth parasite richness in British freshwater fish: saturated or unsaturated parasite communities? *Parasitology* **109**: 175–185.

Kennedy, C. R. and Guégan, J.-F. (1996). The number of niches in intestinal helminth communities of *Anguilla anguilla*: Are there enough spaces in fish for parasites? *Parasitology* **113**: 293–302.

Kennedy, C. R., Laffoley, D. D., Bishop, G., Jones, P., and Taylor, M. (1986a). Communities of parasites of freshwater fish of Jersey Channel Islands UK. *Journal of Fish Biology* **29** : 215–226.

Kennedy, C. R., Bush, A. O., and Aho, J. M. (1986b). Patterns in helminth communities: why are birds and fish different? *Parasitology* **93**: 205–215.

Kennedy, C. R., Hartvigsen, R., and Halvorsen, O. (1991). The importance of fish stocking in the dissemination of parasites throughout a group of reservoirs. *Journal of Fish Biology* **38**: 541–552.

Kermack, W.O. and McKendrick, A.G. (1927). A contribution to the mathematical theory of epidemics. *Proceedings of the Royal Society of London Series A—Mathematical and Physical Sciences* **115**: 700–721.

Kerr, B., Riley, M. A., Feldman, M. W., and Bohannan, B. J. M. (2002). Local dispersal promotes biodiversity in a real-life game of rock-paper-scissors. *Nature* **418** : 171–174.

Keymer, A. E. and Read, A. F. (1991). Behavioural ecology: the impact of parasitism. In: *Parasite–host associations: coexistence or conflict?* (ed. C. A. Toft, A. Aeschlimann, and L. Bolis). Oxford, Oxford University Press.

Keystone, J. S. (2001). Reemergence of malaria: Increasing risks for travelers. *Journal of Travel Medicine* **8**: S42–S47.

Khan, R. A. (1990). Parasitism in marine fish after chronic exposure to petroleum hydrocarbons in the laboratory and to the Exxon Valdez oil spill. *Bulletin of Environmental Contamination and Toxicology* **44**: 759–763.

Khan, R. A. and Thulin, J. (1991). Influence of pollution on parasites of aquatic animals. *Advances in Parasitology* **30**: 201–238.

Killeen, G. F., Fillinger, U., Kiche, I., Gouagna, L. C., and Knols, B. G. (2002). Eradication of *Anopheles gambiae* from Brazil: lessons for malaria control in Africa? *The Lancet Infectious Diseases* **2**: 618–627.

Kinzig, A. P., Pacala, S., and Tilman, D. (2002). *The functional consequences of biodiversity: empirical progress and theoretical extensions*. Princeton, Princeton University Press.

Kiple, K. F. (1993). *The Cambridge world history of human disease*. New York, Cambridge University Press.

Kitron, U. (1987). Malaria, agriculture and development: lessons from past campaigns. *International Journal of Health Services* **17**: 295–326.

Kitron, U., Otieno, L. H., Hungerford, L. L., Odulaja, A., Brigham, W. U., Okello, O. O., Joselyn, M., Mohamed-Ahmed, M. M., and Cook, E. (1996). Spatial analysis of the distribution of tsetse flies in the Lambwe Valley, Kenya, using Landsat TM satellite imagery and GIS. *Journal of Animal Ecology* **65**: 371–380.

Kleshinski, J., Georgiadis, G. M., and Duggan, J. M. (2000). Group C Streptococcal infection in a prosthetic joint. *Southern Medical Journal* **93**: 1217–1220.

Klironomos, J. N., McCune, J., Hart, M., and Neville, J. (2000). The influence of arbuscular mycorrhizae on the relationship between plant diversity and productivity. *Ecology Letters* **3**: 137–141.

Knops, J. M. H., Tilman, D., Haddad, N. M., Naeem, S., Haarstad, J., Ritchie, M. E., Howe, K. M., Reich, P.B., Siemann, E., and Groth, J. (1999). Effects of plant species richness on invasion dynamics, disease outbreaks, and insect abundances and diversity. *Ecology Letters* **2** : 286–293.

Koella, J. C. and Antia, R. (1995). Optimal pattern of replication and transmission for parasites with two stages in their life cycle. *Theoretical Population Biology* **47**: 277–291.

Kohane, M. J. and Parsons, P. A. (1988). Domestication: evolutionary change under stress. *Evolutionary Biology* **23**: 31–48.

Korinek, A.-M. (2000). Les infections nosocomiales en neurochirurgie. *Médecine thérapie* **1**: 75–79.

Kraijjeveld, A. R. and van Alphen, J. J. M. (1995). Geographic variation in encapsulation ability of *Drosophila melanogaster* and evidence for parasitoid-specific components. *Evolutionary Ecology* **9**: 10–17.

Kraaijeveld, A. R., Ferrari, J., and Godfray, H. C. J. (2002). Costs of resistance in insect–parasite and insect–parasitoid interactions. *Parasitology*, **125**: S71–S82.

Krasnov, B. R., Shenbrot, C. I. Khokhlova, I. S., and Degen, A. A. (2004). Relationship between host diversity and parasite diversity: flea assemblages on small mammals. *Journal of Biogeography*, (in press).

Kristoffersen, R. (1991). Occurrence of the digenean *Cryptocotyle lingua* on farmed Arctic charr *Salvelinus alpinus* and periwinkles *Littorina littorea* sampled close to charr farms in northern Norway. *Diseases of Aquatic Organisms* **12**: 59–65.

Kuris, A. M. (1996). Trophic transmission of parasites and host behavior modification in dinosaur predator–prey dynamics. In: *Sixth North American Paleontological Convention*. (ed. Repetski). The Paleontological Society, Washington. Special Publication 8, DC: pp. 226.

Kuris, A. M. (1997). Host behaviour modification: an evolutionary perspective. In: *Parasites and pathogens: effects on host hormones and behaviour* (ed. N. Beckage). New York, Chapman and Hall.

Kuris, A. M. and Lafferty, K. D. (1992). Modelling crustacean fisheries: effects of parasites on management strategies. *Canadian Journal of Fisheries and Aquatic Sciences* **49**: 327–336.

Kuris, A. M. and Lafferty, K. D. (1994). Community structure: larval trematodes in snail hosts. *Annual Review of Ecology and Systematics* **25**: 189–217.

Kuris, A. M., Blau, S. F., Paul, A. J., Shields, J. D., and Wickham, D. E. (1991). Infestation by brood symbionts and their impact on egg mortality of the red king crab, *Paralithodes camtschatica*, in Alaska: geographic and temporal variation. *Canadian Journal of Fisheries and Aquatic Sciences* **48**: 559–568.

Lack, D. (1954). *The natural regulation of animal numbers*. Oxford, Clarendon Press.

Lafferty, K. D. (1992). Foraging on prey that are modified by parasites. *American Naturalist* **140**: 854–867.

Lafferty, K. D. (1993*a*). The marine snail *Cerithidea californica*, matures et smaller sizes where parasitism is high. *Oikos* **68**: 3–11.

Lafferty, K. D. (1993*b*). Effects of parasitic castration on growth, reproduction and population dynamics of the marine snail *Cerithidea californica*. *Marine Ecology Progress Series* **96**: 229–237.

Lafferty, K. D. (1997). Environmental parasitology: What can parasites tell us about human impacts on the environment? *Parasitology Today* **13**: 251–255.

Lafferty, K. D. (1999). The evolution of trophic transmission. *Parasitology Today* **15**: 111–115.

Lafferty, K. D. (2004). Fishing for lobsters indirectly increases epidemics in sea urchins. *Ecological Applications.* **14**: 1566–1573.

Lafferty, K. D. and Gerber, L. (2002). Good medicine for conservation biology: the intersection of epidemiology and conservation theory. *Conservation Biology* **16**: 593–604.

Lafferty, K. D. and Holt, R. D. (2003). How should environmental stress affect the population dynamics of disease? *Ecology Letters* **6**: 797–802.

Lafferty, K. D. and Kuris, A. M. (1993). Mass mortality of abalone *Haliotis cracherodii* on the California Channel Islands: Tests of epidemiological hypotheses. *Marine Ecology Progress Series* **96**: 239–248.

Lafferty, K. D. and Kuris, A. M. (1999). How environmental stress affects the impacts of parasites. *Limnology and Oceanography* **44**: 925–931.

Lafferty, K. D. and Kuris, A. M. (2002). Trophic strategies, animal diversity and body size. *Trends in Ecology and Evolution* **17**: 507–513.

Lafferty, K. D. and Kushner, D. (2000). Population regulation of the purple sea urchin, *Strongylocentrotus purpuratus*, at the California Channel Islands. In: *Fifth California Islands Symposium*, Vol. 99-0038. (ed. D.R. Brown, K.L. Mitchell, K. L., and H.W. Chang). Santa Barbara, CA, Minerals Management Service, pp. 379–381.

Lafferty, K. D. and Page, C. J. (1997). Predation on the endangered tidewater goby, *Eucyclogobius newberryi*, by the introduced African clawed frog, *Xenopus laevis*, with notes on the frog's parasites. *Copeia*: 589–592.

Lafferty, K. D., Sammond, D. T., and Kuris, A. M. (1994). Analysis of larval trematode communities. *Ecology* **75**: 2275–2285.

Lafferty, K. D., Thomas, F., and Poulin, R. (2000). Evolution of host phenotype manipulation by parasites and its consequences. In: *Evolutionary biology of host–Parasite relationships: theory meets reality* (ed. R. Poulin, S. Morand, and A. Skorping). Amsterdam, Elsevier Science, pp. 117–127.

Lambin, X., Krebs, C. J., Moss, R., Stenseth, N. C., and Yoccoz, N. G. (1999). Population cycles and parasitism. *Science* **286**: 2425.

Lancaster, J. and Robertson, A. L. (1995). Microcrustacean prey and macroinvertebrate predators in a stream food web. *Freshwater Biology* **34**: 123–134.

Lancaster, M., Elmore, S., Anderson, J., Swanson, J., Crandell, M., Haramis, L., Grimstad, P., Jones, C., and

Kriton, U. (1999). Invasion of *Aedes albopictus* into an urban La Crosse encephalitis focus in Illinois. *American Journal of Tropical Medicine and Hygiene* **61**: 432–435.

Latham, A. D. M. and Poulin, R. (2003). Spatiotemporal heterogeneity in recruitment of larval parasites to shore crab intermediate hosts: the influence of shorebird definitive hosts. *Canadian Journal of Zoology* **81**: 1282–1291.

Lauckner, G. (1987). Ecological effects of larval trematode infestations on littoral invertebrate populations. *International Journal for Parasitology* **17**: 391–398.

Laurenson, K. M. Gilbert, L., Norman, R. A., Reid, H. W., and Hudson P. J. (2003). Identifying disease reservoirs in complex system: mountain hares as reservoirs of ticks and louping ill virus, pathogens of red grouse. Epidemiology of louping ill. *Journal of Animal Ecology* **72**: 177–186.

Lavrov, D. L., Brown, W. B., and Boore, J. L. (2004). Phylogenetic position of the Pentastomida and (pan)crustacean relationships. *Proceedings of the Royal Society of London, Series B—Biological Science* **271**: 537–544.

Larsen, M. N. and Roepstorff, A. (1999). Seasonal variation in development and survival of *Ascaris suum* and *Trichuris suis* eggs on pastures. *Parasitology* **119**: 209–220.

Lawler, S. P. (1993). Species richness, species composition and population dynamics of protists in experimental microcosms. *Journal of Animal Ecology* **62**: 711–719.

Lawler, S. P. and Morin, P. J. (1993). Food web architecture and population dynamics in laboratory microcosms of protists. *American Naturalist* **141**: 675–686.

Lawrence, J. G. (1999). Gene transfer, speciation, and the evolution of bacterial genomes. *Current Opinion in Microbiology* **2**: 519–23.

Lawton, J. H. (1989). Food webs. In: *Ecological concepts: the contribution of ecology to an understanding of the natural world* (ed. J. M. Cherrett). Oxford, Blackwell Scientific.

Lawton, J. H. (1995). Ecological experiments with model systems. *Science* **269**: 328–331.

Lawton, J. H. (1999). Are there general laws in ecology ? *Oikos* **84**: 177–192.

Lawton, J. H. (2000). *Community ecology in a changing world.* Ecology Institute, 21385 Oldendorf/Luhe, Germany.

Lawton, J. H., Lewinsohn, T. M., and Compton, S. G. (1993). Patterns of diversity for the insect herbivores on bracken. In: *Diversity in ecological communities. Historical and geographical perspectives* (ed. R. E. Ricklefs and D. Schluter). Chicago, The University of Chicago Press, pp. 178–184.

Leaper, R. and Huxham, M. (2002). Size constraints in a real food web: predator, parasite and prey body-size relationships. *Oikos* **99**: 443–456.

Leaper, R. and Raffaelli, D. (1999). Defining the abundance-body size constraint space: data from a real food web. *Ecology Letters* **2**: 191–199.

Lee, R. (2003). Thermal tolerances of deep-sea hydrothermal vent animals from the northeast Pacific. *Biological Bulletin* **205**: 98–101.

Lefcort, H., Aguon, M. Q., Bond, K. A., Chapman, K. R., Chaquette, R., Clark, J., Kornachuk, P., Lang, B. Z., and Martin, J. C. (2002). Indirect effects of heavy metals on parasites may cause shifts in snail species compositions. *Archives of Environmental Contamination and Toxicology* **43**: 34–41.

Legendre, P., Lapointe, F. J., and Casgrain, P. (1994). Modelling brain evolution from behavior: a permutational regression approach. *Evolution* **48**: 1487–1499.

Legendre, S., Clobert, J., Møller, A. P., and Sorci, G. (1999). Demographic stochasticity and social mating system in the process of extinction of small populations: the case of passerines introduced to New Zealand. *American Naturalist* **153**: 449–463.

Lehman, C. L. and Tilman, D. (2000). Biodiversity, stability, and productivity in competitive communities. *American Naturalist* **156**: 534–552.

Leibold, M. A. and Wooton, J. T. (2001). New introduction. In: *Animal ecology* (ed. C. Elton). Chicago, The University of Chicago Press.

Leibold, M. A., Holyoak, M., Mouquet, N., Amarasekare, P., Cahse, J. M. , Hoopes, M. F., Holt, R. D, Shurin, J. B., Law, R., Tilman, D., Loreau, M., and Gonzalez, A. (2004). The metacommunity concept: a framework for multiscale community ecology. *Ecology Letters* **7**: 601–613.

Leigh, E. G. (1970). Natural Selection and Mutability. *American Naturalist* **104**: 301–305.

Lello, J., Boag, B., Fenton, A., Stevenson, I.R., and Hudson, P. J. (2004). Competition and mutualism among the gut helminthes of a mammalian host. *Nature* **428**: 840–844.

Lemaitre, N. and Jarlier, V. (2000). Les infections nosocomiales à mycobactéries. *Médecine thérapie* **1**: 102–107.

Leong, R. T. (1975). Metazoan parasites of fishes of Cold Lake, Alberta: a community analysis, University of Alberta.

Lessios, H. A. (1988). Mass mortality of *Diadema antillarum* in the Caribbean: what have we learned? *Annual Review of Ecology and Systematics* **19**: 371–393.

Levin, S. A. (1992). The problem of pattern and scale in ecology. *Ecology* **73**: 1943–1967.

Levin, B. R. and Bull, J. J. (1994). Short-sighted evolution and the virulence of pathogenic microorganisms. *Trends in Microbiology* **2**: 76–81.

Levin, S. A. (1999). *Fragile dominion: complexity and the commons*. Reading, Perseus Books.

Levin, S. A. and Pimentel, D. (1981). Selection of intermediate rates of increase in parasite–host systems. *American Naturalist* **117**: 308–315.

Levins, R. and Culver, D. (1971). Regional coexistence of species and competition between rare species. *Proceedings of the National Academy of Sciences of the USA* **68**: 1246–1248.

Liautard, J. P. (1997). Les Maladies du Progrès. La Cité **21**.

Likens, G. E. (1992). *The ecosystem approach: its use and abuse*. Oldendorf/Luhe, Ecology Institute.

Lindeman, R. L. (1942). The trophic-dynamic aspect of ecology. *Ecology* **23**: 399–418.

Lindstöm, E. R., Andren, H., Angeltam, P., Cederlund, G., Hornfeldt, B., Jaderberg, L., Lemnell, P. A., Martinsson, B., Skod, K., and Swenson, G. E. (1994). Disease reveals the predator: sarcoptic mange, red fox predation and prey populations. *Ecology* **75**: 1042–1049.

Lipsitch, M., Siller, S., and Nowak, M. A. (1996). The evolution of virulence in pathogens with vertical and horizontal transmission. *Evolution* **50**: 1729–1741.

Lively, C. M. (1999). Migration, virulence, and the geographic mosaic of adaptation by parasites. *American Naturalist* **153**: S34–S47.

Lively, C. M., Craddock, C., and Vrijenhoek, R. J. (1990). Red Queen hypothesis supported by parasitism in sexual and clonal fish. *Nature* **344**: 864–866.

Loehle, C. (1995). Social barriers to pathogen transmission in wild animal populations. *Ecology* **76**: 326–335.

LoGiudice, K., Ostfeld, R. S., Schmidt, K. A., and Keesing, F. (2003). The ecology of infectious disease: effects of host diversity and community composition on Lyme disease risk. *Proceedings of the National Academy of Science of the USA* **100**: 567–571.

Lorber, B. (1997). Listeriosis. *Clinical Infectious Diseases* **24**: 1–11.

Loreau, M. (1995). Consumers as maximizers of matter and energy flow in ecosystems. *American Naturalist* **145**: 22–42.

Loreau, M. (1998a). Ecosystem development explained by competition within and between material cycles. *Proceedings of the Royal Society of London Series B—Biological Sciences* **265**: 33–38.

Loreau, M. (1998b). Biodiversity and ecosystem functioning: a mechanistic model. *Proceedings of the National Academy of Sciences of the USA* **95**: 5632–5636.

Loreau, M. (2001). Microbial diversity, producer–decomposer interactions and ecosystem processes: a theoretical model. *Proceedings of the Royal Society of London Series B—Biological Sciences* **268**: 303–309.

Loreau, M. and Hector, A. (2001). Partitioning selection and complementarity in biodiversity experiments. *Nature* **412**: 72–76.

Loreau, M., Naeem, S., Inchausti, P., Bengtsson, J., Grime, J. P., Hector, A., Hooper, D. U., Huston, M. A., Raffaelli, D., Schmid, B., Tilman, D., and Wardle, D. A. (2001). Biodiversity and ecosystem functioning: current knowledge and future challenges. *Science* **294**: 804–808.

Loreau, M., Naeem, S., and Inchausti, P. (2002). *Biodiversity and ecosystem functioning: synthesis and perspectives.* Oxford, Oxford University Press.

Loreau, M., Mouquet, N. and Gonzalez, A. (2003). Biodiversity as spatial insurance in heterogeneous landscapes. *Proceedings of the National Academy of Sciences of the USA* **100**: 12765–12770.

Lorenz, K. Z. and Tinbergen, N. (1957). Taxis and instinct. In: *Instinctive behavior: the development of a modern concept.* (ed. C.H. Schiller). New York, International Universities Press.

Lotka, A. J. (1925). *Elements of Physical Biology.* Baltimore, Williams & Wilkins.

Lovelock, J. (1995). *The ages of Gaia: a biography of our living earth.* Oxford, Oxford University Press.

Lubchenco, J. (1978). Plant species diversity in a marine intertidal community: importance of herbivore food preference and algal competitive abilities. *American Naturalist* **112**: 23–39.

Lubchenco, J. (1998). Entering the century of the environment: a new social contract for science. *Science* **279**: 491–497.

Lucet, J.-C. (2000). Les infections du site opératoire. *Médecine thérapie* **1**: 80–84.

Lyles, A. M. and Dobson, A. P. (1993). Infectious disease and intensive management: population dynamics, threatened hosts and their parasites. *Journal of Zoo and Wildlife Medicine* **24**: 315–326.

MacArthur, R. H. (1955). Fluctuations of animal populations and a measure of community stability. *Ecology* **36**: 533–536.

MacArthur, R. H. and Wilson, E. O. (1967). *The Theory of Island Biogeography.* Princeton, Princeton University Press.

MacKenzie, K., Williams, H. H., and Williams, B. (1995). Parasites as indicators of water quality and the potential use of helminth transmission in marine pollution studies. *Advances in Parasitology* **35**: 85–114.

Madhavi, R. and Anderson, R. M. (1985). Variability in the susceptibility of the fish host, *Poecilia reticulata,* to infection with *Gyrodactylus bullatarudis* (Monogenea). *Parasitology* **91**: 531–544.

Maender, G. (2002). Parasites as indicators of costal-ecosystem health. *Sound Waves Monthly Newslette—* August 2002.

Manly, B. F. J. (1991). *Randomization, bootstrap and Monte Carlo methods in biology.* 2nd edn. London, Chapman and Hall.

Marcogliese, D. and Cone, D. (2001). Myxozoan communities parasitizing *Notropis hudsonius* (Cyprinidae) at selected localities on the St. Lawrence River, Quebec: possible effects of urban effluents. *Journal of Parasitology* **87**: 951–956.

Marcogliese, D., Ball, M., and Lankester, M. W. (2001). Potential impacts of clearcutting on parasites of minnows in small boreal lakes. *Folia Parasitologica* **48**: 269–274.

Marcogliese, D. J. (2001). Implications of climate change for parasitism of animals in the aquatic environment. *Canadian Journal of Zoology* **79**: 1331–1352.

Marcogliese, D. J. (2002). Food webs and the transmission of parasites to marine fish. *Parasitology* **124**: S83–S99.

Marcogliese, D. J. and Cone, D. K. (1991). Importance of lake characteristics in structuring parasite communities of salmonids from insular Newfoundland. *Canadian Journal of Zoology* **69**: 2962–2967.

Marcogliese, D. J. and Cone, D. K. (1996). On the distribution and abundance of eel parasites in Nova Scotia: influence of pH. *Journal of Parasitology* **82**: 389–399.

Marcogliese, D. J. and Cone, D. K. (1997). Food webs: a plea for parasites. *Trends in Ecology and Evolution* **12**: 320–325.

Martens, P. and McMichael, A. J. (2002). *Environmental change, Climate and health.* Cambridge, Cambridge University Press.

Martens, P., Kovats, R. S., Nijhof, S., de Vries, P., Livermore, M. T. J., Bradley, D. J., Cox, J., and McMichael, A. J. (1999). Climate change and future populations at risk of malaria. *Global Environmental Change, Human and Policy Dimensions* **9**: S89–S107.

Martin, T. E., Møller, A. P., Merino, S., and Clobert, J. (2001). Does clutch size evolve in response to parasites and immunocompetence? *Proceedings of the National Academy of Sciences of the USA* **98**: 2071–2076.

Martinez, N. D. (1991). Artifacts or attributes? Effects of resolution on the Little Rock Lake food web. *Ecological Monographs* **61**: 367–392.

Martinez, N. D. (1992). Constant connectance in community food webs. *American Naturalist* **139**: 1208–1218.

Marx, D. H. (1972). Ectomycorrhizae as biological deterrents to pathogenic root infections. *Annual Review of Phytopathy* **10**: 429–454.

Mas-Coma, S. and Feliu, C. (1984). Helminthofauna from small mammals (insectivores and rodents) on the Pityusic islands. In: *Biogeography and ecology of the Pityusic Islands.* (ed. J. Kuhbier, J. A. Alcover and C. Guerau d'Arellano). Barcelona, Junk Pub, pp. 469–525.

Mas-Coma, S., Funatsu, I.R., and Bargues, M. D. (2001). *Fasciola hepatica* and lymnaeid snails occurring at very high altitude in South America. *Parasitology* **123**: S115–S127.

Mas-Coma, S., Bargues, M. D., Valero, M. A., and Fuentes, M. V. (2003). Adaptation capacities of *Fasciola hepatica* and their relationships with human fascioliasis: from below sea level up to the very high altitude. In: *Taxonomy, ecology and evolution of metazoan parasites Volume II.* (ed. C. Combes and J. Jourdane). Perpignan, Presses Universitaires de Perpignan, pp. 81–123.

Matson, P. A. and Vitousek, P. M. (1987). Cross-system comparisons of soil nitrogen transformations and nitrous oxide flux in tropical forest systems. *Global Biogeochemical Cycles* **1**: 163–170.

Maurer, B. A. (1999). *Untangling ecological complexity. The macroscopic perspective.* Chicago, The University of Chicago Press.

May, R. M. (1972). Will a large complex system be stable? *Nature* **238**: 413–414.

May, R. M. (1973). *Stability and complexity in model ecosystems.* Princeton, Princeton University Press.

May, R. M. (1974). *Stability and complexity in model ecosystems,* 2nd edn. Princeton, Princeton University Press.

May, R. M. and Anderson, R. M. (1978). Regulation and stability of host-parasite population interactions. II. Destabilising processes. *Journal of Animal Ecology* **47**: 249–267.

May, R. M. and Anderson, R. M. (1979). Population Biology of Infectious Diseases. Part II. *Nature* **280**: 455–461.

Maynard Smith, J. and Szathmáry, E. (1995). *The major transitions in evolution.* San Francisco, W. H. Freeman.

McCallum, H. I. and Dobson, A. P. (1995). Detecting disease and parasite threats to endangered species and ecosystems. *Trends in Ecology and Evolution* **10**: 190–194.

McCoy, K. D., Boulinier, T., Chardine, J., Michalakis, Y., and Danchin, E. (1999). Dispersal of the tick *Ixodes uriae* within and among seabird host populations: the need for a population genetic approach. *Journal of Parasitology* **85**: 196–202.

McCoy, K. D., Boulinier, T., Tirard, C., and Michalakis, Y. (2003). Differential host-associated dispersal of the seabird ectoparasite *Ixodes uriae. Evolution* **57**: 288–296.

McGrady-Steed, J., Harris, P. M., and Morin, P. J. (1997). Biodiversity regulates ecosystem predictability. *Nature* **390**: 162–165.

McKendrick, A. G. (1940). The dynamics of crowd infections. *Edinburgh Medical Journal* **47**: 117–136.

McMichael-Phillips, D. F., Lewis, J. W., and Thorndyke, M. C. (1996). The distribution of neuroactive substances within the cercaria of *Sanguinicola inermis. Journal of Helminthology* **70**: 309–317.

McMichael, A. J. (2001). *Human frontiers.* Cambridge, Cambridge University Press.

McMichael, A. J. and Haines, A. (1997). Global climate change: the potential effects on health. *British Medical Journal* **315**: 805–809.

McNamara, J. M. and Houston, N. B. (1992). Evolution of stable levels of vigilance as a function of group size. *Animal Behaviour* **43**: 641–658.

McNaughton, S. J. (1977). Diversity and stability of ecological communities: a comment on the role of empiricism in ecology. *American Naturalist* **111**: 515–525.

McNaughton, S. J. (1992). The propagation of disturbance in savannas through food webs. *Journal of Vegetation Science* **3**: 301–314.

McNaughton, S. J. and Coughenour, M. B. (1981). The cybernetic nature of ecosystems. *American Naturalist* **117**: 985–990.

McNeil, R., Tulio Diaz, M., and Villeneuve, A. (1994). The mystery of shorebird over-summering: a new hypothesis. *Ardea* **82**: 143–152.

McSorley, R. (2003). Adaptations of nematodes to environmental extremes. *Florida Entomologist* **86**: 138–142.

Medley, G. F. (2002). The epidemiological consequences of optimization of the individual host immune response. *Parasitology* **125**: S61–S70.

Memmott, J., Martinez, N. D., and Cohen, J. E. (2000). Predators, parasitoids and pathogens: species richness, trophic generality and body sizes in a natural food web. *Journal of Animal Ecology* **69**: 1–15.

Michalakis, Y. and Hochberg, M. E. (1994). Parasitic effects on host life-history traits: a review. *Parasite* **1**: 191–294.

Michalakis, Y., Olivieri, I., Renaud, F., and Raymond, M. (1992). Pleiotropic action of parasites—how to be good for the host. *Trends in Ecology and Evolution* **7**: 59–62.

Miquel, J., Fons, R., Feliu, C., Marchand, B., Torres, J., and Clara, J. P. (1996). Helminthes parasites des rongeurs Muridés des îles d'Hyères (Var, France): aspects écologiques. *Vie & Milieu* **46**: 219–223.

Minchella, D. J. (1985). Host life-history variation in response to parasitism. *Parasitology* **90**: 205–216.

Minchella, D. J. and Loverde, P. T. (1981). A cost of increased early reproductive effort in the snail *Biomphalaria glabrata. American Naturalist* **118**: 876–881.

Minchella, D. J. and Scott, M. E. (1991). Parasitism: a cryptic determinant of animal community structure. *Trends in Ecology and Evolution* **6**: 250–254.

Mitchell, C. E. (2003). Trophic control of grassland production and biomass. *Ecology Letters* **6**: 147–155.

Mitchell, C. E. and Power, A. G. (2003). Release of invasive plants from fungal and viral pathogens. *Nature* **421**: 625–627.

Mitchell, C. E., Tilman, D. and Groth, J. V. (2002). Effects of grassland species diversity, abundance, and composition on foliar fungal disease. *Ecology* **83**: 1713–1726.

Mode, C. J. (1958). A mathematical model for the co-evolution of obligate parasites and their hosts. *Evolution* **12**: 158–165.

Møller, A. P. and Cassey, P. (2004). On the relationship between T-cell mediated immunity in bird species and the establishment success of introduced populations. *Journal of Animal Ecology*, in press.

Møller, A. P. and Erritzøe, J. (1996). Parasite virulence and host immune defense: Host immune response is related to nest re-use in birds. *Evolution* **50**: 2066–2072.

Møller, A. P. and Erritzøe, J. (2000). Predation against birds with low immunocompetence. *Oecologia* **122**: 500–504.

Møller, A. P. and Erritzoe, J. (2002). Coevolution of host immune defence and parasite-mediated mortality: relative spleen size and mortality in altricial birds. *Oikos* **99**: 95–100.

Møller, A. P. and Erritzoe, J. (2003). Climate, body condition and spleen size in birds. *Oecologia* **442**: 621–626.

Møller, A. P., Merino, S., Brown, C. R., and Robertson, R. J. (2001). Immune defense and host sociality: a comparative study of swallows and martins. *American Naturalist* **158**: 136–145.

Møller, A. P., Martín-Vivaldi, M., and Soler, J. J. (2004a). Parasitism, host immune defence and dispersal. *Journal of evolutionary Biology*, in press.

Møller, A. P., Marzal, A., Navarro, C., and de Lope, F. (2004b). Predation risk, host immune response and parasitism. *Behavioral Ecology*, in press.

Møller, A. P., Martín-Vivaldi, M., and Soler, J. J. (2004c). Parasite effects on host population size: A test using T-cell mediated immunity and abundance-occupancy data for European birds. Unpublished manuscript.

Mollison, D. and Levin, S. (1995). Spatial dynamics of parasitism. In : *Ecology of infectious diseases in natural populations* (ed. B. Grenfell and A. Dobson). Cambridge, Cambridge University Press, pp. 384–398.

Moore, J. (2002). *Parasites and the Behavior of animals*. Oxford, Oxford University Press.

Moore, P.S. and Broome, C.V. (1994). Cerebrospinal meningitis epidemics. *Scientific American* **271**: 38–45.

Moore, J. and Gotelli, N. J. (1990). A phylogenetic perspective on the evolution of altered host behaviours: a critical look at the manipulation hypothesis. In: *Parasitism and host behaviour* (ed. C. J. Barnard and J. M. Behnke). London, Taylor & Francis.

Morand, S. (2000). Wormy world: comparative tests of theoretical hypotheses on parasite species richness. In: *Evolutionary biology of host–parasite relationships: theory meets reality* (ed. R. Poulin, S. Morand, and

A. Skorping). Elsevier, Amsterdam, The Netherlands, pp. 63–79.

Morand, S. and Guégan, J.-F. (2000a). Distribution and abundance of parasite nematodes: ecological specialisation, phylogenetic constraint or simply epidemiology? *Oikos* **88**: 563–573.

Morand, S. and Guégan, J.-F. (2000b). Patterns of endemism in host–parasite associations: lessons from epidemiological models and comparative tests. *Belgian Journal Entomology* **2**: 135–147.

Morand, S. and Poulin, R. (1998). Density, body mass and parasite species richness of terrestrial mammals. *Evolutionary Ecology* **12**: 717–727.

Morand, S. and Sorci, G. (1998). Determinants of life-history evolution in nematodes. *Parasitology Today* **14**: 193–196.

Morand, S., Rohde, K., and Hayward, C. (2002). Order in ectoparasite communities of marine fish is explained by epidemiological processes. *Parasitology* **124**: S57–S63.

Morand, S., Manning, S. D., and Woolhouse, M. E. J. (1996). Parasite–host coevolution and geographic patterns of parasite infectivity and host susceptibility. *Proceedings of the Royal Society of London Series B— Biological Sciences* **263**: 119–128.

Morand, S., Poulin, R., Rohde, K., and Hayward, C. (1999). Aggregation and species coexistence of ectoparasites of marine fishes. *International Journal of Parasitology* **29**: 663–672.

Morand, S., Cribb, T. H., Kulbicki, M., Chauvet, C., Dufour, V., Faliex, E., Galzin, R., Lo, C., Lo-Yat, A., Pichelin, S.P., Rigby, M. C., and Sasal, P. (2000). Determinants of endoparasite species richness of New Caledonian Chaetodontidae. *Parasitology* **121**: 65–73.

Morgan, M. D. and Good R.E. (1988). Stream chemistry in the New Jersey Pinelands: the influence of precipitation and watershed disturbance. *Waters Resources Research* **24**: 1091–1100.

Morin, P. J. (1999). *Community ecology.* Oxford, Blackwell Science.

Morin, P. J. and Lawler, S. P. (1996). Effects of food chain length and omnivory in experiment food webs. In: *Food webs: integration of patterns and dynamics* (ed. G. A. Polis and K. O. Winemiller). New York, Chapman and Hall.

Mork, M. (1996). The effect of kelp in wave damping. *Sarsia* **80**: 323–327.

Morris, J. G. and Potter, M. (1997). Emergence of new pathogens as a function of changes in host susceptibility. *Emerging Infectious Diseases* **3**: 435–441.

Morse, S. S. (1995). Factors in the emergence of Infectious diseases. *Emerging Infectious Diseases* **1**: 7–15.

Morse, S. S. and Schluederberg, A. (1990). Emerging viruses: The evolution of virus and viral diseases. *Journal of Infectious Diseases* **162**: 1–7.

Mouchet, J. and Manguin, S. (1999). Global warming and malaria expansion. *Annales de la Societé Entomologique De France* **35**: 549–555.

Mougeot, F., Redpath, S. M., Moss, R., Matthiopoulos, J., and Hudson, P. J. (2003). Territorial behaviour and population dynamics in red grouse *Lagopus lagopus scoticus*. I. Population experiments. *Journal of Animal Ecology* **72**: 1073–1082.

Mouquet, N. and Loreau, M. (2003). Community patterns in source–sink metacommunities. *American Naturalist* **162**: 544–557.

Mouritsen, K. N. and Poulin, R. (2002). Parasitism, community structure and biodiversity in intertidal ecosystems. *Parasitology* **124**: S101–S117.

Moxon, E. R., Rainey, P. B., Nowak M. A. and Lenski, R. E. (1994). Adaptive evolution of highly mutable loci in pathogenic bacteria. *Current Biology* **4**: 24–33.

Mulder, C. P. H., Koricheva, J., Huss-Danell, K., Högberg, P., and Joshi, J. (1999). Insects affect relationships between plant species richness and ecosystem processes. *Ecology Letters* **2**: 237–246.

Murdoch, W. M., Briggs, C., and Nisbet, R. M. (2003). *Consumer resource dynamics*. Princeton, Princeton University Press.

Murphy, P. M. (2001). Viral exploitation and subversion of the immune system through chemokine mimicry. *Nature Immunology* **2**: 116–122.

Murphy, B. R. and Webster, R. G. (1996). In: *Orthomyxoviruses*. In: *Virology*. (ed. B. N. Fields, D. M. Knipe, P. M. Howley, R. M. Chanock, J. L. Melnick, T. P. Monath et al.). Philadelphia: Lippincott-Raven Publishers, pp. 1397–445.

Murray, T. P., Kay, J. J., Waltner-Toews, D., and Raez-Luna, E. F. (2002). Linking human and ecosystem health on the Amazon frontier. In: *Conservation medicine. Ecological health in practice* (ed. A. A. Aguirre, R. S. Ostfeld, G. M. Tabor, C. House, and M. C. Pearl). Oxford, Oxford University Press, pp. 297–309.

Murray, A. G., O'Callaghan, M., and Jones, B. (2003). A model of spatially evolving herpesvirus epidemics causing mass mortality in Australian pilchard *Sardinops sagax*. *Diseases of Aquatic Organisms* **54**: 1–14.

Myers, R. A. and Worm, B. (2003). Rapid worldwide depletion of predatory fish communities. *Nature* **423**: 280–283.

Naeem, S., Hahn, D. R., and Schuurman, G. (2000). Producer–decomposer co-dependency influences biodiversity effects. *Nature* **403**: 762–764.

Nee, S. (2003). Unveiling prokaryotic diversity. *Trends in Ecology and Evolution* **18**: 62–63.

Nekola, J. C. and White, P. S. (1999). The distance decay of similarity in biogeography and ecology. *Journal of Biogeography* **26**: 867–878.

Newton, I. (1998). *Population limitation in birds*. London, Academic Press.

Nickol, B. B. (1979). *Host–parasite interfaces*. New York, Academic Press.

Nisbet, R. M., Briggs, C. J., Gurney, W. S. C., Murdoch, W. M., and Stewart-Oaten, A. (1993). Two-patch metapopulation dynamics. In: *Patch Dynamics* (ed. S. A. Levin, T. M. Powell, and J. H. Steele). Berlin, Springer-Verlag, pp. 125–135.

Nollen, P. M. (1968). Autoradiographic studies on reproduction in *Philophthalmus megalurus* (Cort, 1914) (Trematoda). *Journal of Parasitology* **54**: 43–48.

Norberg, J. (2004). Biodiversity and ecosystem functioning: A complex adaptive systems approach. *Limnology and Oceanography* **49**: 1269–1279.

Norberg, J., Swaney, D. P., Dushoff, J., Lin, J., Casagrandi, R., and Levin, S. A. (2001). Phenotypic diversity and ecosystem functioning in changing environments: a theoretical framework. *Proceeding of the National Academy of Sciences of the USA* **98**: 11376–11381.

Norrby, E., Sheshberadaran, H., McCullough, K. C., Carpenter, W. C., and Orvell, C. (1985). Is rinderpest virus the archevirus of the *Morbillivirus* genus? *Intervirology* **23**: 2228–2232.

Norris, K. (1999). A trade-off between energy intake and exposure to parasites in oystercatchers feeding on a bivalve mollusc. *Proceedings of the Royal Society of London Series B—Biological Sciences* **266**: 1703–1709.

Nowak, M. (1990). HIV mutation rate. *Nature* **347**: 522.

Nowak, M. A. and May, R. M. (1994). Superinfection and the evolution of parasite virulence. *Proceedings of the Royal Society of London Series B—Biological Sciences* **255**: 81–89.

Nuismer, S. L., Thompson, J. N., and Gomulkiewicz, R. (1999). Gene flow and geographically structured coevolution. *Proceedings of the Royal Society of London Series B—Biological Sciences* **266**: 605–609.

Nuismer, S. L., Thompson, J. N., and Gomulkiewicz, R. (2000). Coevolutionary clines across selection mosaics. *Evolution* **54**: 1102–1115.

Nuismer, S. L., Thompson, J. N., and Gomulkiewicz, R. (2003). Coevolution between hosts and parasites with partially overlapping geographic ranges. *Journal of Evolutionary Biology* **16**: 1337–1345.

Nunn, C. L., Altizer, S., Jones, K. E., and Sechrest, W. (2003). Comparative tests of parasite species richness in primates. *American Naturalist* **162**: 597–614.

Oberdorff, T., Hugueny, B., Compin, A., and Belkessam, D. (1998). Non-interactive fish communities in the coastal streams of north-western France. *Journal of Animal Ecology* **67**: 472–484.

Odling-Smee, F.J., Laland, K.N., and Feldman, M. (2003). *Niche Construction—the neglected process in evolution.* Princeton, NJ, Princeton University Press.

Odum, E. P. (1953). *Fundamentals of ecology* 1971, 3rd edn. (London, Saunders. 1971).

Odum, E.P. (1959). *Fundamentals of ecology.* Philadelphia, WB Saunders Company.

Odum, E. P. (1964). The new ecology. *Bioscience* **14**: 14–16.

Odum, E. P. (1969). The strategy of ecosystem development. *Science* **164**: 262–270.

Oksanen, L. and Oksanen, T. (2000). The logic and realism of the hypothesis of exploitation ecosystems. *American Naturalist* **155**: 703–723.

Ollerenshaw, C. B. (1974). Forecasting liver fluke disease. In: *The Effects of Meteorological Factors upon Parasites.* (ed. A. E. R. Taylor and R. Muller). Symposia of the British Society for Parasitology **12**. Oxford, Blackwell Scientific Publications, pp. 33–52.

Olsen, S. J., Bishop, R., and Brenner, F. W. (2001). The changing epidemiology of *Salmonella*: trends in serotypes isolated from humans in the U.S., 1987–1997. *Journal of Infectious Diseases* **183**: 756–761.

O'Meara, G. F., Evans, L. E., Gettman, L. F., and Cuda, J. P. (1995). Spread of *Aedes albopictus* and decline of *Ae. Aegypti* (Diptera: Culicidae) in Florida. *Journal of Medical Entomology* **32**: 554–562.

Oppliger, A., Vernet, R., and Baez, M. (1999). Parasite local maladaptation in the Canarian lizard *Gallotia galloti* (Reptilia : Lacertidae) parasitized by haemogregarian blood parasites. *Journal of Evolutionary Biology.* **12**: 951–955.

O'Shea, T. (1999). Environmental contaminants and marine mammals. In: *Biology of Marine Mammals* (ed. S. A. Rommel) Washington, DC, Smithsonian Institution Press, pp. 485–564.

Ouedraogo, R. M., Cusson, M., Goettel, M. S., and Brodeur, J. (2003). Inhibition of fungal growth in thermo-regulating locusts, *Locusta migratoria*, infected by the fungus *Metarhizium anisopliae* var *acridum*. *Journal of Invertebrate Pathology* **82**: 103–109.

Overstreet, R. M. (1993). Parasitic diseases of fishes and their relationship with toxicants and other environmental factors. In: *Pathobiology of marine and estuarine organisms* (ed. J. A. Couch and J. W. Fournie). Fla, Boca Raton, CRC Press, pp. 111–156.

Overstreet, R. M. and Howse, H. D. (1977). Some parasites and diseases of estuarine fishes in polluted habitats of Mississippi. *Annals of the New York Academy of Sciences* **298**: 427–462.

Osborne, P. (1985). Some effects of Dutch elm disease on the birds of a Dorset dairy farm. *Journal of Applied Ecology* **22**: 681–691.

Outreman, Y., Bollache, L., Plaistow, S., and Cézilly, F. (2002). Patterns of intermediate host use and levels of association between two conflicting manipulative parasites. *International Journal for Parasitology* **32**: 15–20.

Pace, M. L. and Groffman, P. M. (1998). *Successes, limitations and frontiers in ecosystem science.* New York, Springer.

Packer, A. and Clay, K. (2000). Soil pathogens and spatial patterns of seedling mortality in a temperate tree. *Nature* **404**: 278-281.

Packer, C., Holt, R. D., Hudson, P. J., Lafferty, K. D., and Dobson, A. P. (2003a). Keeping the herds healthy and alert: implications of predator control for infectious disease. *Ecology Letters* **6**: 797–802.

Packer, C. Dobson, A. P. Holt, R. H., and Hudson, P. J. (2003b). Healthy herds and the role of predators in removing parasites from ungulate populations. *Ecology Letters* **6**: 797-802.

Paine, R. T. (1966). Food web complexity and species diversity. *American Naturalist* **100** : 65–75.

Paine, R. T. (1988). Food webs: road maps of interactions or grist for theoretical development? *Ecology* **69**: 1648–1654.

Paperna, I. (1969). Study of an outbreak of schistosomiasis in the newly formed Lake Volta in Ghana. *Zeitschrift fur Tropenmedizin und Parasitologie* **2**: 339–353.

Pappas, P. W. and Uglem, G. L. (1990). *Hymenolepis diminuta* (Cestoda) liberates an inhibitor of proteolytic enzymes during *in vitro* incubation. *Parasitology* **101**: 455–464.

Park, T. (1948). Experimental studies of interspecies competition. I. Competition between populations of the flour beetles, *Tribolium confusum* Duval and *Tribolium castaneum* Herbst. *Ecological Monographs* **18**: 265–308.

Parmesan, C. and Yohe, G. (2003). A globally coherent fingerprint of climate change impacts across natural systems. *Nature* **421**: 37–42.

Pascual, M., Bouma, M. J. and Dobson, A. P. (2002). Cholera and climate: revisiting the quantitative evidence. *Microbes and Infection* **4**: 237–245.

Paterson, A. M. and Gray, R. D. (1997). Host–parasite co-speciation, host switching, and missing the boat. In: *Host–parasite evolution.* (ed. by D. H. Clayton and J. Moore). Oxford University Press, pp. 236–250.

Patten, B. C. and Odum, E. P. (1981). The cybernetic nature of ecosystems. *American Naturalist* **118**: 886–895.

Patz, J. A. (2002). A human disease indicator for the effects of recent global climate change. *Proceedings of the National Academy of Sciences of the USA* **99**: 12506–12508.

Patz, J. A., Epstein, P. R., Burke, T. A., and Balbus, J. M. (1996). Global climate change and emerging infectious disease. *JAMA* **275**: 217–223.

Patz, J. A., Graczyk, T. K., Geller, N., and Vittor, A. Y. (2000). Effects of environmental change on emerging parasitic diseases. *International Journal for Parasitology* **30**: 1395–1405.

Peckarsky, B. L., Fraissinet, P. R., Penton, M. A., and Conklin Jr., D. J. (1990). *Freshwater macroinvertebrates of northeastern north America*. Ithaca, New-York, Cornell University Press.

Pedro-Botet, M. L., Stout, J. E., and Yu, V. L. (2002). Legionnaire's disease contracted from patient homes: the coming of the third plague? *European Journal of Clinical Microbiology and Infectious Diseases* **21**: 699–705.

Peiris, J. S. M., Lai, S. T., Poon, L. L., Yam, L. Y., Lim, W., Nicholls, J., Yee, W. K., Yan, W. W., Cheung, M. T., Cheng, V. C., Chan, K. H., Tsang, D. N., Yung, R. W., Ng, T. K., and Yuen, K. Y. (SARS study group) (2003). Coronavirus as a possible cause of severe acute respiratory syndrome. *Lancet* **361**: 1319–1325.

Pellmyr, O. and Huth, C. J. (1994). Evolutionary stability of mutualism between yuccas and yucca moths. *Nature* **372**: 257–260.

Perkins, S., Cattadori, I. M. , Tagliapietra, V., Rizzoli, A. P., and Hudson P. J. (2003). Empirical evidence for key hosts in persistence of a tick-born disease. *International Journal of Parasitology* **33**: 909–917.

Perry, R. N. (1999). Desiccation survival of parasitic nematodes. *Parasitology* **119**: S19–S30.

Petchey, O. L., Downing, A. L., Mittelbach, G., Persson, L., Steiner, C. F., Warren, P. H., and Woodward, G. (2004). Species loss and the structure and functioning of multitrophic aquatic systems. *Oikos* **104**: 467–478.

Petrushevski, G. K. and Schulman, S. S. (1958). The parasitic diseases of fishes in the natural waters of the USSR. In: *Parasitology of Fishes*. (ed. V.A. Dogiel, G.K. Petrushevski, and Y. I. Polyanski). Leningrad University Press, pp. 299–319.

Pianka, E. R. (1966). Latitudinal gradients in species diversity, a review of concepts. *American Naturalist* **100**: 33–46.

Pickett, S. T. A. and Cadenasso, M. L. (2002). The ecosystem as a multidimensional concept: meaning, model, and metaphor. *Ecosystems* **5**: 1–10.

Pimm, S. L. (1982). *Food Webs*. New York, Chapman and Hall.

Pimm, S. L. (1984). The complexity and stability of ecosystems. *Nature* **307**: 321–326.

Pimm, S. L. and Kitching, R. L. (1988). Food web patterns: trivial flaws or the basis of an active research program? *Ecology* **69**: 1669–1672.

Pimm, S. L and Lawton, J. H. (1977). Number of trophic levels in ecological communities. *Nature* **268**: 329–331.

Pimm, S. L. and Lawton, J. H. (1978). On feeding on more than one trophic level. *Nature* **275**: 542–544.

Plowright, W. (1968). Rinderpest virus. In: *Virology Monographs No 3*. (ed. S. Gard, C. Hallauer, and K. F. Meyer). New York, Springer-Verlag, pp. 1-94.

Plowright, W. (1982). The effects of rinderpest and rinderpest control on wildlife in Africa. *Symposia of the Zoological Society of London* **50**: 1-28.

Plowright, W. and McCullough, B. (1967). Investigations on the incidence of rinderpest virus infection in game animals of N. Tanganyika and S. Kenya 1960/63. *Journal of Hygiene* **65**: 343-358.

Plowright, W. and Taylor, W. P. (1967). Long-term studies of immunity in East African cattle following inoculation with rinderpest culture vaccine. *Researches in Veterinary Science* **8**: 118-128.

Pohley, W. J. (1976). Relationships among three species of *Littorina* and their larval digenea. *Marine Biology* **37**: 179–186.

Poinar, G. Jr. (2002). First fossil record of nematode parasitism of ants; a 40 million year tale. *Parasitology* **125**: 457–459.

Poinar, G. Jr. (2003). Fossil evidence of phorid parasitism (Diptera: Phoridae) by allantonematid nematodes (Tylenchida: Allantonematidae). *Parasitology* **127**: 589–592.

Pojmanska, T., Grabda-Kazubska, B., Kazubski, S. L., Machalska, J., and Niewiadomska, K. (1980). Parasite fauna of five fish species from the Konin lakes complex, artificially heated with thermal effluents, and from Goplo lake. *Acta Parasitologica* **27**: 319–357.

Polak, M. and Starmer, W. T. (1998). Parasite-induced risk of mortality elevates reproductive effort in male *Drosophila*. *Proceeding of the Royal Society of London B Series B—Biological Sciences* **265**: 197–201.

Polis, G. A. (1991). Complex trophic interactions in deserts: an empirical critique of food-web theory. *American Naturalist* **138**: 123–155.

Polis, G. A. (1994). Food webs, trophic cascades and community structure. *Australian Journal of Ecology* **19**: 121–136.

Polis, G. A. (1999). Why are parts of the world green? Multiple factors control productivity and the distribution of biomass. *Oikos* **86**: 3–15.

Polis, G. A. and Strong, D. R. (1996*a*). Food web complexity and community dynamics. *American Naturalist* **147**: 813–846.

Polis, G. A. and Strong, D. (1996*b*). Food web complexity and community dynamics. *American Naturalist* **147**: 813–846.

Polis, G. A. and Winemiller, K. O. (1996). *Food webs: integration of patterns and dynamics*. New York, Chapman and Hall.

Polyanski, Y. I. (1961). Ecology of parasites of marine fishes. In: *Parasitology of fishes*. (ed. V. A. Dogiel, G. K. Petrushevski, and Y. I. Polyanksi). Edinburgh, Oliver & Boyd.

Post, D.M. (2002). The long and short of food-chain length. *Trends in Ecology and Evolution* **17**: 269–277.

Post, W. M., DeAngelis, D. L., and Travis, C.C. (1983). Endemic disease in environments with spatially heterogeneous host populations. *Mathematical Biosciences* **63**: 289–302.

Potts, G. R. (1986). *The partridge. Pesticides, predation and conservation*. London, Collins, pp. 274.

Poulin, R. (1994a). Meta-analysis of parasite-induced behavioural changes. *Animal Behaviour* **48**: 137–146.

Poulin, R. (1994b). The evolution of parasite manipulation of host behaviour: a theoretical analysis. *Parasitology* **109**: S109–S118.

Poulin, R. (1995a). Phylogeny, ecology, and the richness of parasite communities in vertebrates. *Ecological Monographs* **65**: 283–302.

Poulin, R. (1995b). 'Adaptive' changes in the behaviour of parasitized animals: a critical review. *International Journal for Parasitology* **25**: 1371–1383.

Poulin, R. (1996a). Richness, nestedness, and randomness in parasite infracommunity structure. *Oecologia* **105**: 545–551.

Poulin, R. (1996b). How many parasite species are there: are we close to answers? *International Journal for Parasitology* **26**: 1127-1129.

Poulin, R. (1997). Species richness of parasite assemblages: evolutions and patterns. *Annual Review of Ecology and Systematics* **28**: 341–358.

Poulin, R. (1998a). Evolutionary ecology of parasites. From individuals to communities. London, UK, Chapman and Hall Ltd.

Poulin R. (1998b). Comparisons of three estimators of species richness in parasite component communities. *Journal of Parasitology* **84**: 485–490.

Poulin, R. (1999). The functional importance of parasites in animal communities. *International Journal for Parasitology* **29**: 915–920.

Poulin, R. (2000). Manipulation of host behaviour by parasites: a weakening paradigm? *Proceedings of the Royal Society of London Series B—Biological Sciences* **267**: 787–792.

Poulin, R. (2001). Another look at the richness of helminth communities in tropical freshwater fish. *Journal of Biogeography* **28**: 737–743.

Poulin, R. (2003a). Parasites in a global village: food, emerging diseases, and evolution. *Journal of Parasitology* **89** (Suppl): S271–S276.

Poulin, R. (2003b). The decay of similarity with geographical distance in parasite communities of vertebrate hosts. *Journal of Biogeography* **30**: 1609–1615.

Poulin, R. and Guégan, J.-F. (2000). Nestedness, antinestedness, and the relationship between prevalence and intensity in ectoparasite assemblages of marine fish. A spatial model of species coexistence. *International Journal Parasitololy* **30**: 1147–1152.

Poulin, R. and Morand, S. (1999). Geographical distances and the similarity among parasite communities of conspecific host populations. *Parasitology* **119**: 369–374.

Poulin, R. and Morand, S. (2000). The diversity of parasites. *Quarterly Review of Biology* **75**: 277–293.

Poulin, R. and Mouritsen, K. N. (2003). Large-scale determinants of trematode infections in intertidal gastropods. *Marine Ecology Progress Series* **254**: 187–198.

Poulin, R. and Rate, S. R. (2001). Small-scale spatial heterogeneity in infection levels by symbionts of the amphipod *Talorchestia quoyana* (Talitridae). *Marine Ecology-Progress Series* **212**: 211–216.

Poulin, R. and Rohde, K. (1997). Comparing the richness of metazoan ectoparasite communities of marine fishes: controlling for host phylogeny. *Oecologia* **110**: 278–283.

Poulin, R. and Valtonen, E. T. (2001). Nested assemblages resulting from host-size variation: the case of endoparasite communities in fish hosts. *International Journal of Parasitology* **31**: 1194–1204.

Poulin, R. and Valtonen, E. T. (2002). The predictability of helminth community structure in space: a comparison of fish populations from adjacent lakes. *International Journal of Parasitology* **32**: 1235–1243.

Poulin, R., Curtis, M. A., and Rau, M. E. (1992). Effects of *Eubothrium salvelini* (Cestoda) on the behaviour of *Cyclops vernalis* (Copepoda) and its susceptibility to fish predators. *Parasitology* **105**: 265–271.

Poulin, R., Hecker, K., and Thomas, F. (1998). Hosts manipulated by one parasite incur additional costs from infection by another parasite. *Journal of Parasitology* **84**: 1050–1052.

Poulin, R., Morand, S., and Skorping, A. (2000). *Evolutionary biology of host–parasite relationships: Theory meets reality* (ed. R. Poulin, S. Morand, and A. Skorping). Amsterdam, Elsevier Science.

Poveda, G., Rojas, W., Quinones, M. L., Velez, I. D., Mantilla, R. I., Ruiz, D., Zuluaga, J. S., and Rua, G. L. (2001). Coupling between annual and ENSO timescales in the malaria–climate association in Colombia. *Environmental Health Perspectives* **109**: 489–493.

Price, P. W. (1980). *Evolutionary biology of parasites*. Princeton, Princeton University Press.

Price, P. W., Westoby, M., Rice, B., Atsatt, P.R., Fritz, R.S., Thompson, J. N., and Mobley, K. (1986). Parasite mediation in ecological interactions. *Annual Review of Ecology and Systematics* **17**: 487–505.

Price, P. W. (1990). Host populations as resources defining parasite community organization. In: *Parasite communities: patterns and processes* (ed. G. W. Esch, A. O. Bush, and J. M. Aho). London, Chapman and Hall, pp. 21–40.

Price, W. P., Westoby, M., and Rice, B. (1988). Parasite-mediated competition: some predictions and tests. *American Naturalist* **131**: 544–555.

Prins, H. H. T. and van der Jeugd, H. P. (1993). Herbivore population crashes and woodland structure in East Africa. *Journal of Ecology* **81**: 305-314.

Prins, H. H. T. and Weyerhaeuser, F. J. (1987). Epidemics in populations of wild ruminants: anthrax and impala, rinderpest and buffalo in Lake Manyara National Park, Tanzania. *Oikos* **49**: 28-38.

Prior, D. J. and Uglem, G. L. (1979). Behavioural and physiological aspects of swimming in cercariae of the digenetic trematode, *Proterometra macrostoma*. *Journal of Experimental Biology* **83**: 239–247.

Putman, R. J. (1994). *Community ecology*. London, Chapman and Hall.

Raffaelli, D. (2002). From Elton to mathematics and back again. *Science* **296**: 1035-1037.

Rand, D. A., Keeling, M. and Wilson, H. B. (1995). Invasion, stability and evolution to criticality in spatially extended, artificial host-pathogen ecologies. *Proceedings of the Royal Society of London Series B— Biological Sciences* **259**: 55–63.

Rapport, D. J., Costanza, R., Epstein, P. R., Gaudet, C., and Levins, R. (1998). *Ecosystem health*. London, Blackwell Science.

Rapport, D. J., Howard, J., Lannigan, R., McMurtry, R., Jones, D. L., Anjema, C. M., and Bend, J. R. (2002). Introducing ecosystem health into undergraduate medical education. In: *Conservation medicine. Ecological health in practice* (ed. A. A. Aguirre, R. S. Ostfeld, G. M. Tabor, C. House, and M. C. Pearl). Oxford, Oxford University Press, pp. 345–360.

Rauch, E. M., Sayama, H., and Bar-Yam, Y. (2003). Dynamics and genealogy of strains in spatially extended host–pathogen models. *Journal of Theoretical Biology* **221**: 655–664.

Rees, G. (1971). Locomotion of the cercaria of *Parorchis acanthus* Nicoll and the ultrastructure of the tail. *Parasitology* **62**: 489–503.

Regan, H. M. (2001). The currency and tempo of extinction. *American Naturalist* **157**: 1–10.

Reiter, P. (1998). *Aedes albopictus* and the world trade in used tires, 1988–1995: the shape of things to come? *Journal of the American Mosquito Control Association* **14**: 83–94.

Reiter, P. (2000). From Shakespeare to Defoe: malaria in England in the little ice age. *Emerging Infectious Diseases* **6**: 1–11.

Renaud, F. and De Meeüs, T. (1991). A simple model of host–parasite evolutionary relationships. Parasitism: compromise or conflict. *Journal of Theoretical Biology* **152**: 319–327.

Rice, B. and Westoby, M. (1982). Heteroecious rusts as agents of interference competition. *Evolutionary Theory* **6**: 43–52.

Richardson, A. R., Yu, Z., Popovic, T., and Stojiljkovic, I. (2002). Mutator clones of *Neisseria Meningitidis* in Epidemic Serogroup a Disease. *Proceedings of the National Academy of Science of the USA* **99**: 6103–6107.

Richner, H. and Triplet, F. (1999). Ectoparasitism and the trade-off between current and future reproduction. *Oikos* **86**: 535–538.

Ricklefs, R. E. and Schluter, D. (1993). *Species diversity in ecological communities. Historical and geographical perspectives*. Chicago, The University of Chicago Press.

Rigby, M. C. and Moret, Y. (2000). Life-history trade-offs with immune defenses. In: *Evolutionary biology of host–parasite relationships: Theory meets reality* (ed. A. Skorping). Amsterdam, Elsevier Science, pp. 129–142.

Riggs, M. R., Lemly, A. D., and Esch, G. W. (1987). The growth, biomass, and fecundity of *Bothriocephalus acheilognathi* in a North Carolina cooling reservoir. *Journal of Parasitology* **73**: 893–900.

Riley, J. and Henderson, R. J. (1999). Pentastomids and the tetrapod lung. *Parasitology* **119**: S89–S105.

Riley, M. A. and Wertz, J. E. (2002). Bacteriocins: evolution, ecology & application. *Annual Review of Microbiology* **56**: 117–137

Riley, M. A., Goldstone, C. M., and Wertz, J. E. (2003). A phylogenetic approach to assessing the targets of microbial warfare. *Journal of Evolutionary Biology* **16**: 690–697.

Rittig, M. G. and Bogdan, C. (2000). *Leishmania*–host-cell interaction: complexities and alternative views. *Parasitology Today* **16**: 292–297.

Robson, E. M. and Williams, I. C. (1970). Relationships of some species of Digenea with the marine prosobranch *Littorina littorea* (L.) I. The occurrence of larval Digenea in *L. littorea* on the North Yorkshire Coast. *Journal of Helminthology* **44**: 153–68.

Robson, N. D., Cox, A. R. J., McGowan, S. J., Bycroft, B. W., and Salmond, G. P. C. (1997). Bacterial n-acyl-homoserine-lactone dependent signalling and its potential biotechnological applications. *Trends in Biotechnology* **15**: 458–464.

Rodo, X., Pascual, M., Fuchs, G., and Faruque, A. S. G. (2002). ENSO and cholera: a nonstationary link related to climate change? *Proceedings of the National Academy of Sciences of the USA* **99**: 12901–12906.

Rodriguez, M. A. and Hawkins, B. A. (2000). Diversity, function and stability in parasitoid communities. *Ecology Letters* **3**: 35–40.

Rodriguez, D. J. and Torres-Sorando, L. (2001). Models of infectious diseases in spatially heterogeneous environments. *Bulletin of Mathematical Biology* **63**: 547–572.

Roelke-Parker, M. E., Munson, L., Packer, C., Kock, R., Cleaveland, S., Carpenter, M., O'Brien, S. J., Pospischil, A., Hofmann-Lehmann, R., Lutz, H., Mwamengele, G. L. M., Magasa, M. N., Machange, G. A., Summers, B. A., and Appel, M. J. G. (1996). A canine distemper virus epidemic in Serengeti lions (*Panthera leo*). *Nature* **379**: 441–445.

Rogers, D. J. and Randolph, S. E. (2000). The global spread of malaria in a future, warmer world. *Science* **289**: 1763–1766.

Rogers, D. J. and Williams, B. G. (1993). Monitoring trypanosomiasis in space and time. *Parasitology* **106**: 577–592.

Rohani, P., Earn, D. J. and Grenfell, B. T. (1999). Opposite patterns of synchrony in sympatric disease metapopulations. *Science* **286**: 968–971.

Rohde, K. (1982). *Ecology of marine parasites*. Queensland, Australia, University of Queensland Press, St. Lucia.

Rohde, K. (1991). Intra- and inter-specific interactions in low density populations in resource-rich habitat. *Oikos* **60**: 91–104.

Rohde, K. (1992). Latitudinal gradients in species diversity: the search for the primary cause. *Oikos* **65**: 514–527.

Rohde, K. (1998). Is there a fixed number of niches for endoparasites of fish? *International Journal of Parasitology* **25**: 1861–1865.

Rohde, K. (1999). Latitudinal gradients in species diversity and Rapoport's rule revisited: a review of recent work and what can parasites teach us about the causes of the gradients? *Ecography* **22**: 593–613.

Rohde, K. (2001). Spatial scaling laws may not apply to most animal species. *Oikos* **93**: 499–504.

Rohde, K. and Heap, M. (1998). Latitudinal differences in species and community richness and in community structure of metazoan endo- and ectoparasites of marine teleost fish. *International Journal for Parasitology* **28**: 461–474.

Rohde, K., Hayward, C., Heap, M., and Gosper, D. (1994). A tropical assemblage of ectoparasites: gill and head parasites of *Lethrinus miniatus* (Teleostei Lethrinidae). *International Journal of Parasitology* **24**: 1031–1053.

Rohde, K., Hayward, C., and Heap, M. (1995). Aspects of the ecology of metazoan ectoparasites of marine fishes. *International Journal of Parasitology* **25**: 945–970.

Rohde, K., Worthen, W.B., Heap, M., Hugueny, B., and Guégan, J. F. (1998). Nestedness in assemblages of metazoan ecto- and endoparasites of marine fish. *International Journal for Parasitology* **28**: 543–549.

Romi, R., Pierdominici, G., Severini, C., Tamburro, A., Cocchi, M., Menichetti, D., Pili, E. and Marchi, A. (1997). Status of malaria vectors in Italy. *Journal of Medical Entomology* **34**: 262–271.

Root, T. L., Price, J. T., Hall, K. R., Schneiders, S. H., Rosenzweig, C., and Pounds, J. A. (2003). Fingerprints of global warming on wild animals and plants. *Nature* **421**: 57–60.

Rosenzweig, M. L. (1995). *Species diversity in space and time*. Cambridge, Cambridge University Press.

Rothschild, M. (1962). Changes in behaviour in the intermediate hosts of trematodes. *Nature* **193**: 1312.

Rothschild, L. J. and Mancinelli, R. L. (2001). Life in extreme environments. *Nature* **409**: 1092–1101.

Rottem, S. and Naot, Y. (1998). Subversion and exploitation of host cells by mycoplasmas. *Trends in Microbiology* **6**: 436–440.

Roughgarden, J. and Feldman, M. (1975). Species packing and predation pressure. *Ecology* **56**: 489–492.

Rousset, F., Thomas, F., De Meeüs, T., and Renaud, F. (1996). Inference of parasite-induced host mortality from distributions of parasite loads. *Ecology* **77**: 2203–2211.

Roy, J., Saugier, B. and Mooney, H. A. (2001). *Terrestrial global productivity*. San Diego, CA, Academic Press.

Ruiz, G. M., Fofonoff, P., Carlton, J. T., Wonham, M. J., and Hines, A. H. (2000). Invasions of coastal marine communities in North America: apparent patterns, processes, and biases. *Annual Review of Ecology and Systematics* **31**: 481–531.

Rushton, S. P., Lurz, P. W. W., Fuller, R., and Garson, P. J. (1997). Modelling the distribution of the red and grey squirrel at the landscape scale: a combined GIS and population dynamics approach. *Journal of Applied Ecology* **34**: 1137–1154.

Rykiel, E. (1997). Ecosystem science for the twenty-first century. *Bioscience* **47**: 705–707.

Sabria, M. and Yu, V. L. (2002). Hospital-acquired Legionellosis: Solution of preventable infection. *The Lancet Infectious Diseases* **2**: 368–373.

Salgado-Maldonado, G. and Kennedy, C. R. (1997). Richness and similarty of helminth communities in the tropical cichlid fish *Cichlasoma urophthalmus* from the Yucatan Peninsula, Mexico. *Parasitology* **114**: 581–590.

Samper, C. (2003). The Millennium ecosystem Assessment: science and policy for sustainable development. *Bioscience* **53**: 1148–1149.

Sanders, M. J. (1966). Parasitic castration of scallop *Pecten alba* (Tate) by a Bucephalid Trematode. *Nature* **212**: 307.

Sanders, M. J. and Lester, R. J. G. (1981). Further observations on a bucephalid trematode infection in scallops (*Pecten alba*) in Port-Phillip Bay, Victoria. *Australian Journal of Marine and Freshwater Research* **32**: 475–478.

Sangster, N. C. and Dobson, R. J. (2002). Anthelmintic resistance. In: *The Biology of Nematodes* (ed. by D.L. Lee). London, Taylor and Francis, pp. 531–567.

Sankurathri, C. and Holmes, J. (1976). Effects of thermal effluents on parasites and commensals of *Physa gyrina* Say (Mollusca: Gastropoda) and their interactions at Lake Wabamun, Alberta. *Canadian Journal of Zoology* **54**: 1742–1753.

Sasaki, A. and Iwasa, Y. (1991). Optimal growth schedule of pathogens within a host: switching between lytic and latent cycles. *Theoretical Population Biology* **39**: 201–239.

Sattenspiel, L. and Dietz, K. (1995). A structured epidemic model incorporating geographic-mobility among regions. *Mathematical Biosciences* **128**: 71–91.

Sattenspiel, L. and Simon, C. P. (1988). The spread and persistence of infectious diseases in structured populations. *Mathematical Biosciences* **90**: 341–366.

Schad, G. A. (1963). Niche diversification in a parasite species flock. *Nature* **198**: 404–406.

Schardp (2001). Epichloë festucae and related mutualistic symbionts of grasses. *Fungal genetics and biology* **33**: 69–82.

Scheckler, W. E., Brimhall, D., Buck, A.S., Farr, B. M., Friedman, C., and Garibaldi, R. A. (1998). Requirements for infrastructure and essential activities of infection control and epidemiology. *Infectious Control of Hospital Epidemiology* **19**: 114–124.

Schell, S. C. (1970). *How to know the trematodes*. Dubuque, Iowa, William C. Brown Company Publishers.

Schindler, D. W. (1998). Replication versus realism: the need for ecosystem-scale experiments. *Ecosystems* **1**: 323–334.

Schindler, D. W. (2001). The cumulative effects of climate warming and other human stresses on Canadian freshwaters in the new millenium. *Canadian Journal of Fisheries and Aquatic Sciences* **58**: 18–29.

Schlein, L. (1998). Hunting down the last of the poliovirus. *Science* **279**:168.

Schulze, E. D. and Mooney, H. A. (1993). *Biodiversity and ecosystem function*. Berlin, Springer.

Schwarz, R. H. (1999). IUD-associated infections: causes and prevention. *Contemporary Ob/Gyn* **9**: 103–113.

Schuitemaker, H., Koot, M., Kootstra, N. A., Dercksen, M. W., De Goede, R. E., Van Steenwijk, R. P., Lange, J. M., Schattenkerk, J. K., Miedema, F., and Tersmette, M. (1992). Biological phenotype of human immunodeficiency virus type 1 clones at different stages of infection: progression of disease is associated with a shift from monocytotropic to T-cell-tropic virus population. *Journal of Virology* **66**: 1354–60.

Scott, M. E. (1985). Dynamics of challenge infections of *Gyrodactylus bullatarudis* Turnbull (Monogenea) on guppies, *Poecilia reticulata* (Peters). *Journal of Fish Diseases* **8**: 495–503.

Scott, M. E. (1987). Regulation of mouse colony abundance by *Heligomosomoides polygyrus* (Nematoda). *Parasitology* **95**: 111–129.

Scott, M. E. (1988). The impact of infection and disease on animal populations: implications for conservation biology. *Conservation Biology* **2**: 40–56.

Scott, M. E. and Anderson, R. M. (1984). The population dynamics of *Gyrodactylus bullatarudis* (Monogenea) within laboratory populations of the fish host *Poecilia reticulata*. *Parasitology* **89**: 159–194.

Scott, M. E. and Dobson, A. (1989). The role of parasites in regulating host abundance. *Parasitology Today* **5**: 176–183.

Service, M. W. (1991). Agricultural development and arthropod-borne disease: A review. *Revista de Saude Publica* **25**: 165–178.

Settle, W. H. and Wilson, L. T. (1990). Invasion by the variegated leafhopper and biotic interactions, parasitism, competition and apparent competition. *Ecology* **71**: 1461–1470.

Severini, C., Menegon, M., Gradoni, L. and Majori, G. (2002). Use of the *Plasmodium vivax* merozoite surface protein 1 gene sequence analysis in the investigation of an introduced malaria case in Italy. *Acta Tropica* **84**: 151–157.

Seytre, B. and Schaffer, M. (2004). Histoire de l'éradication de la poliomyélite. *Presse Universtaire de France* (PUF), 224 p.

Shaw, D. J. and Dobson, A. P. (1995). Patterns of macroparasite abundance and aggregation in wildlife populations: a quantitative review. *Parasitology* **111**: S111–S133.

Shaw, D. J., Grenfell, B. T and Dobson, A. P. (1998). Patterns of parasite aggregation and the negative binomial distribution **117**: 597–610.

Shope, R. (1991). Global climate change and infectious diseases. *Environmental Health Perspectives* **96**: 171–174.

Shroyer, D.A. (1986). *Aedes albopictus* and arboviruses: a concise review of the literature. *Journal of the American Mosquito Control Association* **2**: 424–428.

Shurin, J. B. and Hallen, E. G. (2001). Effects of competition, predation, and dispersal on species richness at local and regional scales. *American Naturalist* **158**: 624–637.

Shurin, J. B., Havel, J. E., Leibold, M. A., and Pinel-Alloul, B. (2000). Local and regional zooplankton species richness: a scale-independent test for saturation. *Ecology* **81**: 3062–3073.

Siddall, R. and Clers, S. D. (1994). Effect of sewage sludge on the miracidium and cercaria of *Zoogonoides viviparus* (Trematoda: Digenea). *Helminthologia (Bratislava)* **31**: 143–153.

Silver, H. M. (1998). Listeriosis during pregnacy. *Obstetric Gynecology Surveys* **53**: 737–740.

Sinclair, A. R. E. (1979a). Dynamics of the Serengeti ecosystem. In: *Serengeti: dynamics of an ecosystem* (ed. A. R. E. Sinclair and M. Norton-Griffiths). Chicago, University of Chicago Press, pp. 1–30.

Sinclair, A. R. E. (1979b). The eruption of the ruminants. In: *Serengeti: dynamics of an ecosystem* (ed. A. R. E. Sinclair and M. Norton-Griffiths). Chicago, University of Chicago Press: 82-103.

Simberloff, D. (1980). A succession of paradigms in ecology: essentialism to materialism and probabilism. *Synthese* **43**: 3–39.

Simberloff, D. and Moore, J. (1997). Community ecology of parasites and free-living organisms: host-parasite evolution. In: *Host–parasite evolution: general principles and avian models* (ed. Clayton D. H. and Moore J.). Oxford, Oxford University Press, 174–197.

Simon, N. (1962). *Between the sunlight and the thunder. The wildlife of Kenya*. London, Collins.

Skorping, A., Read, A. F., and Keymer, A. E. (1991). Life history covariation in intestinal nematodes of mammals. *Oikos* **60**: 365-372.

Skorping, A. and Hogstedt, G. (2001). Trophic cascades: a role for parasites? *Oikos* **94**: 191–192.

Slatkin, M. (1974). Competition and regional coexistence. *Ecology* **55**: 128–134.

Smith, L. D. S. and Sugiyama, H. (1988). *Botulism: the organism, its toxin, the disease*, 2nd edn. Springfield, Thomas.

Smith, N. F. (2001). Spatial heterogeneity in recruitment of larval trematodes to snail intermediate hosts. *Oecologia* **127**: 115–122.

Smith, N. J. H. (1981). Colonization lessons from a tropical forest. *Science* **214**: 755–761.

Soler, J., Møller A. P. and Soler M. (1998). Mafia behavior and the evolution of facultative virulence. *Journal of Theoretical Biology* **191**: 267–277.

Solis-Soto, J. M. and De Jong-Brink, M. (1994). Immunocytochemical study on biologically active neurosubstances in daughter sporocysts and cercariae of *Trichobilharzia ocellata* and *Schistosoma mansoni*. *Parasitology* **108**: 301–311.

Solis-Soto, J. M. and De Jong-Brink, M. (1995). An immunocytochemistry study comparing the occurrence of neuroactive substances in the nervous system of cercariae and metacercariae of the eye fluke *Diplostomum spathacum*. *Parasitology Research* **81**: 553–559.

Sorci, G., Massot, M., and Clobert, J. (1994). Maternal parasite load increases sprint speed and philopatry in female offspring of the common lizard. *American Naturalist* **144**: 153–164.

Sorci, G., Clobert, J., and Michalakis, Y. (1996). Cost of reproduction and cost of parasitism in the common lizard *Lacerta vivipara*. *Oikos* **76**: 121–130.

Sousa, W. P. and Gleason, M. (1989). Does parasitic infection compromise host survival under extreme environmental conditions? The case for *Cerithidea californica* (Gastropoda : Prosobranchia). *Oecologia* **80**: 456–464.

Southgate, V. R. (1997). Shistosomiasis in the Senegal River Basin: before and after the construction of dams at Diama Senegal and Manantali, Mali and future prospects. *Journal of the Helminthological Society of Washington* **71**: 125–132.

Spinage, C. A. (1962). Rinderpest and faunal distribution patterns. *African Wild Life* **16**: 55–60.

Sprent, J. F. A. (1992). Parasites lost. *International Journal for Parasitology* **22**: 139–151.

Srivastava, D. S. (1999). Using local–regional richness plots to test for species saturation: pitfalls and potentials. *Journal of Animal Ecology* **68**: 1–16.

Srividya, A., Michael, E., Palaniyandi, M., Pani, S. P., and Das, P. K. (2002). A geostatistical analysis of the geographic distribution of lymphatic filaris prevalence in southern India. *American Journal of Tropical Medecine and Hygiene* **67**: 480–489.

Stadnichenko, A. P., Ivanenko, L. D., Gorchenko, I. S., Grabinskaya, O. V., Osadchuk, L. A., and Sergeichuk, S. A. (1995). The effect of different concentrations of nickel sulphate on the horn snail (Mollusca: Bulinidae) infected with the trematode *Cotylurus cornutus* (Strigeidae). *Parazitologiya (St Petersburg)* **29**: 112–116.

Stanhope, M. J., Brown, J. R., and Amrine-Madsen A. (2004). Evidence from the evolutionary analysis of nucleotide sequences for a recombinant history of SARS-CoV. *Infection, Genetics and Evolution*, **4**: 15–19.

Stanko M., Miklisova D., Goüy de Bellocq J., and Morand S. (2002). Mammal density and patterns of ectoparasite species richness and abundance. *Oecologia* **131**: 289–295.

Steinberg, P. D., Estes, J. A., and Winter, F. C. (1995). Evolutionary consequences of food chain length in kelp forest communities. *Proceedings of the National Academy of Sciences of the USA* **92**: 8145–8148.

Stevens, T. (1996). *The importance of spatial heterogeneity and recruitment in organisms with complex life cycles: an analysis of digenetic trematodes in a salt marsh community.* Santa Barbara, University of California.

Stewart, B. S. and Yochem, P. K. (2000). Community ecology of California Channel Islands pinnipeds. In: *Fifth California Islands Symposium*, Vol. 99–0038 (ed D. R. Brown, K. L. Mitchell, and H. W. Chang). Santa Barbara, California, Minerals Management Service, pp. 413–420.

Stiven, A. E. (1964). Experimental studies on the host parasite system hydra and *Hydramoeba hydroxena* (Entz.). II. The components of a single epidemic. *Ecological Monographs* **34**: 119–142.

Stock, T. M. and Holmes, J. C. (1988). Functional relationships and microhabitat distributions of enteric helminths of grebes (Podicipedidae): the evidence for interactive communities. *Journal of Parasitology* **74**: 214–227.

Straight, T. M. and Martin, G. J. (2002). Brucellosis. *Current Treatment Options in Infectious Diseases* **4**: 447–456.

Strong, D. R. and Levin D. A. (1975). Species richness of the parasitic fungi on British trees. *Proceedings of the National Academy of Sciences of the USA* **72**: 2116–2119.

Strong, D. R., Lawton, J. H., and Southwood, T. R. E. (1985). *Insects on plants: community patterns and mechanisms.* London, Blackwell Science.

Sukhdeo, M. V. K. and Mettrick, D. F. (1987). Parasite behaviour: understanding platyhelminth responses. *Advances in Parasitology* **26**: 74–144.

Sukhdeo, M. V. K. and Sukhdeo, S. C. (1994). Optimal habitat selection by helminths within the host environment. *Parasitology* **109**: S41–S54.

Sukhdeo, M. V. K. and Sukhdeo, S. C. (2002). Fixed behaviours and migration in parasitic flatworms. *International Journal for Parasitology* **32**: 329–342.

Sukhdeo, M. V. K. and Sukhdeo, S. C. (2004). Trematode behaviours and the perceptual worlds of parasites. *Canadian Journal of Zoology*, in press.

Sures, B., Taraschewski, H., and Rydlo, M. (1997). Intestinal fish parasites as heavy metal bioindicators: A comparison between *Acanthocephalus lucii* (Palaeacanthocephala) and the zebra mussel, *Dreissena polymorpha. Bulletin of Environmental Contamination and Toxicology* **59**: 14–21.

Swart, R. L., Ross, P. S., Vedder, L. J., Timmerman, H. H., Heisterkamp, S., Loveren, H. V., Vos, J. G., Reijnders, P. J. H. and Osterhaus, A. D. M. E. (1994). Impairment of immune function in harbor seals (*Phoca vitulina*) feeding on fish from polluted waters. *Ambio* **23**: 155–159.

Swift, J. (1734). *On poetry: a rapsody.* London, J. Higgonson.

Swinton, J. , Harwood, J., Grenfell B. T., and Harwood, J. (1998). Persistence thresholds for phocine distemper virus infection in harbour seal *Phoca vitulina* metapopulations. *Journal of Animal Ecology* **67**: 54–68.

Taddei, F., Radman, M., Maynard-Smith, J., Toupance, B., Gouyon, P. H., and Godelle, B. (1997). Role of mutator alleles in adaptive evolution. *Nature* **387**: 700–702.

Talbot, L. M. and Talbot, M. H. (1963). The wildebeest in western Maasailand. *Wildlife Monograph* **12**: 8-88.

Tansley, A. G. (1935). The use and abuse of vegetational concepts and terms. *Ecology* **16**: 284–307.

Tavares-Cromar, A. F. and Dudley Williams, D. (1996). The importance of temporal resolution in food web analysis: evidence from a detritus-based food stream. *Ecological Monographs* **66**: 91–113.

Taylor, D. R. (2002). Ecology and evolution of chestnut blight fungus. In : *Adaptive Dynamics of Infectious Diseases* (ed. U. Dieckmann, J. A. J. Metz, M. W. Sabelis, and K. Sigmund). Cambridge University Press, pp. 286–296.

Taylor, P. D. and Frank, S. A. (1996). How to make a kin selection model. *Journal of Theoretical Biology* **180**: 27–37.

Tegner, M. J. and Levin, L. A. (1983). Spiny lobsters and sea urchins: analysis of a predator-prey interaction. *Journal of Experimental Marine Biology and Ecology* **73**: 125–150.

Temple, S. A. (1986). Do predators always capture substandard individuals disproportionately from prey populations? *Ecology* **68**: 669–674.

Tersmette, M., Gruters, R. A., De Wolf, F., De Goede, R. E., Lange, J. M., Schellekens, P. T., Goudsmit, J., Huisman, H. G., and Miedema, F. (1989). Evidence for a role of virulent human immunodeficiency virus (HIV) variants in the pathogenesis of acquired immunodeficiency syndrome: studies on sequential HIV isolates. *Journal of Virology* **63**: 2118–2125.

Tezuka, Y. (1989). The C:N:P ratio of phytoplankton determines the relative amounts of dissolved inorganic nitrogen and phosphorus released during aerobic decomposition. *Hydrobiologia* **173**: 55–62.

Thébault, E. and Loreau, M. (2003). Food-web constraints on biodiversity-ecosystem functioning relationships. *Proceedings of the National Academy of Sciences of the USA* **100**: 14949–14954.

Thirgood, S. J., Redpath, S. M., Haydon, D. T., Rothery, P, Newton I., and Hudson, P. J. (2000). Habitat loss and raptor predation; disentangling long- and short-term causes of red grouse declines. *Proceedings of the Royal Society of London Series B—Biological Sciences* **267**: 651–656.

Thomas, C. D., Cameron, A., Green, R. E., Bakkenes, M., Beaumont, L. J., Collingham, Y. C., Erasmus, B. F. N., Ferreira de Siqueira, M., Grainger, A., Hannah, L., Hughes, L., Huntley, B., van Jaarsveld, A. S., Midgley, G. F., Miles, L., Ortega-Huerta, M. A., Peterson, A. T., Phillips, O. L., and Williams, S. E. (2004). Extinction risk from climate change. *Nature* **427**: 145–148.

Thomas, F. and Renaud, F. (2001). *Microphallus papilloro-bustus* (Trematoda): a review of its effects in lagoon ecosystems. *Revue d'Ecologie Terre et Vie* **56**: 147–156.

Thomas, F., Adamo, S., and Moore, J. (2005). Parasitic manipulation : where are we and where should we go ? *Behavioural Processes*, in press.

Thomas, F., Fauchier, J., and Lafferty, K. (2002). Conflict of interest between a nematode and a trematode in an amphipod host: test of the 'sabotage' hypothesis. *Behavioral Ecology and Sociobiology* **51**: 296–301.

Thomas, F., Renaud, F., Rousset, F., Cézilly, F., and DeMeeüs, T. (1995). Differential mortality of two closely related host species induced by one parasite. *Proceedings of the Royal Society of London Series B—Biological Sciences* **260**: 349–352.

Thomas, F., Mete, K., Helluy, S., Santalla, F., Verneau, O., De Meeûs, T., Cézilly, F., and Renaud, F. (1997). Hitch-hiker parasites or how to benefit from the strategy of another parasite. *Evolution* **51**: 1316–1318.

Thomas, F., Renaud, F., De Meeüs, T., and Poulin, R. (1998a). Manipulation of host behaviour by parasites: ecosystem engineering in the intertidal zone ? *Proceedings of the Royal Society of London Series B—Biological Sciences* **265**: 1091–1096.

Thomas, F., Renaud, F., and Poulin, R. (1998b). Exploitation of manipulators: 'hitch-hiking' as a parasite transmission strategy. *Animal Behaviour* **56**: 199–206.

Thomas, F., Poulin, R., de Meeüs, T., Guégan, J.F., and Renaud, F. (1999a). Parasites and ecosystem engineering: what roles could they play ? *Oikos* **84**: 167–171.

Thomas, F., Oget, E., Gente, P., Desmots, D., and Renaud, F. (1999b). Assortative pairing with respect to parasite load in the beetle *Timarcha maritima* (Chrysomelidae). *Journal of Evolutionary Biology* **12**: 385–390.

Thomas, F., Guégan, J. F., Michalakis, Y., and Renaud, F. (2000a). Parasites and host life-history traits: implications for community ecology and species co-existence. *International Journal for Parasitology* **30**: 669–674.

Thomas, F., Poulin, R., Guégan, J. F., Michalakis, Y., and Renaud, F. (2000b). Are there pros as well as cons of being parasitized? *Parasitology today* **16**: 533–536.

Thomas, M. B. and Blanford, S. (2003). Thermal biology in insect-parasite interactions. *Trends in Ecology and Evolution* **18**: 344–350.

Thompson, J. N. (1982). *Interaction and coevolution*. New York, John Wiley & Sons.

Thompson, J. N. (1994). *The coevolutionary process*. Chicago, The University of Chicago Press.

Thompson, J. N. (1999). The evolution of species interactions. *Science* **284**: 2116–2118.

Thompson, R. M. and Townsend, C. R. (2000). Is resolution the solution? The effect of taxonomic resolution on the calculated properties of three stream food webs. *Freshwater Biology* **44**: 413–422.

Thompson, J. N., Nuismer, S. L., and Gomulkiewicz, R. (2002). Coevolution and maladaptation. *Integrative and Comparative Biology* **42**: 381–387.

Thoney, D. A. (1993). Community ecology of the parasites of adult spot, *Leiostomus xanthurus*, and Atlantic croaker, *Micropogonias undulatus* (Sciaenidae) in the Cape Hatteras Region. *Journal of Fish Biology* **39**: 515–534.

Thorns, C. J. (2000). Bacterial food-borne zoonoses. *Revue Scientifique et Technique—Office International des Epizooties* **19**: 226–239.

Thrall, P. H. and Burdon, J. H. (2003). Evolution of virulence in a plant host–pathogen metapopulation. *Science* **299**: 1735–1737.

Tierney, J. F., Huntingford, F. A., and Crompton, D. W. T. (1993). The relationship between infectivity of *Schistocephalus solidus* (Cestoda) and anti-predator behaviour of its intermediate host, the three-spined stickleback, *Gasterosterus aculeatus*. *Animal Behaviour* **46**: 603–605.

Tilman, D. (1988). *Plant strategies and the dynamics and function of plant communities*. Princeton, Princeton University Press.

Tilman, D. (1996). Biodiversity: population versus ecosystem stability. *Ecology* **77**: 350–363.

Tilman D. (1997). Community invasibility, recruitment limitation, and grassland biodiversity. *Ecology* **78**: 81–92.

Tilman, D. and Kareiva, P. (1997). *Spatial ecology*. Princeton, Princeton University Press.

Tilman, D., Lehman, C., and Bristow, C. E. (1998). Diversity–stability relationships: statistical inevitability or ecological consequence? *American Naturalist* **151**: 277–282.

Tilman, D., Wedin, D., and Knops, J. (1996). Productivity and sustainability influenced by biodiversity in grassland ecosystems. *Nature* **379**: 718–720.

Tilman, D., Knops, J., Wedin, D., Reich, P., Ritchie, M., and Siemann, E. (1997a). The influence of functional diversity and composition on ecosystem processes. *Science* **277**: 1300–1302.

Tilman, D., Lehman, C. and Thompson, K. (1997b). Plant diversity and ecosystem productivity: theoretical considerations. *Proceedings of the National Academy of Sciences of the USA* **94**: 1857–1861.

Tilman, D., Reich, P. B., Knops, J., Wedin, D., Mielke, T., and Lehman, C. (2001). Diversity and productivity in a long-term grassland experiment. *Science* **294**: 843–845.

Tinsley, R. C. (1983). Ovoviviparity in platyhelminth life-cycles. *Parasitology* **86**: 161–196.

Tinsley, R. C. (1990a). The influence of parasite infection on mating success in spadefoot toads, *Scaphiopus couchii*. *American Zoologist* **30**: 313–324.

Tinsley, R. C. (1990*b*). Host behaviour and opportunism in parasite life cycles. In : *Parasitism and Host Behaviour.* (ed. C. J. Barnard and J.M. Behnke). London, Taylor and Francis, pp. 158–192.

Tinsley, R. C. (1995). Parasitic disease in amphibians: control by the regulation of worm burdens. *Parasitology* **111**: S153–178.

Tinsley, R. C. (1996). Parasites of *Xenopus*. In: *The biology of xenopus* (ed. R. C. Tinsley and H. R. Kobel). Oxford, Oxford University Press, pp. 233–261.

Tinsley, R. C. (1999*a*). Overview: extreme environments. In: *Parasite adaptation to environmental constraints* (ed. R. C. Tinsley). Cambridge University Press, pp. S1–S6.

Tinsley, R. C. (1999*b*). Parasite adaptation to extreme conditions in a desert environment. *Parasitology* **119**: S31–S56.

Tinsley, R. C. (2003). Polystomatid monogeneans and anuran amphibians: an evolutionary arms race leading to parasite extinctions? In: *Taxonomy, ecology and evolution of metazoan parasites, Volume II* (ed. by C. Combes and J. Jourdane). Perpignan, Presses Universitaires de Perpignan, pp. 259–285.

Tinsley, R. C. (2004). Platyhelminth parasite reproduction: some general principles derived from monogeneans. *Canadian Journal of Zoology*, **82**: 270–291.

Tinsley, R. C. and Jackson, J. A. (1998). Speciation of *Protopolystoma* Bychowsky, 1957 (Monogenea: Polystomatidae) in hosts of the genus *Xenopus* (Anura: Pipidae). *Systematic Parasitology* **40**: 93–141.

Tinsley, R. C. and Jackson, J. A. (2002). Host factors limiting monogenean infections: a case study. *International Journal for Parasitology* **32**: 353–365.

Tinsley, R. C., Loumont, C., and Kobel, H. R. (1996). Geographical distribution and ecology. In: *The biology of xenopus.* (ed. R. C. Tinsley and H. R. Kobel). Oxford, Oxford University Press, pp. 35–59.

Tocque, K. and Tinsley, R. C. (1991*a*). The influence of desert temperature cycles on the reproductive biology of *Pseudodiplorchis americanus* (Monogenea). *Parasitology* **103**: 111–120.

Tocque, K. and Tinsley, R. C. (1991*b*). Asymmetric reproductive output by the monogenean *Pseudodiplorchis americanus*. *Parasitology* **102**: 213–220.

Tocque, K. and Tinsley, R. C. (1994). Survival of *Pseudodiplorchis americanus* (Monogenea) under controlled environmental conditions. *Parasitology* **108**: 185–194.

Toft, C. A. (1986). Communities of parasites with parasitic life-styles. In: *Community Ecology* (ed. J. M. Diamond and T. J. Case). New York, Harper and Row, pp. 445–463.

Toft, C. A. (1991). An ecological perspective: the population and community consequences of parasitism. In:

Parasite–host associations: coexistence or conflict? (ed. C. A. Toft, A. Aeschlimann, and L. Bolis). Oxford, Oxford University Press.

Toft, C. A. and Karter, A. J. (1990). Parasite–host coevolution. *Trends in Ecology and Evolution* **5**: 326–329.

Tokeshi, M. (1999). *Species coexistence. Ecological and evolutionary perspectives*. Oxford, Blackwell Science.

Tompkins, D. M., Greenman, J. V., and Hudson, P. J. (2000*a*). Differential impact of a shared nematode parasite on two gamebird hosts: implications for apparent competition. *Parasitology* **122**: 187–193.

Tompkins, D. M., Greenman, J. V., Robertson, P. A., and Hudson, P. J. (2000*b*). The role of shared parasites in the exclusion of wildlife hosts: *Heterakis gallinarum* in the ring-necked pheasant and the grey partridge. *Journal of Animal Ecology* **69**: 829–841.

Tompkins, D. M., Draycott, R. A. H., and Hudson, P. J. (2000*c*). Field evidence for apparent competition mediated via the shared parasites of two gamebird species. *Ecology Letters*, **3**: 10–14.

Tompkins, D. M., Greenman, J. V., and Hudson, P. J. (2001). Differential impact of a shared nematode parasite on two gamebird hosts: implications for apparent competition. *Parasitology* **122**: 187–193.

Tompkins, D. M., Sainsbury, A. W., Nettleton, P., Buxton, D., and Gurnell, J. (2002). Parapox causes a deleterious disease in red squirrels associated with UK population declines. *Proceedings of the Royal Society of London Series B—Biological Sciences* **269**: 529–533.

Tompkins, D. M., White, A. R., and Boots, M. (2003). Ecological replacement of native red squirrels by invasive greys driven by disease. *Ecology Letters* **6**: 189–196.

Torchin, M. E., Lafferty, K. D., and Kuris, A. M. (1996). Infestation of an introduced host, the European green crab, *Carcinus maenas*, by a symbiotic nemertean egg predator, *Carcinonemertes epalt*. *Journal of Parasitology* **82**: 449–453.

Torchin, M. E. and Mitchell, C. E. (2004). Parasites, pathogens, and invasions by plants and animals. *Frontiers in Ecology and the Environment* **2**: 183–190.

Torchin, M. E., Lafferty, K. D., and Kuris, A. M. (2001). Release from parasites as natural enemies: increased performance of a globally introduced marine crab. *Biological Invasions* **3**: 333–345.

Torchin, M. E., Lafferty, K. D., and Kuris, A. M. (2002). Parasites and marine invasions. *Parasitology* **124**: S137–S151.

Torchin, M. E., Lafferty, K. D., Dobson, A. P., McKenzie, V. J., and Kuris, A. M. (2003). Introduced species and their missing parasites. *Nature* **421**: 628–630.

Tortorella, D., Gewurz, B. E., Furman, M. H., Schust, D. J., and Ploegh, H. L. (2000). Viral subversion of the immune system. *Annual Review of Immunology* **18**: 861–926.

Townsend, C. R., Thompson, R. M., McIntosh, A. R., Kilroy, C., Edwards, E., and Scarsbook, M.R. (1998). Disturbance, resource supply, and food-web architecture in streams. *Ecology Letters* **1**: 200–209.

Turner, M. G., Gardner, R. H., and O'Neill, R. V. (2001). *Landscape ecology in theory and practice: pattern and process*. New York, Springer.

Uglem, G. L. and Just, J. J. (1983). Trypsin inhibition by tapeworms: antienzyme secretion or pH adjustment? *Science* **220**: 79–81.

Uglem, G. L. and Prior, D. J. (1983). Control of swimming in cercariae of *Proterometra macrostoma* (Digenea). *Journal of Parasitology* **69**: 866–870.

Upeniece, I. (1999). Fossil record of parasitic helminths in fishes. *Proceedings of the 5th International Symposium on Fish Parasites*. Institute of Parasitology, Academy of Sciences of the Czech Republic, p. 154.

Valtonen, E. T., Holmes, J. C., and Koskivaara, M. (1997). Eutrophication, pollution and fragmentation: effects on parasite communities in roach (*Rutilus rutilus*) and perch (*Perca fluviatilis*) in four lakes in central Finland. *Canadian Journal of Fisheries and Aquatic Sciences* **54**: 572–585.

van Baalen, M. (2002). Contact network and the evolution of virulence. In : *Adaptive dynamics of infectious disease: in pursuit of virulence management* (ed. U. Dieckmann, A. J. Metz, M. W. Sabelis, and K. Sigmund). Cambridge, Cambridge University Press, pp. 85–103.

van Baalen, M. and Jansen, V. A. A. (2001). Dangerous liaisons: the ecology of private interest and common good. *Oikos* **95**: 211–224.

van Baalen, M. and Sabelis, M. W. (1995). The dynamics of multiple infection and the evolution of virulence. *American Naturalist* **146**: 881–910.

Vance, R. R. (1978). Predation and resource partitioning in one predator-two prey model communities. *American Naturalist* **112**: 797–813.

van der Heijden, M. G. A., Klironomos, J. N., Ursic, M., Moutoglis, P., Streitwolf-Engel, R., Boller, T., Wiemken, A., and Sanders, I. R. (1998). Mycorrhizal fungal diversity determines plant biodiversity, ecosystem variability and productivity. *Nature* **396**: 69–72.

van der Putten, W. H. and Van der Stoel, C. D. (1998). Plant parasitic nematodes and spatio-temporal variation in natural vegetation. *Applied Soil Ecology* **10**: 253–262.

van der Putten, W. H., Van Dijk, C., and Peters, B. A. M. (1993). Plant-specific soil-borne diseases contribute to succession in foredune vegetation. *Nature* **362**: 53-56.

van Loveren, H., Ross, P. S., Osterhaus, A., and Vos, J. G. (2000). Contaminant induced immunosuppression and mass mortalities among harbor seals. *Toxicology Letters* **112**: 319–324.

van Ruijven, J. and Berendse, F. (2003). Positive effects of plant species diversity on productivity in the absence of legumes. *Ecology Letters* **6**: 170–175.

Venables, W. N. and Ripley, B. D. (1999). *Modern Applied Statistics with S-Plus*, 3rd edn. NY. Springer Verlag,

Verneau, O., Bentz, S., Sinnappah, N. D., du Preez, L., Whittington, I., and Combes, C. (2002). A view of early vertebrate evolution inferred from the phylogeny of polystome parasites (Monogenea: Polystomatidae). *Proceedings of the Royal Society of London Series B—Biological Sciences* **269**: 535–543.

Vidal-Martinez, V. M. and Poulin, R. (2003). Spatial and temporal repeatability in parasite community structure of tropical fish hosts. *Parasitology* **127**: 387–398.

Viney, M. (2002). Environmental control of nematode life cycles. In: *The behavioural ecology of parasites* (ed. E. E. Lewis, J. F. Campbell, and M. V.K. Sukhdeo). Wallingford, CABI Publishing, pp. 111–128.

Vitousek, P. M., Mooney, H. A., Lubchenco, J. and Melillo, J. M. (1997). Human domination of Earth's ecosystems. *Science* **277**: 494–499.

von Holst, D. (1999). The concept of stress and its relevance for animal behavior. In : *Stress and behavior* (ed. A. P. Møller, M. Milinski, and P. J. B. Slater). San Diego CA, Academic Press, pp. 1–132.

von Humboldt, A. (1808). Ansichten der Natur mit wissenschaftlichen Erlauterungen. J. G. Cotta, Tübingen, Germany.

von Uexküll, J. (1934). Streifzge durch die Unwelten von Tieren und Menschen (A stroll through the worlds of animal and men). In *Instinctive behavior* (ed. C. H. Schiller). New York, International Universities Press, pp. 5–80.

Wadman, M. (1996). Xenotransplantation trials 'should proceed but need guidelines'. *Nature* **382**: 197.

Wakelin, D. (1996). *Immunity to parasites*. Cambridge, Cambridge University Press.

Walker, B. H. (1992). Biodiversity and ecological redundancy. *Conservation Biology* **6**: 18–23.

Walker, B., Kinzig, A., and Langridge, J. (1999). Plant attribute diversity, resilience, and ecosystem function: the nature and significance of dominant and minor species. *Ecosystems* **2**: 95–113.

Ward, J. R. and Lafferty, K. D. (2004). The elusive baseline of marine disease: Are diseases in ocean ecosystems increasing? *Public Library of Science*, in press.

Warner, R.E. (1968). The role of introduced diseases in the extinction of the endemic Hawaiian avifauna. *Condor* **70**: 101–120.

Wasserman, S. and Faust, K. (1994). *Social network analysis: methods and applications*. Cambridge, Cambridge University Press.

Watson, H., Lee, D. L., and Hudson, P. J. (1987). The effect of *Trichostrongylus tenuis* on the caecal mucosa of young, old and anthelmintic treated wild red grouse *Lagopus lagopus scoticus*. *Parasitology* **94**: 405–411.

Watts, D. J. and Strogatz, S. H. (1998). Collective dynamics of small world networks. *Nature* **393**: 440–442.

Webster, B. L., Southgate, V. R., and Tchuem Tchuenté, L. A. (1999). Mating interactions between *Schistosoma haematobium* and *S. mansoni*. *Journal of Helminthology* **73**: 351–356.

Webster, B. L., Southgate, V. R., and Tchuem Tchuenté, L. A. (2003). On *Schistosoma haematobium*, *S. intercalatum* and occurrences of their natural hybridization in south west Cameroon. In : *Taxonomy, ecology and evolution of metazoan parasites Volume II* (ed. C. Combes and J. Jourdane). Perpignan, Presses Universitaires de Perpignan, pp. 319–337.

Webster, R. G. (1997). Predictions for future human influenza pandemics. *The Journal of Infectious Diseases* **176** (Suppl 1): S14–S19.

Weiher, E. and Keddy, P. (1999). *Ecological assembly rules. Perspectives, advances, retreats*. Cambridge, Cambridge University Press.

Weinbauer, M. G. and Rassoulzadegan, F. (2004). Are viruses driving microbial diversification and diversity? *Environmental Microbiology* **6**: 1–11.

Wenzel, R. P. (1985). Nosocomial infections, diagnosis related groups, and study on the efficacy of nosocomial infection control. Economic implications for hospitals under the prospective payment system. *American Journal of Medicine* **78**: 3–7.

Werner, E. E. and Peacor, S. D. (2003). A review of trait-mediated indirect interactions in ecological communities. *Ecology* **84**: 1083–1100.

West, S. A. and Buckling, A. (2003). Cooperation, virulence and siderophore production in bacterial parasites. *Proceedings of the Royal Society of London Series B—Biological Sciences* **270**: 37–44.

Wharton, D. A. (1999). Parasites and low temperatures. *Parasitology* **119**: S7–S17.

Wharton, D. A. (2002). Nematode survival strategies. In: *The biology of nematodes*. (ed. D. L. Lee). London, Taylor and Francis, pp. 389–411.

Whitfield, P. J., Anderson, R. M., and Bundy, D. A. P. (1977). Experimental investigations on the behaviour of the cercariae of an extoparasitic digenean *Transversotrema patialense*: general activity patterns. *Parasitology* **75**: 9–30.

Whittaker, R. J. (1998). *Island biogeography. Ecology, evolution, and conservation*. Oxford, Oxford University Press.

Whittaker, R. H. and Woodwell, G. M. (1972). Evolution of natural communities. In : *Ecosystem structure and function* (ed. J. A. Wiens). Corvallis, Oregon State University Press, pp. 137–168.

Whittington, I. (1996). Benedeniine capsalid monogeneans from Australian fishes: pathogenic species, site-specificity and camouflage. *Journal of Helminthology* **70**: 177–184.

Wickler, W. (1976). Evolution-oriented ethology, kin selection, and altruistic parasites. *Zeitschrift für Tierpsychologie* **42**: 206–214.

Williams Jr., E. H. and Bunkley-Willimas, L. (1990). The worldwide coral reef bleaching cycle and related sources of coral mortality. *Atoll Research Bulletin* **335**: 1–71.

Williams, P., Camara, M., Hartman, A., Swift, S., Milton, D., Hope, V. J., Winzer, K., Middleton, B., Pritchard, D. I., and Bycroft, B. W. (2000). Quorum-sensing and the population-dependent control of virulence. *Philosophical Transactions of the Royal Society of London Series B—Biological Sciences* **355**: 667–680.

Williams, H. H., MacKenzie, K., and McCarthy, A. M. (1992). Parasites as biological indicators of the population biology, migrations, diet, and phylogenetics of fish. *Reviews in Fish Biology and Fisheries* **2**: 144–176.

Williamson, M. H. (1957). An elementary theory of interspecific competition. *Nature* **180**: 422–425.

Willig, M. R., Kaufman, D. M., and Stevens, R. D. (2003). Latitudinal gradients of biodiversity: pattern, process, scale, and synthesis. *Annual Review of Ecology and Systematics* **34**: 273–309.

Wilson, D. S. (1980). *The natural selection of populations and communities*. Menlo, Park Benjamin/Cummings Publ. Co.

Wilson, H. B., Holt, R. D., and Hassell, M. P. (1998). Persistence and area effects in a stochastic tritrophic model. *American Naturalist* **151**: 587–595.

Wilson, K. and Grenfell, B. T. (1997). Generalized linear modelling for parasitologists. *Parasitology Today* **13**: 33–38.

Wilson, K., Bjornstad, O. N., Dobson, A. P., Merler, S., Poglayen, G., Randolph, S. E., Read, A. F., and Skorping, A. (2002). Heterogeneities in macroparasite infections: patterns and processes. In : *The ecology of wildlife diseases* (ed. P. J. Hudson, A. Rizzoli, B. T. Grenfell, H. Heeterbeek, and A. P. Dobson). Oxford, Oxford University Press, pp. 102–118.

Winemiller, K. O. and Polis, G. A. (1996). Food webs: what can they tell us about the world? In: *Food webs: integration of patterns and dynamics* (ed. G. A. Polis GA and K. O. Winemiller). New York, Chapman and Hall.

Wood, S. R., Kirkham, J., Marsh, P. D., Shore, R. C., Nattress, B., and Robinson, C. (2000). Architecture of intact natural human plaque biofilms studied by confocal laser scanning microscopy. *Journal of Dental Research* **79**: 21–27.

Woodroffe, R. (1999). Managing threats to wild mammals. *Animal Conservation* **2**: 185–193.

Wootton, J. T. (1994). The nature and consequences of indirect effects in ecological communities. *Annual Review of Ecology and Systematics* **25**: 443–466.

Worthen, B. and Rohde, K. (1996). Nested subsets analyses of colonization-dominated communities: metazoan ectoparasites of marine fishes. *Oikos* **75**: 471–478.

Wright, D. H., Patterson, B. D., Mikkelson, G. M., Cutler, A., and Atmar, W. (1998). A comparative analysis of nested subset patterns of species composition. *Oecologia* **113**: 1–20.

Wright, S. J. (2002). Plant diversity in tropical forests: a review of mechanisms of species coexistence. *Oecologia* **130**: 1–14.

Yachi, S. and Loreau, M. (1999). Biodiversity and ecosystem productivity in a fluctuating environment: the insurance hypothesis. *Proceedings of the National Academy of Science of the USA* **96**: 1463–1468.

Yan, G., Stevens, L., Goodnight, C. J., and Schall, J. J. (1998). Effects of a tapeworm parasite on the competition of *Tribolium* beetles. *Ecology* **79**: 1093–1103.

Yang, E. V. and Glaser, R. (2002). Stress-induced immunomodulation and the implications for health. *International Immunopharmacology* **2**: 315–324.

Yasuda, J., Shortridge, K. F., and Shizimu, Y. and Kida, H. (1991). Molecular evidence for a role of domestic ducks in the introduction of avian H3 influenza viruses to pigs in southern china, where the A/Hong Kong/68 (H3N2) strain emerged. *Journal of General Virology* **72**: 2007–2010.

Yodzis, P. (1996). Food webs and perturbation experiments: theory and practice. In: *Food webs: integration of patterns and dynamics* (ed. G. A. Polis and K. O. Winemiller). New York, Chapman and Hall.

Young, P. C. (1972). Relationship between presence of larval anisakine nematodes in cod and marine mammals in British home waters. *Journal of Applied Ecology* **9**: 459–485.

Young, O. R. (2003). Fish farming in the Arctic. *Polar Environmental Times* **3**: 5.

Yue, C., Coles, G., and Blake, N. (2003). Multiresistant nematodes on a Devon farm. *Veterinary Record* **153**: 604.

Zampella, R. A. (1994). Characterization of surface water quality along a watershed disturbance gradient. *Water Resources Bulletin* **30**: 605–611.

Zampella, R. A. and Bunnell J. F. (1998). Use of reference-site fish assemblages to assess aquatic degradation in pinelands streams. *Ecological Applications* **8**: 645–658.

Zampella, R. A. and Laidig K. J. (1997). Effect of watershed disturbance on pinelands stream vegetation. *Journal of the Torrey Botanical Society* **124**: 52–66.

Zampella, R. A., Bunnell, J. F., Laidig, K. J., and Dow, C. L. (2001). The Mullica River basin: a Report to the Pinelands Commission on the status of the landscape and *selected aquatic and wetland resources*. New Lisbon, Long-term Environmental-monitoring Program, Pinelands Commission.

Zholdasova, I. (1997). Sturgeons and the Aral Sea ecological catastrophe. *Environmental Biology of Fishes* **48**: 373–380.

Zhu, Y., Chen, H., Fan, J., Wang, Y., Li, Y., Chen, J., Fan, J., Yang, S., Hu, L. , Leung, H., Mew, T. W., Teng, P. S., Wang, Z., and Mundt, C. C. (2000). Genetic diversity and disease control in rice. *Nature* **406**: 718–722.

Zuwarski, T. H., Mousley, A., Mair, G. R., Brennan, G. P., Maule, A. G., Gelmar, M., and Halton, D. W. (2001). Immunomicroscopical observations on the nervous system of adult *Eudiplozoon nipponicum* (Monogenea: Diplozoidae). *International Journal for Parasitology* **31**: 783–792.

Index